普通高等教育系列教材

Fundamentals of Manufacturing Technology

机械制造技术基础

（英汉双语对照）

任小中　任乃飞　王红军　编著

王贵成（中文）　钟庆伦（英文）　主审

机械工业出版社

本书是按照我国高等教育要与国际接轨，培养国际性复合型人才的要求，结合作者近年来在"质量工程"建设方面的实践与成果编著的英汉双语教材。这是我国第一部以中、英文形式编写的"机械制造技术基础"双语教材。

本书由机械类专业的多门技术基础课的核心内容综合而成。本着"重基础、精内容、强实践"的原则，吸取国外同类教材的特点，以机械制造技术为主线，重视基本概念、基本理论和基本方法的学习，并通过相关实践教学环节的训练，理论联系实际，培养学生科技创新和工程实践的能力。

全书共有 7 章，主要内容包括绪论、金属切削原理、机床与刀具、机械加工工艺规程的制订、机床夹具设计原理、机械加工质量分析与控制、机械装配技术基础、先进制造技术简介等。

本书配套齐全，不仅有配套的电子教案和电子课件（教师版），还有对应的双语习题集（学生版）和主要习题解答（教师版），均由机械工业出版社出版或提供。

本书内容综合性强、编写体系新颖，可作为高等院校机械类专业和近机械类专业本、专科生的教材或教学参考书，特别适合作为同类课程的双语教材，也可供机械制造业工程技术人员参考。

图书在版编目（CIP）数据

机械制造技术基础：Fundamentals of Manufacturing Technology：英汉对照/任小中，任乃飞，王红军编著. —北京：机械工业出版社，2014.7（2024.2 重印）
普通高等教育系列教材
ISBN 978-7-111-46890-5

Ⅰ.①机… Ⅱ.①任…②任…③王… Ⅲ.①机械制造工艺—双语教学—高等学校—教材—英、汉 Ⅳ.①TH16

中国版本图书馆 CIP 数据核字（2014）第 116397 号

机械工业出版社（北京市百万庄大街 22 号　邮政编码 100037）
策划编辑：刘小慧　责任编辑：刘小慧　王勇哲　安桂芳　王晓艳
版式设计：霍永明　责任校对：刘怡丹
封面设计：张　静　责任印制：郜　敏
北京富资园科技发展有限公司印刷
2024 年 2 月第 1 版第 7 次印刷
184mm×260mm·26.25 印张·636 千字
标准书号：ISBN 978-7-111-46890-5
定价：57.00 元

电话服务　　　　　　　　　　网络服务
客服电话：010-88361066　　机 工 官 网：www.cmpbook.com
　　　　　010-88379833　　机 工 官 博：weibo.com/cmp1952
　　　　　010-68326294　　金　书　网：www.golden-book.com
封底无防伪标均为盗版　　　　机工教育服务网：www.cmpedu.com

Preface

With the globalization of the economy and the increase of international communication, the society requires higher foreign language proficiency and stronger international competitive capacity from talents. A significant measure of China's campus pedagogy to be in line with the international education is to teach bilingually, which could enhance students' foreign language level and open up their global perspective while imparting subject matter knowledge, and which is an effective way of cultivating high-quality international talents.

The basis of bilingual teaching lies in bilingual textbooks that are crucial to ensure the teaching quality. Original English textbooks introduced have their own advantages, but they do not correspond with the domestic teaching program and curriculum system. Therefore, combining the achievements and experiences acquired in the construction of "quality engineering", we have compiled, both in English and Chinese, this bilingual textbook, which is the first book on "Fundamentals of Manufacturing Technology" in China, to provide a better teaching resource for bilingual teaching and to improve bilingual teaching quality.

As a basic specialty course, its primary contents have inherited the essence of the traditional course. Some contents of such courses as "Metal Cutting Principle and Cutting Tools", "Machine Tools", "Machine Manufacturing Technology" and "Fixture Design", etc. have been synthesized, optimized and integrated, with an eye to the requirements of the 21st century for personnel training, together with the teaching achievements and experiences accumulated for many years by the authors, into this textbook, which reflects our compiling ideas to "emphasize the foundation, to refine the contents and to strengthen the practice". The rudimentary knowledge, principal theories and basic approaches of manufacturing technology are taken as the plot line of the book and both traditional and advanced manufacturing technologies and equipment are introduced. Thus the fundamental hierarchy of manufacturing technology is constructed in this textbook that possesses the following features:

1) Integral knowledge hierarchy. The book forms a complete fundamental hierarchy of manufacturing technology, covering the basic theory and knowledge of metal cutting principle and cutting tools, machine tools, equipment and machining quality control and so on.

2) New compiling system and wide application. Written in both English and Chinese, the book is fit for the students in bilingual classes as well as those in ordinary ones. Its main purpose is to help students understand better the knowledge hierarchy and contents, learn the technical terms and knowledge in machine manufacturing field, and lay the foundation for international communication in this field.

3) Proper arrangement and prominent key points. Based on the sufficiency-scale principle and in accordance with the mechanical engineering student training program, the contents to be mastered are discussed in detail, and the descriptive contents are simplified to thin the bilingual book.

4) Consistency, accuracy and better readability. The biggest challenge to domestic authors of bilingual books is not the word-for-word translation but the accuracy of English wording and the consistency of English sense with Chinese meaning. Therefore, professional vocabularies, descriptive manners, frequently-used grammars and simple and easy sentence patterns of the same kind of foreign textbooks are used to the greatest extent in this book so as to make it more readable.

5) Combination of theory with practice and concentration on practicality. While the metal-cutting theories, mechanical structures of the typical equipment and the manufacturing methods are introduced, how to apply the fundamental knowledge into production rationally and to build up the ability to analyze and solve practical problems are paid attention to in this book.

6) Complete supporting materials. The book is equipped with such supporting materials as the electronic teaching plan and courseware (for teachers only), the bilingual workbook (for students) and the keys to most exercises (for teachers' reference). The electronic courseware is open for the teachers to add, delete or recompose its contents according to their own will to satisfy various kinds of individualized teaching requirements.

The book has been compiled jointly by Prof. Ren Xiaozhong of Henan University of Science & Technology, Prof. Ren Naifei of Jiangsu University and Prof. Wang Hongjun of Beijing Information Science & Technology University. Besides, Xu Huili and Zhang Zhiwen (Henan University of Science & Technology), Chang Yunpeng (Luoyang Institute of Science and Technology) have also taken part in the compilation work. The specific division of the writing tasks is as follows:

Chinese manuscript: Ren Xiaozhong (Introduction, Chapter 4 and Chapter 7); Ren Naifei (Section 2.1 to 2.4 of Chapter 2); Wang Hongjun (Section 1.1 to 1.3 of Chapter 1, Section 3.4 to 3.6 of Chapter 3); Xu Huili (Section 1.4 of Chapter 1, Section 2.5 of Chapter 2, Section 3.1 to 3.3, 3.7 of Chapter 3, Chapter 5 and Chapter 6).

English manuscript: Ren Xiaozhong (Introduction, Chapter 4); Ren Naifei (Section 2.1 to 2.4 of Chapter 2); Wang Hongjun (Section 1.1 to 1.3 of Chapter 1, Section 3.4 to 3.6 of Chapter 3); Zhang Zhiwen (Section 1.4 of Chapter 1, Section 2.5 of Chapter 2, Section 3.1 to 3.3, 3.7 of Chapter 3, Chapter 5 and Chapter 6); Chang Yunpeng (Chapter 7).

Ren Xiaozhong is also in charge of the overall editing and compiling work of the entire book.

The Chinese manuscript of the book has been thoroughly reviewed and revised by Prof. Wang Guicheng of Jiangsu University and that of the English by Prof. Zhong Qinglun of Henan University of Science & Technology. They have offered valuable advice and suggestions and made some corrections. Here we extend our heartfelt thanks for their significant contributions.

Some textbooks published at home and abroad were used as reference during the compilation, for which we express most cordially thanks to the authors. At the same time, we also announce our sincerely acknowledgement to all those who have provided their help and kindness for the publication of the book.

The book has proudly acquired special financial support from the Textbook Publishing Fund of Henan University of Science & Technology. We gratefully announce our sincerely acknowledgement.

Due to various limitations, there may be some improper contents or even mistakes in the first edition. Criticisms and corrections from all experts and readers are respectfully welcome and invited so that the flaws and errors can be corrected in future editions of this book.

Ren Xiaozhong

前　言

随着经济全球化和国际交流活动的日益频繁，社会对人才外语水平和国际竞争能力的要求越来越高。双语教学是我国高等教育与国际教育接轨的一项重要举措。双语教学可以在传授学科知识的同时，提高学生的外语水平，开拓学生的国际视野，是培养高素质国际性人才的有效途径。

双语教材是开展双语教学的基础，合适的双语教材是保证双语教学质量的关键。尽管引进的原版教材有其优势，但它们与国内教学大纲和教学体系不相适应。为此，我们结合近年来在"质量工程"建设方面取得的成果与经验编写了这本英汉双语教材。这是我国第一部以中、英文形式编著的"机械制造技术基础"双语教材，旨在为双语教学提供优质的教学资源，提高双语教学质量。

本书针对专业基础课程的特点，以"重基础、精内容、强实践"作为编写指导思想，继承传统内容的精华，融入作者多年积累的教学成果和经验，着眼于 21 世纪对人才培养的要求，对原金属切削原理与刀具、金属切削机床、机械制造工艺学与夹具设计等课程的主要内容进行了综合与优化。全书以机械制造技术的基础知识、基本理论和基本方法为主线，在传承传统工艺与制造装备技术的同时，也介绍了一些先进制造技术，较系统地构建了机械制造技术的基础体系。本书具有以下特色：

1）知识体系完整。本书涵盖了金属切削原理、机床与工艺装备、机械制造工艺以及加工质量控制等各方面的基本理论和知识，形成了完整的机械制造技术基础体系。

2）编写体系新颖，适用范围广。教材用英汉双语编写，不仅适合于双语班的学生，也适合于非双语班的学生。其主要目的是通过英文内容的学习，使学生更好地理解英文教材的体系和内容，学习机械制造领域的专业术语和知识，为在本专业领域内进行国际交流奠定基础。

3）重点突出，编排得当。为了不增加篇幅，依据机械类专业人才培养大纲，本着"够用为度"的原则，对要求掌握的内容进行详述，对叙述性的内容进行简化。

4）中英文内容的一致性、准确性和可读性强。这是对作者的最大挑战。并非一味追求严格地按中文内容翻译，而是注重中英文基本内容的一致性和英文词义的准确性。尽量采用国外同类教材中的专业词汇和描述方式，尽量采用常用的语法和简单易懂的句子，使内容易读、易懂。

5）理论联系实际，注重实用。在介绍金属切削理论、典型装备的机械结构以及机械制造方法的同时，注重讲述在实际生产中如何合理运用这些基本知识，强调培养分析和解决实际问题的能力。

6）配套齐全。本教材配有电子教案和电子课件（教师版），还有对应的双语习题集（学生版）和主要习题解答（教师版）。其中电子课件为开放式课件，任课教师可根据各自情况自行增、删或改编，以满足个性化的教学要求。

本书由河南科技大学任小中教授、江苏大学任乃飞教授和北京信息科技大学王红军教授共同编著。参加本书编写的还有许惠丽、张志文（河南科技大学）、常云朋（洛阳理工学院）。具体编写分工如下：

中文部分：任小中（绪论，第4章，第7章）；任乃飞（第2章的2.1~2.4节）；王红军（第1章的1.1~1.3节，第3章的3.4~3.6节）；许惠丽（第1章的1.4节，第2章的2.5节，第3章的3.1~3.3，3.7节，第5章，第6章）。

英文部分：任小中（绪论，第4章）；任乃飞（第2章的2.1~2.4节）；王红军（第1章的1.1~1.3节，第3章的3.4~3.6节）；张志文（第1章的1.4节，第2章的2.5节，第3章的3.1~3.3，3.7节，第5章，第6章）；常云朋（第7章）。

全书由任小中负责统稿。本书由江苏大学王贵成教授和河南科技大学钟庆伦教授分别担任中文部分和英文部分的主审，两位主审分别对教材进行了仔细的审阅，提出了很多宝贵的建议和意见，并对其中一些内容分别进行了订正，在此表示由衷的感谢！

本书参考了国内外出版的一些教材，谨此向有关作者表示诚挚的谢意！并向所有关心和帮助本书出版的人表示感谢！

本书得到了河南科技大学教材出版基金的资助，在此表示衷心的感谢！

由于编者所及资料和水平有限，书中难免有错漏和不当之处，敬请各位专家和广大读者批评指正。

编　者

CONTENTS 目录

Preface 前言
Introduction 绪论 ·· 1
  0.1 Manufacturing industry and its position in national economy 制造业及其在国民经济中的地位 ········ 2
  0.2 Development of machine manufacturing technology 机械制造技术的发展 ····················· 4
  0.3 The research objects and the purpose of the book 本书的研究对象和目的 ······················ 6
  0.4 Requirements for learners 学习要求 ·· 8

Chapter 1 Principles of Metal Cutting 金属切削原理 ··· 11
  1.1 Introduction 概述 ·· 12
  1.2 Basic knowledge of metal cutting 金属切削的基本知识 ··· 12
    1.2.1 Cutting motions and machining variables 切削运动与切削用量 ································ 12
    1.2.2 Geometry of cutting tools 刀具的几何形状 ·· 14
    1.2.3 Undeformed chip dimensions 切削层参数 ·· 22
    1.2.4 Cutting tool materials 刀具材料 ·· 22
  1.3 Basic theory of the metal cutting process 金属切削过程的基本理论 ····································· 28
    1.3.1 Deformation in the metal cutting process 金属切削过程中的变形 ······························· 28
    1.3.2 Cutting forces and cutting power 切削力与切削功率 ·· 32
    1.3.3 Cutting heat and cutting temperature 切削热与切削温度 ·· 38
    1.3.4 Tool wear and tool life 刀具磨损与刀具寿命 ··· 40
  1.4 Applications of basic theory in the cutting process 金属切削基本规律的应用 ························· 46
    1.4.1 Control of chips 切屑的控制 ·· 48
    1.4.2 Machinability 切削加工性 ··· 50
    1.4.3 Cutting fluids 切削液 ··· 52
    1.4.4 Proper choice of tool geometric parameters 刀具几何参数的合理选择 ························· 56
    1.4.5 Proper choice of cutting variables 切削用量的合理选择 ·· 62

Chapter 2 Machine Tools and Cutting Tools 机床与刀具 ··· 71
  2.1 Introduction 概述 ·· 72
    2.1.1 Classification and model of machine tools 机床的分类及型号 ································· 72
    2.1.2 Surface generating methods and kinematic analysis of machine tools 表面的形成方法与机床
           运动分析 ··· 76
  2.2 Lathes and lathe cutters 车床与车刀 ·· 80
    2.2.1 Basic contents of turning 车削的基本内容 ·· 80
    2.2.2 Constructional features of a center lathe (CA6140) CA6140 卧式车床的结构特征 ············· 82
    2.2.3 Lathe cutters 车刀 ··· 108

2.3 Grinding wheels and grinding machines 砂轮与磨床 ······ 110
   2.3.1 Introduction 概述 ······ 110
   2.3.2 Grinding wheels 砂轮 ······ 110
   2.3.3 Grinding machines 磨床 ······ 114
2.4 Gear cutting machines and cutting tools 齿轮加工机床及其刀具 ······ 118
   2.4.1 Gear cutting methods 齿轮加工方法 ······ 118
   2.4.2 Gear cutting machines 齿轮加工机床 ······ 120
   2.4.3 Gear cutting tools 齿轮刀具 ······ 130
2.5 Other machine tools and cutting tools 其他类型的机床与刀具 ······ 132
   2.5.1 Hole-making machine tools and cutting tools 孔加工机床与刀具 ······ 132
   2.5.2 Milling machines and milling cutters 铣床与铣刀 ······ 142

Chapter 3 Machining Process Planning 机械加工工艺规程的制订 ······ 149

3.1 Basic concepts 基本概念 ······ 150
   3.1.1 Production course and process course of the machine 机器的生产过程和工艺过程 ······ 150
   3.1.2 Elements of process course 工艺过程的组成 ······ 150
   3.1.3 Production program and production type 生产纲领和生产类型 ······ 154
3.2 Procedure of process planning 工艺规程制订的步骤 ······ 156
   3.2.1 Process rule and its functions 工艺规程及其作用 ······ 156
   3.2.2 Original information required in process planning 制订工艺规程所需的原始资料 ······ 158
   3.2.3 Procedure of process planning 工艺规程制订的步骤 ······ 158
3.3 Selection of location datum 定位基准的选择 ······ 160
   3.3.1 Datum and its classification 基准及其分类 ······ 160
   3.3.2 Selection of location datum 定位基准的选择 ······ 162
3.4 Drawing up the process route 工艺路线的拟订 ······ 170
   3.4.1 Selection of surface machining methods 表面加工方法的选择 ······ 170
   3.4.2 The division of machining phases 加工阶段的划分 ······ 174
   3.4.3 Process concentration and dispersal 工序集中与工序分散 ······ 174
   3.4.4 Arrangement of operation sequence 工序顺序的安排 ······ 176
3.5 Determination of machining allowance 加工余量的确定 ······ 178
   3.5.1 Concept of machining allowance 加工余量的概念 ······ 178
   3.5.2 Factors affecting machining allowance 影响加工余量的因素 ······ 182
   3.5.3 Methods to determine machining allowance 确定加工余量的方法 ······ 182
3.6 Determination of operation dimensions and their tolerances 工序尺寸及其公差的确定 ······ 184
   3.6.1 Introduction 概述 ······ 184
   3.6.2 Dimension chain 尺寸链 ······ 186
   3.6.3 Calculation formulas of dimension chains 尺寸链的基本计算公式 ······ 188
   3.6.4 Examples of calculating process dimension chains 工艺尺寸链计算举例 ······ 192
3.7 Economic analysis of technological processes 工艺过程的经济分析 ······ 198
   3.7.1 Determination of time rating 时间定额的确定 ······ 198
   3.7.2 Economic analysis of process plans 工艺方案的经济性分析 ······ 200

Chapter 4 Design Principles of Machine Tool Fixtures 机床夹具设计原理 ········ 205
4.1 Introduction to machine tool fixtures 机床夹具概述 ········ 206
4.1.1 Introduction to fixtures 夹具概述 ········ 206
4.1.2 Classification of fixtures 夹具的分类 ········ 206
4.1.3 Functions and elements of fixtures 夹具的功用和组成 ········ 208
4.2 Location of the workpiece in a fixture 工件在夹具中的定位 ········ 212
4.2.1 Location principles 工件定位原理 ········ 212
4.2.2 Locators 定位元件 ········ 216
4.2.3 Analysis and calculation of location errors 定位误差的分析与计算 ········ 222
4.3 Clamping of the workpiece 工件的夹紧 ········ 234
4.3.1 Elements and requirements of the clamping device 夹紧装置的组成和要求 ········ 234
4.3.2 Principles of determining the clamping force 夹紧力的确定原则 ········ 236
4.3.3 Several commonly used clamping mechanisms 几种常用的夹紧机构 ········ 242
4.4 Other devices in machine tool fixtures 机床夹具的其他装置 ········ 254
4.4.1 Indexing device 分度装置 ········ 254
4.4.2 Tool guiding element 刀具引导元件 ········ 256
4.4.3 Tool aligning device 对刀装置 ········ 260
4.4.4 Connecting elements 连接元件 ········ 260
4.5 Methods to design special-purpose fixtures 专用夹具的设计方法 ········ 262
4.5.1 Basic requirements for the design of special-purpose fixtures 专用夹具设计的基本要求 ········ 262
4.5.2 Approaches and procedures for the design of special-purpose fixtures 专用夹具设计的方法和步骤 ········ 262
4.5.3 A case of special-purpose fixture design 专用夹具设计案例 ········ 264

Chapter 5 Analysis and Control of Machining Quality 机械加工质量分析与控制 ········ 271
5.1 Introduction 概述 ········ 272
5.1.1 Introduction to machining quality 机械加工质量概述 ········ 272
5.1.2 Basic concepts of machining accuracy 机械加工精度的基本概念 ········ 272
5.1.3 Basic concepts of machined surface quality 加工表面质量的基本概念 ········ 274
5.2 Factors affecting machining accuracy 影响加工精度的因素 ········ 278
5.2.1 Machining principle errors 加工原理误差 ········ 280
5.2.2 Errors of machine tools 机床误差 ········ 282
5.2.3 Errors caused by elastic deformation of process systems 工艺系统弹性变形引起的加工误差 ········ 292
5.2.4 Errors caused by thermal deformation of process systems 工艺系统热变形引起的加工误差 ········ 302
5.2.5 Errors caused by inner stress of the workpiece 工件内应力引起的误差 ········ 306
5.3 Statistic analysis of machining errors 加工误差的统计分析 ········ 308
5.3.1 Categories of machining errors 加工误差的分类 ········ 308
5.3.2 Distribution curve method 分布曲线分析法 ········ 310
5.3.3 Point diagram method 点图分析法 ········ 318
5.4 Factors influencing surface quality 影响加工表面质量的因素 ········ 320
5.4.1 Factors influencing surface roughness 影响表面粗糙度的因素 ········ 320
5.4.2 Factors influencing physical and mechanical properties of surface layer 影响表面层物理力学性能的因素 ········ 324

5.5 Vibrations in machining processes 机械加工过程中的振动 ································ 326
   5.5.1 Introduction 概述 ··· 326
   5.5.2 Forced vibration in machining processes 机械加工过程中的强迫振动 ········· 326
   5.5.3 Self-excited vibration in machining processes 机械加工过程中的自激振动 ····· 328
   5.5.4 Ways to control vibrations in machining processes 控制机械加工振动的方法 ··· 332

Chapter 6 Fundamentals of Machine Assembly Technology 机械装配技术基础 ············ 337
  6.1 Introduction 概述 ··· 338
     6.1.1 Concept of assembly accuracy 装配精度的概念 ································· 338
     6.1.2 Factors affecting assembly accuracy 影响装配精度的因素 ····················· 340
     6.1.3 Relation of part accuracy to assembly accuracy 零件精度与装配精度的关系 ··· 340
  6.2 Assembly methods 装配方法 ·· 342
     6.2.1 Assembly dimension chain 装配尺寸链 ·· 342
     6.2.2 Interchangeable assembly method 互换装配法 ·································· 344
     6.2.3 Selective assembly method 选择装配法 ·· 352
     6.2.4 Individual fitting assembly method 修配装配法 ································ 354
     6.2.5 Adjustment assembly method 调整装配法 ······································ 358
  6.3 Assembly process planning 装配工艺规程的制订 ·· 362
     6.3.1 Major contents of assembly process regulation 装配工艺规程的主要内容 ···· 362
     6.3.2 Principles and original information required in assembly process planning 制订装配工艺规程的基本原则及所需要的原始资料 ······························· 362
     6.3.3 Approaches and Procedures to assembly process planning 制订装配工艺规程的方法与步骤 ······ 364

Chapter 7 Brief Introduction to Advanced Manufacturing Technology 先进制造技术简介 ··········· 369
  7.1 Introduction 概述 ··· 370
     7.1.1 Definition of advanced manufacturing technology 先进制造技术的定义 ······· 370
     7.1.2 Characteristics of advanced manufacturing technology 先进制造技术的特点 ··· 370
  7.2 Advanced machining technology 先进加工技术 ··· 370
     7.2.1 Ultra-precision machining technology 超精密加工技术 ························ 370
     7.2.2 Nanofabrication technology 纳米加工技术 ······································ 374
     7.2.3 High-speed machining technology 高速加工技术 ······························· 380
     7.2.4 Modern non-traditional machining technology 现代特种加工技术 ············· 386
     7.2.5 Rapid prototyping & manufacturing technology 快速原型制造技术 ··········· 398

References 参考文献 ·· 408

Introduction

绪 论

0.1 Manufacturing industry and its position in national economy

1. The concept of manufacturing industry

The wealth of a nation depends on its ability to retrieve natural resources and to manufacture goods. There are rich nations and poor nations, and rich people and poor people. However, the bottom line for creating national wealth is still to rely on the ability to manufacture.

The word "manufacture" originated from two Latin roots "manu", meaning by hand, and "facere", meaning to make. This means that for hundreds of years, manufacturing was done manually. With the development of society and the progress of manufacturing technology, the connotation of manufacturing shows a significant historical trend. Since the Industrial Revolution, machinery has played an increasingly important role. If you look up the dictionary you may find that the definition of manufacturing is "making of articles by physical labor or machinery, especially on a large scale". With machine tools, humans can produce goods faster and better.

In short, manufacturing means the whole procedure by which people, according to their purpose and applying their knowledge and skills, make original materials into valuable products, and put them into market by means of manual or available objective tools and facilities.

Manufacturing industry is the social production department whose task is to provide production materials for national economic departments and daily consumer goods for whole society. As we know, there are different kinds of machines, tools and instruments in different factories, and the machines, or tools, or instruments are composed of many workparts with different shapes and sizes. The industry to produce various workparts and assemble them into tools, instruments and machines is called machine manufacturing industry.

2. Importance of manufacturing industry

With a general survey of the world, all economically powerful countries have their own developed manufacturing industry which has performed meritorious deeds for their economic boom. The importance of manufacturing industry can be listed below:

1) Manufacturing industry is the mainstay industry of national economy and the engine of economic growth. In the developed countries, the manufacturing industry has created about 60% social wealth and 45% national economy income. In the United States, about 68% wealth comes from the manufacturing industry. In Japan, 49% GNP comes from the manufacturing industry. In China, the production value of the manufacturing industry takes up about 45% of the total industrial output value.

2) Manufacturing industry is the basic carrier to realize the industrialization of high technologies. Take America for instance. Companies in manufacturing industry have covered all researches and developments in America industries, and provided most of technical innovations used in manufacturing industry. The most of technological advancements to promote the long-term economical growth of America have come from the manufacturing industry. It is found by surveying industrialization history that numerous science and technology achievements are conceived in the development of manufacturing industry. At the same time, the manufacturing industry also provides scientific and technological means. A large number of high technologies arisen in the 20th century, such as nuclear technology, space technology, information technology, biomedical technology etc. were all produced and converted into productive forces of scale. Its direct effectiveness was that many high-tech

0.1 制造业及其在国民经济中的地位

1. 制造业的概念

一个国家的财富取决于其拥有自然资源和制造商品的能力。世界上有穷国和富国、穷人和富人，但创造国民财富的根本仍然依赖于制造能力。

制造一词来源于拉丁语词根 manu（手）和 facere（做）。这说明几百年来，制造一直是靠手工完成的。随着社会的发展和制造技术的进步，制造也在顺应历史潮流有着更深层次的内涵。自第一次工业革命以来，机器发挥着越来越重要的作用。不妨查查字典，你会发现制造的定义是"利用人力或机器大规模制作物品"。人类使用机床可以把商品做得既快又好。

总之，制造是指人们根据自己的意图，运用掌握的知识和技能，利用手工或一切可以利用的工具和设备把原材料制成有价值的产品，并把这些产品投放市场的整个过程的总称。

制造业是为国民经济各部门提供生产原料和为全社会提供日常消费品的社会生产部门。制造业涉及国民经济各个行业。人们知道，在不同的工厂使用着各种各样的机器、工具和仪器，而这些机器、工具和仪器是由许多具有不同尺寸和形状的零件组成的。生产各种各样的零件并把它们装配成工具、仪器和机器的行业成为机械制造业。

2. 制造业的重要性

纵观世界各国，任何一个经济强大的国家，无不具有发达的制造业。许多国家的经济腾飞，制造业功不可没。制造业的重要性具体表现在以下几个方面：

1）制造业是国民经济的支柱产业和经济增长的发动机。在发达国家中，制造业创造了约60%的社会财富、约45%的国民经济收入。其中美国68%的财富来源于制造业，日本49%的国民生产总值来源于制造业。我国制造业产值占工业总产值的比例为45%。

2）制造业是高技术产业化的基本载体。以美国为例，制造业企业几乎囊括了美国产业的全部研究和开发，提供了制造业内外所用的大部分技术创新，使得美国长期经济增长的大部分技术进步都来源于制造业。纵观工业化历史，众多的科技成果都孕育于制造业的发展之中。制造业也是科技手段的提供者，科学技术与制造业相伴成长。如20世纪兴起的核技术、空间技术、信息技术、生物医学技术等高新技术无一不是通过制造业的发展而产生并转化为规模生产力的。其直接结果是导致如集成电路、计算机、移动通信设备、国际互联网、机器人、核电站、航天飞机等产品相继问世，并由此形成了制造业中的高新技术产业。

3）制造业是吸纳劳动就业的重要途径。在工业国家中，约有1/4的人口从事各种形式的制造活动。在我国，制造业吸引了一半的城市就业人口，农村剩余劳动力的转移也有近一半流入了制造业。

products, such as IC, computer, mobile communications equipment, Internet, robot, nuclear power station and space shuttle, etc. came out one after another, thereby, generating the high technology industries in manufacturing industry.

3) Manufacturing industry is the key industry to recruit labor employment. In industrialized countries, the people worked at manufacturing activities in various forms take up 1/4 of employers in whole country. In China, one half of employed population of a city works at the manufacturing industry and about half of surplus labors in countryside transfers into manufacturing industry.

4) Manufacturing industry is the main force in international trade. In recent years, the growth rate of international trade is nearly two times more than that of the world economy. As the primary products have lower technology content, and its competitiveness in international market is getting weaker and weaker, countries of the world are enlarging the export of finished goods by all means to increase its competitiveness and added value in international market. The exports of finished goods in America, Britain, France, Germany, Japan, Korea and Singapore have taken up above 90% of all exports. China's exports in the manufacturing industry have kept over 80% and created about 3/4 foreign exchange earnings since 1990s.

5) Manufacturing industry is an important assurance of national security. Modern wars have come into the time of high-tech warfare. The competition in armaments is just the competition in manufacturing technology to a large extent. Without the excellent equipments and powerful equipment manufacturing industry, any country would have no safety not only in military and political affairs, but in economical and cultural activities.

Machine manufacturing industry is the important component of manufacturing industry. It takes on the dual tasks to provide consumer goods for users and various technical equipments for national economic departments. Machine manufacturing industry is the important foundation of national industry system and the important part of national economy. The production level and economic benefit of national economic departments depend largely on the technical performance, quality and reliability of the equipments supplied by machine manufacturing industry.

0.2 Development of machine manufacturing technology

1. History of machine manufacturing technology

The earliest human's manufacturing activities could go back to the Stone Age. At that time, people made use to provide stones to make laboring tools which were used to hunt up natural resources for existence and survival. With the advent of the Bronze Age, and later the Iron Age, some primal manufacturing activities, such as spinning, smelting, forging, etc., began to come forth in order to meet the needs of natural economy based on agriculture.

The word "lathe", for instance, has a romantic root. It derives from the word "lath". It is said that the earliest lathe was named as "tree lathe". The lathe was operated by two people: one turning the bar being machined by means of a flexible tree branch (the lath) and a rope, the other holding a piece of shell or gallet as a cutting tool and moving along the bar. This was a relative crude turning process from a current view of achievable machining accuracy. Existence of some form of crude machine tools can be traced to as early as 700 B.C. In 1668, the horse-powered milling machine and pedal grinding machine were presented in China. In 1775, a British man named John Wilkinson in-

4）制造业是国际贸易的主力军。近年来，国际贸易的增长速度比世界经济的增长速度高约两倍。由于初级产品的技术含量低，在国际市场的竞争力越来越弱，各国都千方百计扩大制成品的出口，以提高国际竞争力和附加价值。美、英、法、德、日等国家的制成品出口占全部出口比例的90%以上。20世纪90年代以后，我国制造业的出口一直维持在80%以上，创造了接近3/4的外汇收入。

5）制造业是国家安全的重要保障。现代战争已进入"高技术战争"的时代，武器装备的较量在很大意义上就是制造技术水平的较量。没有精良的装备，没有强大的装备制造业，一个国家不仅不会有军事和政治上的安全，而且经济和文化上的安全也会受到威胁。

机械制造业是制造业的重要组成部分。它肩负着直接为用户提供消费品和为国民经济各部门提供各种技术装备的双重任务。机械制造业是国家工业体系的重要基础和国民经济的重要组成部分。国民经济各部门的生产水平和经济效益，在很大程度上取决于机械制造业所提供装备的技术性能、质量和可靠性。

0.2 机械制造技术的发展

1. 机械制造技术的历史

人类最早的制造活动可以追溯到石器时代。当时，人类利用天然石料制作劳动工具，用其猎取自然资源为生。随着青铜器以及后来的铁器时代的到来，为了满足以农业为主的自然经济的需要，出现了如纺织、冶炼、锻造等较为原始的制造活动。

例如，车床这个词具有一个传奇的来源。它是由"lath"这个词派生出来的。据说最早的车床称为"树车床"。该车床由两个人操作，一人利用一根柔韧的树枝和一根绳子转动被加工的棒料，另一个人则手持一个坚硬的贝壳或碎石片作为刀具沿着棒料移动。从当前能达到的加工精度角度来看，这是一种相当简易和粗糙的车削方法。早在公元前700年就出现了一些简易机床。1668年，我国就出现了马拉铣床和脚踏磨床。1775年，一位名叫John Wilkinson的人发明了一台镗床。该项发明为瓦特蒸汽机的制造扫清了障碍。后来就出现了由Henry Maudsley研制的第一台丝杠车床。1817年，一位名叫Roberts的人发明了龙门刨床。接着，John Nasmyth大约在1840年研制了钻床。1845年，Stephen Fitch设计了世界上第一台转塔车床。全自动的转塔车床是由Christopher Spencer在1869年发明的，这是首款利用凸轮控制刀具进给的自动车床，因而可使大部分加工任务自动完成。Christopher Spencer也因研制了多轴车床而享誉世界。第一台磨床是由美国人于1864年开发出来的。于是，几乎所有通用机床均已研制成功。

由Eli Whitney发明的零件互换性为制造做出了卓越的贡献。运用互换性零件的制造理

vented a boring machine. This invention paved the way for James Watt's steam engine. This was followed by Henry Maudsley's engine lathe equipped with leadscrew in 1797. Later, a planer was invented by Roberts in 1817. The drill press was the next machine tool to be developed around 1840 by John Nasmyth. Stephen Fitch designed the first turret lathe in 1845. A completely automatic turret lathe was invented by Christopher Spencer in 1869. This was the first kind of automatic lathe using cams to feed the cutting tool in and out of the workpiece, thereby automating most of the machining tasks. Christopher Spencer is also credited with the development of a multi-spindle lathe. Finally the first grinding machine was developed in America around 1864. Thus, this completed the development of almost all general-purpose machine tools.

Parts interchangeability, invented by Eli Whitney, brought another major improvement to manufacturing. Using the interchangeable parts concept, the individual subcomponents were produced with strict uniformity. In this case, any combination of them would fit together properly. It also meant that assembly could proceed by employing relatively unskilled labors. By combining fixtures and gauges developed for interchangeable manufacture, the concept of production line and mass production became a reality in the 20th century. Henry Ford refined and developed the use of assembly line for the major component manufacture of his automobile. Mass production and scientific management of manufacturing helped to produce more, better, and less expensive goods.

2. Development of cutting tool materials

After 1860, iron and steel had become primary structure material. As it was difficult to cut, new tool materials were needed to develop. Frederick Taylor co-invented high speed steel (HSS) with Maunsel White in 1898, which allowed a four times increase in cutting speed in the basic production process of turning, milling, and drilling. In 1907, carbide alloy was developed for the first time in German, thus having cutting speed increase by 4 to 20 times. Up to now, different cutting tool materials, such as carbon tool steel, HSS, carbide, ceramics, artificial diamond and Cubic Boron Nitride (CBN) and so on, have been developed.

3. Development of machining quality

With the progress and development of manufacturing technology, the machining accuracy is improved continuously. In 1775, Wilkinson's boring machine was used to machine the cylinder in Watt's steam engine. Its machining error in cylinder boring was 1 mm. That was not a simple thing at that time. Later on, the dimension error of machine parts reached 0.01mm in 1850. At the beginning of 20th century, as the micrometer and optic comparator whose measuring precision can reach 0.001mm were invented, the machining accuracy was improved gradually to μm stage (10^{-6}m). The highest accuracy is 1μm in 1940's. At the end of 20th century, the precision machining meant that the machining error $\leqslant 0.1\mu$m, and the surface roughness $Ra \leqslant 0.01\mu$m, and the super-precision machining meant that the machining error $\leqslant 0.01\mu$m, and the surface roughness $Ra \leqslant 0.001\mu$m.

0.3 The research objects and the purpose of the book

"*Fundamentals of Manufacturing Technology*" is written mainly for the undergraduates who major in the specialities related to mechanical engineering. It is completed by integrating and reorganizing the contents in metal cutting principle and cutting tools, machine tools, and machine manufac-

念使每一个零件都做得严格一致。此时，零件之间都可实现正确配合。这就意味着即使雇佣基本上没什么技能的工人照样可以完成装配工作。为互换性制造研制的夹具与量具的有机结合使得生产线和大量生产在20世纪成为了现实。Henry Ford 将装配线在其汽车主要零部件生产上的应用进一步发扬壮大。大量生产与制造的科学管理有助于生产出更多、更好、更便宜的商品。

2. 刀具材料的发展

1860年后，钢铁成为主要的结构材料。由于其难以切削，需要研制新的刀具材料。Frederick Taylor 与 Maunsel White 在1898年共同发明了高速钢（high speed steel, HSS）。这种钢的切削速度在一般的车削、铣削和钻削过程中可提高4倍。1907年，硬质合金首先在德国问世，使得切削速度可提高4~20倍。如今，已研制出了各种各样的刀具材料，如碳素工具钢、高速钢、硬质合金、陶瓷、人造金刚石、立方氮化硼等。

3. 加工质量的发展

随着制造技术的进步与发展，加工精度也在不断地提高。1775年，由 Wilkinson 发明的用于加工瓦特蒸汽机气缸的镗床，其加工误差为1mm。但在当时这也是了不起的事情。后来到了1850年，机械零件的尺寸误差达到0.01mm。20世纪初，由于能够测量0.001mm的千分尺和光学比较仪的问世，使加工精度逐渐向 μm 级过渡，成为机械加工精度发展的转折点。20世纪40年代的最高加工精度是 $1\mu m$。20世纪末，精密加工是指其加工误差$\leq 0.1\mu m$，表面粗糙度 $Ra \leq 0.01\mu m$；超精密加工是指其加工误差$\leq 0.01\mu m$ 而表面粗糙度 $Ra \leq 0.001\mu m$。

0.3 本书的研究对象和目的

《机械制造技术基础》这本书主要是为主修机械工程相关专业的本科学生而编著的。它是把金属切削原理与刀具、金属切削机床、机械制造工艺与夹具设计等内容经过有机整合而成的。本书的主要研究对象如下：

1) 机械加工工艺。包括金属切削原理、机械制造装备以及工艺规程编制。
2) 机械加工质量控制。
3) 机械装配工艺。
4) 先进制造技术简介。

本书的目的是使学生了解金属切削原理和各种机床与刀具的应用；让学生掌握不同零件的加工方法以及加工和装配工艺；培养学生具有工艺和夹具的设计能力，以及分析和解决在机械制造和装配中遇到的质量问题。

turing technology and fixtures. The primary objects studied in this book are as follows:

1) Machining technology. Machining technology includes metal cutting principle, manufacturing equipment and process planning and so on.

2) Machining quality control.

3) Machine assembly technology.

4) Introduction to advanced manufacturing technology.

The purposes of this book are for students to learn the metal cutting principle and the applications of various machine tools and cutting tools; to master the machining methods for different workparts, and the machining and assembling processes; to have the abilities to design process planning and fixtures, and to analyze and solve the quality problems met in machine manufacturing and assembly process.

0.4 Requirements for learners

"Fundamentals of Manufacturing Technology" is a main technical fundamental course of mechanical engineering in college and universities. In order to realize the cultivation goal of this course, specific requirements are put forward as follows:

1) To master the basic theories and rules of metal cutting process, and have the ability to select rational machining method according to technical requirement.

2) To learn the basic concepts and structure about machining equipment (i.e. machine tools, cutting tools, and fixtures), and understand how to select machine tools and cutting tools, and have the preliminary ability to design jigs and fixtures.

3) To know the factors affecting on machining quality, and have the preliminary abilities to analyze and solve the specific technological problems.

4) To learn and master the basic principle and knowledge about process planning, and have the preliminary abilities to draw up the technological rule.

5) To learn about the advanced manufacturing technologies.

In summary, mechanical manufacturing technology is an engineering technology with high integrity and strong practicality. It involves various contents and broad knowledge and has a close relation to production practice. As the key points summarized in this book are the general principles and rules involved in manufacturing activities, they have a considerable flexibility. In view of these characteristics, while reading this book, the learners should also pay attention to the learning and accumulating of practice knowledge. If you combine the course learning with some practice teaching links such as production practice, curriculum experiment, course project, you will find it easy to understand the basic concepts in manufacturing technology. Only when you are good at finding, analyzing, summarizing, and applying, can you improve the ability to analyze and solve the practical problems in machine manufacturing process.

0.4 学习要求

"机械制造技术基础"是大学机械工程专业的主要技术基础课。为了实现本门课程的培养目标,特提出如下具体要求:

1) 掌握金属切削过程的基本理论和规律,并能根据工艺要求选择合理的加工方法。

2) 了解常用机械加工装备(机床、刀具、机床夹具)的基本概念和结构,懂得机床、刀具的选用,并具有设计夹具的初步能力。

3) 掌握影响机械加工质量(加工精度和表面质量)的因素,具有分析和解决具体工艺问题的初步能力。

4) 了解和掌握制订机械加工工艺规程和机器装配工艺规程的基本原理和基本知识,具有制订工艺规程的初步能力。

5) 对先进制造技术有一定的了解。

总之,机械制造技术是一门综合性、实践性很强的工程技术。它涉及的内容繁多,知识面广,且与生产实际联系密切。由于本书总结的要点是机械制造活动的一般原理和规律,因此在生产实践中具有很大的灵活性。鉴于机械制造技术的这些特点,在学习本课程时,要注意实践知识的学习和积累。只要把这门课程与生产实习、课程实验、课程设计等实践性教学环节相结合,就易弄懂机械制造技术的基本概念。只有善于发现、分析、总结和应用,才能提高分析和解决机械制造过程中实际问题的能力。

Chapter 1

Principles of Metal Cutting

第 1 章

金属切削原理

1.1 Introduction

There are many metal machining methods in machine building industry. Metal cutting means that the cutting tool is used for removing the redundant material from workpiece. Its purpose is to obtain the finished workpiece with desired size, shape and surface quality. Therefore, metal cutting is one of the most widely used manufacturing processes.

The principle of metal cutting is a science to study the basic rules in metal cutting process. The cutting regularity and its control methods can be found so as to establish the necessary theoretical foundation for solving some technical problems in cutting process. The main goals of studying the metal cutting principle are to solve the technology problems produced in cutting processes and to find the effective ways to increase cutting efficiency and to improve machining quality, and to reduce machining cost.

1.2 Basic knowledge of metal cutting

1.2.1 Cutting motions and machining variables

1. Cutting motions

Metal cutting is a kind of mechanical machining process to obtain the finished part with a certain dimension, form accuracy and surface quality by using the cutting tool to remove the excess material from the blank. In turning process, the rotation of workpiece and the longitudinal movement of cutting tool generate the external cylindrical surface, as shown in Fig. 1-1. There are three constantly changing surfaces on the workpiece, which are the surface to be machined, machining surface and machined surface respectively.

As shown in Fig. 1-1, during the metal cutting process, there is a relative motion between the workpiece and the cutting tool. Cutting motions, according to their function in metal cutting, can be divided into main motion and feed motion.

(1) Main motion A main motion is a basic motion required to remove the surplus materials from the workpiece. Its main features are: The movement velocity is the highest of all; the consumed power is the largest; there is only one main motion in cutting motions.

(2) Feed motion A Feed motion is a motion which brings new metal into cutting process constantly so as to machine the whole surface of a workpiece. Its main features are: The feed velocity is smaller than the velocity of main motion; the consumed power is very small; there could be one or more feed motions among cutting motions; a feed motion can be continuous or disconnected.

(3) Resultant cutting motion A resultant cutting motion is the motion composed of main motion and feed motion. The main motion and the feed motion engaged in cutting simultaneously in most of cutting process. The direction of instantaneous resultant of cutting motions in a given point of the cutting edge relative to the workpiece is called resultant cutting motion direction, and the velocity is called resultant speed. Its magnitude and direction can be expressed by the vector v_e, that is:

$$v_e = v_c + v_f \tag{1-1}$$

1.1 概述

在机械制造工业中有许多金属加工方法。金属切削加工指的是利用切削刀具从工件表面上切除多余的金属，从而获得具有一定尺寸、几何形状和表面质量要求的零件。因此，金属切削加工是得到最广泛使用的机械制造方法之一。

金属切削原理是研究金属切削加工过程基本规律的一门科学，从而找到切削规律及其控制方法，以便为解决切削过程中产生的有关工艺问题奠定必要的理论基础。研究金属切削原理的主要目的就是要解决切削过程中产生的工艺问题，并寻求提高切削效率、改善加工质量和降低加工成本的有效途径。

1.2 金属切削的基本知识

1.2.1 切削运动与切削用量

1. 切削运动

金属切削加工是利用刀具切去工件毛坯上多余的金属层，以获得具有一定的表面精度和表面质量的机械零件的机械加工方法。如图 1-1 所示，零件车削加工时，工件旋转和刀具的轴向进给运动形成零件的外圆表面。在切削过程中，工件上通常存在三个不断变化的切削表面：待加工表面、过渡表面和已加工表面。

如图 1-1 所示，金属切削过程是工件和刀具相互运动的过程。切削运动根据其功用不同，可分为主运动和进给运动。

（1）主运动 主运动是从工件表面上切除多余金属所必需的最基本运动。主运动的速度最高、所消耗的功率最大。在切削运动中，主运动只有一个。

（2）进给运动 进给运动是不断地把待切金属投入切削过程，从而加工出全部已加工表面的运动。进给运动一般速度较低，消耗的功率较少，可以由一个或多个运动组成。它可以是间歇的，也可以是连续的。

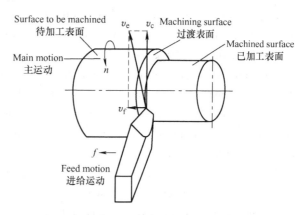

Fig. 1-1　Cutting motions and workpiece surface
切削运动与工件表面

（3）合成切削运动 合成切削运动是由主运动和进给运动合成的运动。大多数切削加工中，主运动和进给运动同时进行。刀具切削刃上选定点相对工件的瞬时合成运动方向称合成切削运动方向，其速度称合成切削速度，大小和方向可以用矢量 v_e 表示

$$v_e = v_c + v_f \tag{1-1}$$

其中，v_c 是主运动速度，v_f 进给运动速度。

where, v_c stands for the velocity of main motion, and v_f stands for the feed rate.

2. Cutting variables

Cutting variables consist of cutting speed, feed rate and depth of cut, as shown in Fig. 1-2.

(1) Cutting speed v_c The cutting speed means the instantaneous speed of the main motion in a given point of the cutting edge relative to workpiece, generally expressed in meters per second or m/min. For external turning, the cutting speed may be expressed by the following formula:

$$v_c = n\pi d_w/1000 \tag{1-2}$$

where, d_w—diameter of the work to be machined (mm);

n—rotary speed of the work (or tool) (r/s or r/min).

(2) Feed rate f The feed rate may be defined as the small relative movement per cycle (per revolution or per stroke) of the cutting tool in a direction usually normal to the cutting speed. The feed rate is measured in millimeter per revolution (mm/r).

(3) Depth of cut a_p The depth of cut is the normal distance between the unmachined surface and the machined surface, measured in mm. For external longitudinal turning, the depth of cut can be calculated as the following:

$$a_p = (d_w - d_m)/2 \tag{1-3}$$

where, d_w—diameter of workpiece surface to be machined;

d_m—diameter of machined workpiece surface.

1.2.2 Geometry of cutting tools

The tool geometry refers to the tool angles, the shape of the tool face and the form of the cutting edges. The optimum tool geometry depends on the workpiece material, machining variables (v, f, a_p), the material of the tool point and the type of cutting.

1. Elements of cutting tool

There are many kinds of metal cutting tools, as shown in Fig. 1-3, such as lathe tool, milling cutter, drill, reamer, etc. Even though they have different shapes, they have a common geometry in cutting part, i.e. "wedge part". Lathe tool is the simplest single-point cutting tool, which is the typical of other cutting tools. The principles which apply to the lathe tool point also apply to other types of cutting tools.

A turning tool consists of the shank and the working part called cutting point (or tool bits), as shown in Fig. 1-4. The structural elements of a turning tool can be named as follows: rake face 1, leading flank 5, trailing flank 4, leading cutting edge 6, trailing cutting edge 2 and tool nose 3.

2. Reference system for designating tool angles

In order to determine the space position of a rake face, flanks and cutting edges, the reference system should be established firstly, which is made up of a group of coordinate and measurement planes used to define and stipulate cutting tool angles. Based on the types of the cutting motion, the reference systems can be classified into two types: one is the reference system for marking the cutting tool angle on tool drawing, the other is the reference system for indicating the tool angles under cutting conditions. The former is determined by the direction of the main motion, and the latter is determined by the direction of the resultant cutting motion.

(1) Reference system with orthogonal plane The reference system consists of three planes as

2. 切削用量

切削用量包括切削速度 v_c、进给量 f 和背吃刀量（切削深度）a_p，如图 1-2 所示。

（1）切削速度 v_c　切削刃选定点相对于工件的主运动的瞬时速度，单位为 m/s 或 m/min。车削外圆的计算公式如下

$$v_c = n\pi d_w/1000 \qquad (1\text{-}2)$$

式中　d_w——工件待加工表面直径（mm）；

　　　n——工件转速（r/s 或 r/min）。

（2）进给量 f　工件或刀具每转一转时，两者沿进给方向的相对位移，单位为 mm/r。

（3）背吃刀量 a_p　工件上已加工表面和待加工表面间的垂直距离，单位为 mm，车削外圆时

$$a_p = (d_w - d_m)/2 \qquad (1\text{-}3)$$

式中　d_w——待加工表面直径（mm）；

　　　d_m——已加工表面直径（mm）。

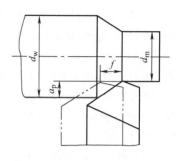

Fig. 1-2　Feed and depth of cut
进给量与切削深度

1.2.2　刀具的几何形状

刀具的几何形状指的是刀具角度、前面的形状以及切削刃的形状。最佳的刀具几何形状取决于被加工工件材料、切削用量（v_c, f, a_p）、刀具材料以及切削类型。

1. 切削刀具的组成

金属切削刀具的种类很多，如图 1-3 所示，如车刀、铣刀、钻头、铰刀等。即使它们形状各异，但它们切削部分的几何形状具有共同特征，即"楔部"。车刀是最简单的单刃刀具，它是其他刀具的典型代表。适用于车刀的一些原理同样适用于其他类型的刀具。

车刀由刀柄和刀头组成，如图 1-4 所示。车刀切削部分的结构要素包括前面 1、主后面 5、副后面 4、主切削刃 6、副切削刃 2 和刀尖 3。

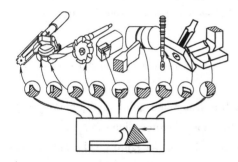

Fig. 1-3　Geometry of cutting point
各种刀具切削部分的几何形状

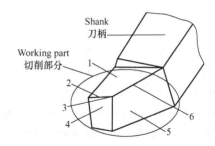

Fig. 1-4　Turning tool 车刀
1—Rake face 前面　2—Trailing cutting edge 副切削刃
3—Tool nose 刀尖　4—Trailing flank 副后面
5—Leading flank 主后面　6—Leading cutting edge 主切削刃

2. 刀具角度的参考系

为了确定刀具切削部分各表面和切削刃的空间位置，需要建立能够定义和规定刀具角度

shown in Fig. 1-5.

1) Reference plane P_r. P_r is a plane which passes through a designated point on the cutting edge and is perpendicular to the direction of the main motion.

2) Cutting edge plane P_s. P_s is a plane which passes through a designated point on the cutting edge, and is tangential to the cutting surface and perpendicular to the reference plane.

3) Orthogonal plane P_o. P_o is a plane which passes through a designated point on the cutting edge, and is perpendicular to the projection of cutting edge upon P_r and the cutting edge plane P_s simultaneously.

(2) Reference system with normal plane As shown in Fig. 1-5, the reference system with normal plane consists of the reference plane P_r, cutting edge plane P_s and normal plane P_n, where P_r and P_s are the same as that in orthogonal plane system. The normal plane P_n is a plane which passes through a designated point on the cutting edge and is perpendicular to the cutting edge.

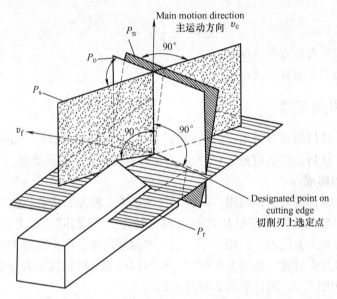

Fig. 1-5 Reference system with orthogonal plane and Reference system with normal plane
正交平面与法平面参考系

(3) Reference system with feed and back sections The reference system consists of three planes (P_r, P_f, P_p), as shown in Fig. 1-6.

1) Feed section P_f. P_f is a plane which passes through a designated point on the cutting edge, and is parallel to the feed direction and perpendicular to the tool reference plane P_r.

2) Back section P_p. P_p is a plane which passes through a designated point on the cutting edge and is perpendicular to the tool reference plane P_r and the feed section P_f.

3. Tool angles designated in orthogonal plane reference system

The designated angles of the cutting tool refer to the angles marked on the design drawing of the cutting tool, which are used for manufacturing and sharpening cutting tools. There are 6 main cutting tool angles. Take the turning tool for example; its designated angles in orthogonal plane reference system are shown in Fig. 1-7.

的参考坐标系，而该参考坐标系是由一组坐标测量平面构成的。按构成参考系时所依据的切削运动的差异，参考系分为刀具标注角度参考系和刀具工作角度参考系。前者由主运动方向确定，后者由合成切削运动方向确定。

（1）正交平面参考系　如图1-5所示，正交平面参考系由以下三个平面组成：

1）基面P_r。基面是指过切削刃上选定点垂直于主运动方向的平面。

2）切削平面P_s。切削平面是指过切削刃上选定点与切削刃相切并垂直于基面的平面。

3）正交平面P_o。正交平面是指过切削刃上选定点并同时垂直于切削平面与基面的平面。

（2）法平面参考系　如图1-5所示，法平面参考系由基面P_r、切削平面P_s和法平面P_n组成。其中基面P_r和切削平面P_s与正交平面参考系的相同，法平面P_n是指过切削刃上选定点并垂直于切削刃的平面。

（3）假定工作平面参考系　如图1-6所示，假定工作平面参考系由P_r、P_f、P_p三个平面组成。

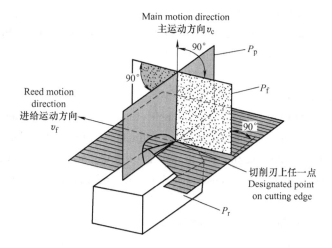

Fig. 1-6　Longitudinal-transverse system 假定工作平面参考系

1）假定工作平面P_f。过切削刃上选定点平行于假定进给方向并垂直于基面P_r的平面。

2）背平面P_p。过切削刃上选定点同时垂直于基面P_r和假定工作平面P_f的平面。

3. 在正交平面参考系内刀具的标注角度

刀具的标注角度指的是在刀具设计图上标注的角度，用于制造和刃磨刀具。刀具的主要标注角度有6个，以车刀为例，在正交平面参考系内刀具的标注角度如图1-7所示。

（1）在正交平面内标注的角度

前角γ_o——在正交平面内测量的前面与基面之间的夹角。

后角α_o——在正交平面内测量的后面与切削平面之间的夹角。

楔角β_o——在正交平面内测量的前面与后面之间的夹角。由图1-7可知

$$\beta_o = 90° - (\gamma_o + \alpha_o) \tag{1-4}$$

（2）在切削平面内标注的角度

刃倾角λ_s——在切削平面内测量的主切削刃与基面之间的夹角。

（3）在基面P_r内测量的角度

(1) Angles measured in orthogonal plane P_o

Rake angle γ_o—included angle between the rake face and the reference plane.

Clearance angle α_o—included angle between the flank and the cutting edge plane.

Wedge angle β_o—included angle between the flank and the rake face. It can be seen from Fig. 1-7 that

$$\beta_o = 90° - (\gamma_o + \alpha_o) \tag{1-4}$$

(2) Angle measured in cutting edge plane P_s

Cutting edge inclination angle or back rake angle λ_s—included angle between the cutting edge and the reference plane.

(3) Angles measured in reference plane P_r

Major cutting edge angle κ_r—included angle between the projection of the cutting edge on the reference plane and the feed direction.

Minor cutting edge angle κ_r'—included angle between the projection of the trailing cutting edge on the reference plane and the feed direction.

Nose angle ε_r—included angle between the projections of leading and trailing cutting edges on the reference plane. It can be seen from Fig. 1-7 that

$$\varepsilon_r = 180° - (\kappa_r + \kappa_r') \tag{1-5}$$

4. Tool working angles

The tool working angles are the real cutting angles under working state. Because the setting position and feed movement of the cutting tool have an influence on the tool angles, working angles are not equal to the angles indicated on the tool drawing. In order to enable tool working angles to be optimal in cutting process, it is very necessary to know the relationship between the working angle and the designated angle.

(1) Influence of transverse feed on working angles As shown in Fig. 1-8, when the cutting tool feeds transversely to cut the workpiece, the moving track of the designated point on the cutting edge relative to the workpiece is an Archimedes spiral line. Therefore, the working reference plane P_{re} and working cutting plane P_{se} turn a angle μ relative to P_r and P_s respectively, causing the variations of both rake and clearance angles. The working rake and clearance angles can be calculated as the following in the reference system (P_{re}, P_{se}, P_{oe}):

$$\gamma_{oe} = \gamma_o + \mu \tag{1-6}$$

$$\alpha_{oe} = \alpha_o - \mu \tag{1-7}$$

$$\mu = \arctan(f / \pi d) \tag{1-8}$$

where, γ_{oe}, α_{oe}—working rake angle and working clearance angle;

f—feed (mm/r);

d—the diameter of the workpiece at the cutting point (mm);

μ—included angle between the main motion direction and the resultant motion direction.

It is known from Eq. (1-8) that μ increases with the increase of the feed and the decrease of the work diameter. With the cutter getting near the center, the value μ would increase severely, and

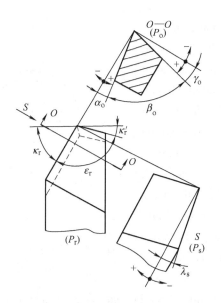

Fig. 1-7 Turning tool angles designated in orthogonal plane reference system
车刀在正交平面参考系内标注的角度

主偏角 κ_r——主切削刃在基面上的投影与进给方向的夹角。
副偏角 κ_r'——副切削刃在基面上的投影与进给方向的夹角。
刀尖角 ε_r——主切削刃与副切削刃之间的夹角。由图 1-7 可知

$$\varepsilon_r = 180° - (\kappa_r + \kappa_r') \tag{1-5}$$

4. 刀具的工作角度

刀具的工作角度是刀具在工作时的实际切削角度。由于刀具的安装位置和刀具的进给运动对刀具角度的影响,刀具的工作角度和刀具的标注角度不相等,为了确保加工过程中获得合理的工作角度,有必要了解两者之间的关系。

(1) 横向进给运动对工作角度的影响 如图 1-8 所示,当车刀横向进给切削工件时,切削刃相对工件的运动轨迹为一平面阿基米德螺旋线。此时,工作基面 P_{re} 和工作切削平面 P_{se} 相对于 P_r 和 P_s 转动一个 μ 角,从而引起刀具的前角和后角发生变化。其计算公式如下

$$\gamma_{oe} = \gamma_o + \mu \tag{1-6}$$

$$\alpha_{oe} = \alpha_o - \mu \tag{1-7}$$

$$\mu = \arctan(f/\pi d) \tag{1-8}$$

式中 γ_{oe}、α_{oe}——工作前角和工作后角;
  f——进给量(mm/r);
  d——切削点处工件的直径(mm);
  μ——正交平面内 P_{re} 和 P_r 之间的夹角,即主运动方向与合成运动方向的夹角。

由式 (1-8) 可知,当进给量增大时,μ 值增大;当瞬时直径 d 减小时,μ 值也增大。因此,车削至接近工件中心时,μ 值增长很快,工作后角将由正变负,导致工件被挤断。

(2) 纵向进给对工作角度的影响 图 1-9 所示为纵车外圆车刀的工作角度。在考虑纵向进给运动时,切削刃相对于工件表面的运动轨迹为螺旋线。此时,基面 P_r 和切削平面 P_s 就

the clearance angle would become negative.

(2) Influence of longitudinal feed on working angles In cylindrical turning shown in Fig. 1-9, considering the longitudinal feed movement of the cutting tool, the moving track of the designated point on the cutting edge relative to the workpiece is an Archimedes spiral line. Both the reference plane P_r and the cutting plane P_s would tilt the angle μ relative to the reference system for designating tool angles. The working rake angle γ_{oe} increases, and the working clearance angle α_{oe} decreases.

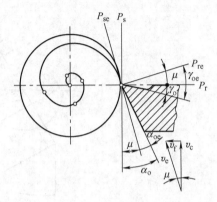

Fig. 1-8 Influence of transverse feed on working angles
横向进给对工作角度的影响

Working angles in the longitudinal section can be expressed as follows:

$$\gamma_{fe} = \gamma_f + \mu_f \tag{1-9}$$

$$\alpha_{fe} = \alpha_f - \mu_f \tag{1-10}$$

$$\mu_f = \arctan(f/\pi d_w) \tag{1-11}$$

where, γ_{fe}—working rake angle in the longitudinal section;

α_{fe}—working clearance angle in the longitudinal section;

d_w—the diameter of the unmachined workpiece (mm);

μ_f—angle between the main motion direction and the resultant motion direction.

In orthogonal plane, the working angles γ_{oe}, α_{oe} are the same as that expressed in Eq. (1-6) and Eq. (1-7), and

$$\mu = \arctan(f \sin\kappa_r / \pi d_w) \tag{1-12}$$

where, μ—included angle between P_{se} and P_s in the orthogonal plane.

(3) Influence of tool set height on working angles With the tool nose above the centre height or below the center height (see Fig. 1-10), the changing of the cutting speed direction would result in the position changing of the reference plane P_r and the cutting edge plane P_s^*. The relationship between the working angles and the designed angles can be calculated as follows:

$$\gamma_{pe} = \gamma_p \pm \theta_p \tag{1-13}$$

$$\alpha_{pe} = \alpha_p \mp \theta_p \tag{1-14}$$

$$\tan\theta_p = \frac{h}{\sqrt{\left(\dfrac{d}{2}\right)^2 - h^2}} \tag{1-15}$$

where, γ_{pe}—working rake angle in the transverse section;

α_{pe}—working clearance angle in the transverse section;

θ_p—included angle between P_r and P_{re};

h—the deviation of tool nose from the workpiece center (mm);

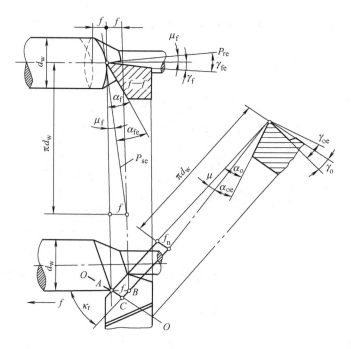

Fig. 1-9 Influence of longitudinal feed on working angles 纵向进给对工作角度的影响

会在空间偏转一个 μ 角，从而使刀具的工作前角 γ_{oe} 增大，工作后角 α_{oe} 减小。

在纵向平面内的工作角度为

$$\gamma_{fe} = \gamma_f + \mu_f \tag{1-9}$$

$$\alpha_{fe} = \alpha_f - \mu_f \tag{1-10}$$

$$\mu_f = \arctan(f/\pi d_w) \tag{1-11}$$

式中　γ_{fe}——纵向平面内的工作前角；

α_{fe}——纵向工作平面内的工作后角；

d_w——工件待加工表面直径（mm）；

μ_f——主运动方向与合成运动方向的夹角。

在正交平面内，工作角度 γ_{oe}、α_{oe} 与式（1-6）、式（1-7）相同，而

$$\mu = \arctan(f\sin\kappa_r/\pi d_w) \tag{1-12}$$

式中　μ——正交平面内 P_{se} 和 P_s 之间的夹角。

（3）刀具安装对工作角度的影响　如图 1-10 所示，当刀尖安装高于或低于工件中心时，则此时的切削速度方向发生变化，引起基面和切削平面的位置改变。此时工作角度与标注角度的换算关系如下

$$\gamma_{pe} = \gamma_p \pm \theta_p \tag{1-13}$$

$$\alpha_{pe} = \alpha_p \mp \theta_p \tag{1-14}$$

$$\tan\theta_p = \frac{h}{\sqrt{\left(\dfrac{d}{2}\right)^2 - h^2}} \tag{1-15}$$

d—the diameter of the work piece at the designed point of the cutting edge(mm).

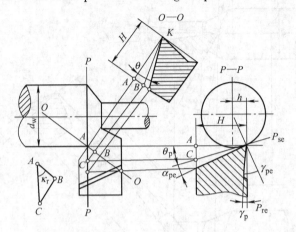

Fig. 1-10 Influence of tool set height on working angles 刀尖安装高度对工作角度的影响

1.2.3 Undeformed chip dimensions

The undeformed chip means the layer of workpiece material removed by the cutting edge in a single cutting action (i.e. one way cutting in shaping, or cutting a circle of transition surface in turning).

The shape and the dimension of the undeformed chip affect directly the load on the cutting tool. For the sake of the calculation, the shape and the dimension of undeformed chip are generally measured in the reference plane P_r of the cutting tool, as shown in Fig. 1-11. The dimensions of the undeformed chip are named as undeformed chip dimensions.

Undeformed chip thickness h_D—measured perpendicular to the cutting surface. In longitudinal turning, there is

$$h_D = f \sin\kappa_r \tag{1-16}$$

Undeformed chip width b_D—measured parallel to the cutting surface. In longitudinal turning,

$$b_D = a_p / \sin\kappa_r \tag{1-17}$$

Cross-section area of undeformed chip A_D—measured in the reference plane P_r.

$$A_D = h_D \cdot b_D = f \cdot a_p \tag{1-18}$$

1.2.4 Cutting tool materials

The performances of cutting tool materials have an important influence on the productivity, tool life, and machining quality of the workpiece. The cutting performances of cutting tools depend on the material, geometric shape, and structure of cutting tool to a large extent. Therefore, we should pay more attention to the selection and application of cutting tool materials.

1. Properties of tool materials

(1) High hardness

(2) High wear-resistance

(3) Sufficient strength and toughness

式中　r_{pe}——背平面内的工作前角；
　　　α_{pe}——背平面内的工作后角；
　　　θ_p——背平面内P_r与P_{re}的夹角；
　　　h——刀尖高于或低于工件中心的数值（mm）；
　　　d——工件切削刃上选定点处直径（mm）。

1.2.3　切削层参数

切削层为切削部分切过工件的一个单程所切除的工件材料层。切削层的尺寸与形状直接影响着刀具承受的负荷。为简化计算，切削层的尺寸与形状规定在刀具基面中测量。切削层的尺寸称为切削层参数。

现以外圆车削为例来说明切削层参数的定义。外圆车削时，工件转一转，主切削刃移动一个进给量f所切除的金属层称为切削层，如图1-11所示。当主、副切削刃为直线，且$\lambda_s = 0°$时，切削层公称截面为平行四边形。

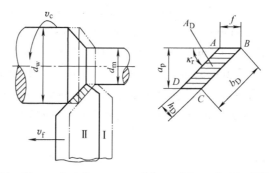

Fig. 1-11　Geometrical dimensions of the undeformed chip 切削层参数

切削层公称厚度h_D——垂直于过渡表面测量的切削层尺寸。外圆纵车时，
$$h_D = f \sin\kappa_r \tag{1-16}$$
切削层公称宽度b_D——沿着过渡表面测量的切削层尺寸。外圆纵车时，
$$b_D = a_p / \sin\kappa_r \tag{1-17}$$
切削层公称横截面积A_D——切削层在基面内的面积。
$$A_D = h_D \cdot b_D = f \cdot a_p \tag{1-18}$$

1.2.4　刀具材料

刀具材料性能对生产率、刀具寿命和零件表面质量影响巨大。刀具的切削性能很大程度上取决于刀具材料、刀具几何形状和刀具结构。因此，应当重视刀具材料的正确选择和合理使用。

1. 刀具材料应具备的性能

（1）高的硬度
（2）高的耐磨性
（3）足够的强度和韧性
（4）高的耐热性

(4) High heat resistance

(5) Good thermal conductivity and processing properties

2. High-speed steel

High-speed steel (HSS) is a kind of alloy tool steel with W, Mo, Cr, V and other alloy elements. It has good bending strength, and its hardness under normal temperature is 62-65HRC. The high heat resistance is over 600℃. High-speed steel can be used for manufacturing complex tools in shape, such as drills, form cutters, broaches, gear cutting tools and so on.

According to chemical composition, high-speed steels are generally divided into two types: the tungsten type (T series) and the molybdenum type (M series). High-speed steels can be divided into plain high-speed steels and the high-speed steels with high performance.

(1) Plain high-speed steels Plain high-speed steels have good processing property, and are used for conventional machining of general engineering materials. Commonly used plain high-speed steels are W18Cr4V, W6Mo5Cr4V2, W14CrVMnRE and W9Mo3Cr4V.

(2) High-speed steels with high performance High-speed steels are mainly used for machining high temperature alloy, titanium alloy and stainless steel and other hard-to-machine materials. Cobalt HSS (W2Mo9Cr4VCo8) has high hardness (70HRC) and hot hardness, but it is very expensive. The aluminum HSS (W6Mo5Cr4V2Al) is a new type of high-speed steel which is developed by China. Its hot hardness, bending strength, and impact toughness are similar to that of W2Mo9Cr4VCo8. The HSS with high vanadium content (W6Mo5Cr4V3) is usually used to machine high-strength steels.

3. Cemented carbides

The best thing to have happened for metal cutting is the invention of cemented carbides around 1926 in Germany. They are produced by the cold compaction of metallic carbide (WC, TiC, TaC, NbC) powder in metallic bonding materials (Co, Mo, Ni), followed by liquid-phase sintering. Cemented carbide is one of the uppermost tool materials. The vast majority of lathe tools, face milling cutters, parts of end milling cutters, drills and reamers are made of cemented carbides. But the cemented carbides can't be widely used for making complex tools.

The hardness of cemented carbides is around 89-94HRA. The carbide has a good wear-resistance, and its cutting temperature can reach up to 800-1000℃. Therefore, the cutting speed of cemented carbides is 4-10 times higher than that of high-speed steels. Its tool life can increase dozens of times. But the cemented carbides have lower bending strength, poor toughness, worse impact toughness, shock and vibration.

According to GB/T 18376.1-2008, cemented carbides can be divided into three categories:

(1) P series (YT types) The typical symbols of P series carbides are P01, P10, P20, P30, P40 and so on. The number in the symbols stands for the contents of TiC and Co. The larger the number is, the less the TiC contents, the more the Co contents, then, the lower the wear-resistance and the higher the toughness of the carbide. P series are suitable for machining ferrous metals with long chips, such as steels, malleable cast irons and so on.

(2) M series (YW types) The typical symbols of M series carbides are M10, M20, M30,

(5) 良好的导热性和工艺性

2. 高速钢

高速钢是加入了 W、Mo、Cr、V 等合金元素的高合金工具钢。高速钢的抗弯强度较好，常温硬度为 62~65HRC。耐热性可达 600℃，可以制造刃形复杂的刀具。如钻头、成形车刀、拉刀和齿轮刀具等。

按照高速钢的化学成分，高速钢通常分为两大类：钨类（钨系）和钼类（钼系）。按切削性能，又有普通高速钢和高性能高速钢之分。

(1) 普通高速钢　普通高速钢的工艺性好，切削性能可满足一般工程材料的常规加工。常用的品种有 W18Cr4V、W6Mo5Cr4V2、W14CrVMnRE 和 W9Mo3Cr4V。

(2) 高性能高速钢　此类高速钢主要用于高温合金、钛合金、不锈钢等难加工材料的切削加工。钴高速钢、铝高速钢以及高钒高速钢等属于此类高速钢。钴高速钢（W2Mo9Cr4VCo8）具有高硬度（70HRC）和热硬性，但价格很贵。铝高速钢（W6Mo5Cr4V2Al）是我国自主研制的新型高速钢，价格低廉，其高硬度、抗弯强度、冲击韧度均与 W2Mo9Cr4VCo8 相当。高钒高速钢（W6Mo5Cr4V3）一般用于切削高强度钢。

3. 硬质合金

德国于 1926 年发明了硬质合金。硬质合金是由金属碳化物（WC、TiC、TaC、NbC 等）粉末和金属粘结剂（Co、Mo、Ni 等）经过冷挤压和液相烧结制成的。它是当今最主要的刀具材料之一。绝大部分车刀、面铣刀和部分立铣刀、深孔钻、铰刀等均已采用硬质合金制造，但用于复杂刀具尚受到很大限制。

硬质合金的硬度为 89~94HRA，相当于 71~76HRC，耐磨性好，耐热性可达 800~1000℃。因此，硬质合金比高速钢的切削速度高 4~10 倍。刀具使用寿命可提高几十倍，但其抗弯强度低、韧性差、怕冲击和振动。

根据 GB/T 18376.1—2008，常用的硬质合金可分为三类：

(1) P 系列（YT 类）　常用牌号有 P01、P10、P20、P30、P40 等，牌号中的数字越大，则 TiC 的含量越少，Co 的含量越多，其耐磨性越低而韧性越高。P 系列适合加工具有长切屑的钢铁材料，如钢、可锻铸铁等。

(2) M 系列（YW 类）　常用牌号有 M10、M20、M30、M40。牌号中的数字越大，其耐磨性越低而韧性越高。M 系列不仅适合加工钢铁材料，还适合于加工非铁（有色）金属和非金属材料，甚至还可以加工耐热钢和不锈钢等。

(3) K 系列（YG 类）　常用牌号有 K01、K10、K20、K30、K40 等，K 类硬质合金与钢的粘结温度较低，其抗弯强度与韧性比 P 类高。K 系列适合加工具有短切屑的钢铁材料，如灰铸铁、甚至淬火钢、非铁（有色）金属和非金属材料等。

以上三类硬质合金中，主要成分是 WC，故统称为 WC 基硬质合金。近年来还出现了许多新的硬质合金品种。以镍和钼为粘结相的 TiC 基硬质合金（也称金属陶瓷），适合于碳素钢、合金钢的半精加工和精加工。通过添加 Cr_2O_3 获得的超细晶粒硬质合金（粒度小于 0.5~1.0μm）具有优良的耐磨性，适用于加工冷硬铸铁、淬硬钢、不锈钢、高温合金等难加工材料。

硬质合金牌号主要根据工件材料和切削加工的类型进行选择。常用的硬质合金牌号及用途见表 1-1。

and M40. The larger the number is, the lower the wear resistance, and the higher the toughness of it. M series are suitable for machining not only ferrous metals, but nonferrous and non-metallic materials, and even heat-resistant steels and stainless steels.

(3) K series (YG types) The typical symbols of K series carbides are K01, K10, K20, K30, K40 and so on. K series carbides are not suitable for machining steels because they are easy to adhesive to steel. But its bending strength and toughness are higher than that of P series carbides. K series are suitable for machining ferrous metals with short chips, such as grey cast irons, and even hardened steels, and nonferrous metals and non-metallic materials.

All of the above three kinds of cemented carbides take WC as their main composition, so they are called WC matrix carbide alloys. A lot of new cemented carbides have come in to use in recent years. TiC matrix carbide alloys (called metal ceramics), taking TiC as its main composition and Ni or Mo as its bonder, are used for finishing and semi-finishing carbon steels and alloy steels. Ultra fine grained cemented carbides (grain size is less than 0.5-1.0μm) can be obtained by adding Cr_2O_3 to ordinary carbide alloys. They have very good wear-resistance and are suited to machine not only chilled cast iron and hardened steel, but stainless steel, high temperature alloy and other hard-to-cut materials.

The kind of cemented carbide material is selected according to workpiece materials and the types of machining. The commonly used cemented carbides and its applications are listed in Table 1-1.

4. Other tool materials

(1) Ceramics Ceramics have very high hardness (HRA91-95), chemical stability, anti-adhesiveness, heat resistance and wear resistance, and have less affinity to the workpiece. They can withstand cutting temperatures as high as 1200℃. But they are brittle, have poor bending strength and impact toughness.

Table 1-1 Common used cemented carbides and its applications 常用的硬质合金牌号及用途

Designation 牌号	Direction of increase in characteristics 性能变化方向	Application 用途
P01	Wear resistance, Cutting speed 耐磨性、切削速度 ↑ Toughness, Feed 韧性、进给量 ↓	Steel, finish cutting 钢：精车加工
P10		Steel, turning, milling, high cutting speed, with small or medium chip sections 钢：高速、中、小截面条件下的车削、铣削
P20		Steel, turning, milling, medium cutting speeds and with medium chip sections 钢：中等切速、中等切屑截面条件下的车削、铣削
P30		Steel, turning, milling, medium or low cutting speeds 钢：中速或低速条件下的车削、铣削

(续)

Designaition 牌号	Direction of increase in characteristics 性能变化方向	Application 用途
M10	Wear resistance, Cutting speed 耐磨性、切削速度 ↑ / Toughness, Feed 韧性、进给量 ↓	Steel, manganese steel, grey cast iron and alloy cast iron, turning, medium or high cutting speeds 钢、锰钢、灰铸铁和合金铸铁：中速或高速条件下的车削
M20		Steel, manganese steel, gray cast iron, turning, medium cutting speeds 钢、锰钢、灰铸铁：中速条件下的车削
M30		Steels, austenitic steel, grey cast iron, high temperature alloys, turning milling, medium cutting speeds 钢、奥氏体钢、灰铸铁、高温合金：中速条件下的车削、铣削
K01	Wear resistance, Cutting speed 耐磨性、切削速度 ↑ / Toughness, Feed 韧性、进给量 ↓	Hardened steel, high-silicon aluminum alloy, titanium alloy, turning, finish-turning, boring, milling 冷硬铸铁、高硅铝合金、钛合金：车削、精车、镗加工和铣削加工
K10		Cast, iron more than 220HBW, malleable cast iron, silicon-aluminum alloy, copper alloy, turning, drilling, boring, milling 硬度高于220HBW 的灰铸铁、可锻铸铁、硅铝合金、铜合金：车削、钻削、镗加工和铣削加工
K20		Grey cast iron, nonferrous metal, turning, boring, milling 灰铸铁、有色金属：车削、镗削和铣削
K30		Grey cast irons with low hardness, low tensile steel, turning, milling 低硬度灰铸铁、低强度钢：车削和铣削

4. 其他刀具材料

（1）陶瓷　陶瓷有很高的硬度和耐磨性，耐热性可达1200℃以上，常温硬度达91～95HRA，化学稳定性好，但最大弱点是抗弯强度低，韧性差。

（2）金刚石　金刚石分天然和人造两种。金刚石是目前最硬的物质（10000HV），金刚石刀具既能胜任陶瓷、硬质合金等高硬度非金属材料的切削加工，又可切削其他非铁（有色）金属及其合金。

（3）立方氮化硼（CBN）　立方氮化硼是由六方氮化硼在高温高压下加入催化剂转变而成的。它的特点是：硬度为8000～9000HV，耐磨性好、热稳定性高，可耐1300～1500℃的高温，具有良好的导热性和较低的摩擦因数。

(2) Diamond Diamonds are divided into natural and artificial diamonds. Diamond is the hardest material (10000HV) up to now. A diamond tool is qualified with machining not only high hardness and wear resistant materials such as carbide alloy, ceramics, etc., but also non-ferrous metals.

(3) Cubic boron nitride (CBN) The hardness of CBN is HV8000-9000. It has good wear-resistance and high thermal stability. It can withstand the cutting temperature in the range of 1300-1500℃, and has a good thermal conductivity and low friction coefficient.

1.3 Basic theory of the metal cutting process

Metal cutting is a process in which the cutting tool is used to remove the excess metal in the form of chips from a rough blank so as to generate finished surface. In metal cutting process, a series of physical phenomena, such as built-up edge, cutting force, cutting heat and temperature, surface hardening and tool wear and so on, would occur. These phenomena have direct influence on the machining quality, productivity and economic benefit. The study of these phenomena and its varying rule has important significance to understand the technology characteristics of mechanical processing methods, and to guarantee the machining quality, reduce the production cost and improve productivity.

1.3.1 Deformation in the metal cutting process

1. Formation of chip and deformation zones

When metal is compressed, the internal stress and strain of metal materials are generated along the direction inclined 45° to the stress plane. The shear stress increases as the load increasing. When the shear stress reaches the material yield limit, the metal is sheared and slips along the direction of 45° and ultimately fractures. Therefore, metal cutting is a process during which the workpiece cutting layer under the extrusion of the tool produces plastic deformation and chips. Now we take the plastic material as an example to analyze the chip formation and deformation during metal cutting.

A lot of experiments and theoretical analyses showed that the chip generating process in metal cutting process is the deformation process of metal in the cutting layer. When the tool and the workpiece begin to contact, the workpiece is extruded by the cutting edge and the rake face, then elastic deformation takes place. With the cutting going on, the extrusion effect increases and gradually the shearing stress in the material reaches to yield strength τ_s, then the cutting layer begins to slip along the direction of the maximum shearing stress, the plastic deformation takes place. Line OA is called the initial sliding line, as shown in Fig. 1-12. Take the point P in cutting layer as an example to analyze the formation of typically continuous chips. When the point P in the cutting layer approaches gradually to the cutting edge and reaches the position 1, the shearing stress in the material at this point reaches to the yield strength. As the point 1 moves forward and slides along line OA simultaneously, finally it will flow to point 2 because of resultant movement. The distance between 2′ and 2 is the slippage. With the slide going on, the shear stress will increase gradually until it comes up to the maximum shear stress at point 4. The metal layer being cut is segregated from parent work material,

1.3 金属切削过程的基本理论

金属切削过程就是利用切削刀具从毛坯上切除多余的金属使之变为切屑并产生光洁表面的过程。在金属切削过程中，发生了一系列的物理现象，如积屑瘤、切削力、切削热、表面硬化和刀具的磨损等。这些现象对切削加工质量、生产率和经济效益等都有直接的影响。研究这些现象及其变化规律，对于认识各种机械加工方法的工艺特点，保证加工质量，降低生产成本和提高生产率，都具有十分重要的意义。

1.3.1 金属切削过程中的变形

1. 切屑的形成及变形区的划分

金属材料受压其内部产生应力应变，大约与受力方向成 45°的斜面内，切应力随载荷增大而逐渐增大，当切应力达到材料的屈服极限时，金属即沿着 45°方向产生剪切滑移，最终导致破坏。因此，金属的切削过程，实质上是工件切削层在刀具的挤压下，产生以剪切滑移为主的塑性变形，从而形成切屑的过程。现以塑性材料为例说明切屑的形成及切削过程中的变形情况。

大量的切削实验和理论分析表明，金属切削过程中的切屑形成过程就是切削层金属的变形过程。如图 1-12 所示，在刀具和工件开始接触的最初瞬间，切削刃和前面在接触点挤压工件，使工件内部产生应力和弹性变形。随着切削过程的继续，切削刃和前面对工件材料的挤压作用不断增加，使工件材料内部的应力和变形逐渐增大，当应力达到材料屈服强度 τ_s 时，被切削层的金属开始沿切应力最大的方向滑移，产生塑性变形，图 1-12 中的 OA 面就代表"始滑移面"。以被切削层中的 P 点为例，当 P 到达点 1 位置时，由于 OA 面上的切应力达到材料的屈服强度，则点 1 在向前移动的同时也沿 OA 面滑移，其合成运动将使点 1 流动到点 2。点 2′到点 2 之间的距离就是它的滑移量。随着滑移的产生，切应力将逐渐增加。当 P 移动到点 4 的位置时，应力和变形都达到了最大值，被切削层与本体金属分离，从而形成切屑沿前面流出。OM 线代表"终滑移面"。整个切屑形成过程经历了挤压、滑移、挤裂、切离四个阶段。

在 OA 到 OM 之间的区域称为第一变形区（Ⅰ）——主要剪切变形区，主要呈现塑性变形。实验证明，第一变形区的厚度随切削速度增大而变薄。在一般速度下，第一变形区的厚度仅为 0.02~0.2mm。因此可用一个平面 OO' 表示第一变形区（Ⅰ），如图 1-13 所示。

当切屑沿前面流出时，如果前面与切屑底层金属的摩擦相当大，使切屑底层金属又一次产生塑性变形，即第二次变形。第二变形区（Ⅱ）又称第二剪切变形区。它实际上是前面上的内摩擦区。切屑底层与前面之间的强烈摩擦，对切削力、切削热、积屑瘤的形成与消失，以及对刀具的磨损等都有直接影响。

第三变形区（Ⅲ）是已加工表面上的挤压与摩擦区。已加工表面受到切削刃钝圆部分和后面的挤压与摩擦，产生变形与回弹，造成已加工表面的纤维化和加工硬化。第三变形区对已加工表面的质量和刀具后面的磨损都有很大的影响。

and flows away along the tool face. Therefore, line OM is termed as final sliding line. The whole process of chip formation goes through four stages, that is, extruding, sliding, cracking and segregating.

The area between the line OA and the line OM is called the first deformation zone (Ⅰ)—primary shear deformation zone, showing plastic deformation. Experiment shows that with the increase of the cutting speed, the thickness of the undeformed chip in zone Ⅰ becomes thinner. In ordinary cutting speed range, the width of deformation zone Ⅰ is very small (about 0.2-0.02mm). Therefore it can be represented approximately by the shear plane OO' shown in Fig. 1-13.

When the deformed metal (called chip) flows over the tool face, if the friction between the tool face and the underside of the chip (deformed material) is considerable, then the chip gets further deformed, which is termed as secondary deformation. The second deformation zone (Ⅱ) is called secondary shear deformation zone. Actually it is an inner friction deformation zone on the tool face. The sever friction in this region has direct influence on the formation and disappear of the cutting forces, cutting heat, built-up edge and tool wear.

The third deformation zone (Ⅲ) is called extrusion and friction zone on the finished surface. As the machined surface withstands the extrusion and friction from the dull round cutting edge and flank, it would produce not only plastic deformation but elastic deformation, causing the machined surface fibering and hardening. Zone Ⅲ has a significant impact on the quality of machined surface and the wear of tool flank.

2. Chip types

Based on the properties of materials to be cut and the cutting conditions, usually there are four types of chips, as shown in Fig. 1-14.

(1) Ribbon chip   As shown in Fig. 1-14a, the ribbon chip appears to be smooth and continuous. They are formed easily when the plastic metals, such as carbon steel, alloy steel, copper alloy and aluminum alloy and so on, are cut with smaller cutting thickness at high cutting speed by means of the cutter with larger rake angle.

(2) Cracked chip   As shown in Fig. 1-14b, the cracked chip has clearly defined stair-like sides and a smooth undersurface. It is formed easily when the plastic metals are cut under the cutting conditions of lower cutting speed, larger undeformed chip thickness and smaller rake angle.

(3) Unit chip   It is formed easily when the metals with poor plasticity are cut under the cutting conditions of lower cutting speed, larger undeformed chip thickness, and smaller rake angle. As the shear stress in the shear plane is more than the ultimate strength of the material to be cut, the chips are generated in the form of single segments (namely, unit chips), as shown in Fig. 1-14c.

(4) Splintering chip   When brittle materials such as cast iron, brass copper are cut, the material directly ahead of the tool is compressed and has not undergone plastic deformation but breaks into irregular pieces, as shown in Fig. 1-14d. The harder and more brittle the material to be cut, the larger the undeformed chip thickness, the easier the splintering chip is produced.

3. Built-up edge (BUE)

When metals are cut at a medium cutting speed, it usually brings about a phenomenon that a

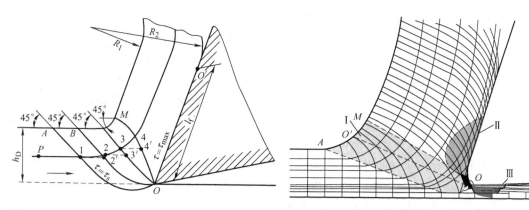

Fig. 1-12　Shearing slide in the first region of deformation
第一变形区金属的滑移

Fig. 1-13　Three deformation zones 三个变形区

2. 切屑的类型

在切削加工中，由于工件材料不同，通常产生四种类型的切屑。如图 1-14 所示。

（1）带状切屑　如图 1-14a 所示，带状切屑连续不断呈带状。采用较高的切削速度、较小的切削厚度和前角较大的刀具，切削塑性较好的金属材料时，易形成带状切屑。

（2）节状切屑　如图 1-14b 所示，节状切屑的外表面呈锯齿状并带有裂纹，但底部仍然相连。采用较低的切削速度、较大的切削厚度和前角较小的刀具，切削中等硬度的塑性材料时，易形成节状切屑。

（3）单元切屑　当采用极低的切削速度，大的切削厚度，小的前角，切削塑性较差的材料时易形成单元切屑。切削层金属在塑性变形过程中，剪切面上产生的切应力超过材料的强度极限，切屑沿剪切面完全断开，形成形状相似而又互相分离的屑块，如图 1-14c 所示。

（4）崩碎切屑　切削脆性材料时，由于材料的塑性小，抗拉强度低，切削层金属在产生弹性变形后，几乎不产生塑性变形而突然崩裂，形成形状极不规则的碎块，如图 1-14d 所示。工件材料越是硬脆，切削厚度越大时，越容易形成崩碎切屑。

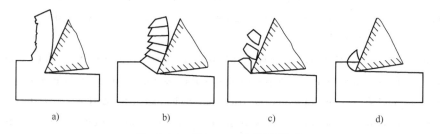

Fig. 1-14　Chip types 切屑的类型
a）Ribbon chip 带状切屑　b）Cracked chip 节状切屑
c）Unit chip 单元切屑　d）Splintering chip 崩碎切屑

3. 积屑瘤

在以中速切削金属材料时，经常在前面上靠刃口处粘接一小块很硬的金属楔块，并代替切削刃进行切削，如图 1-15 所示。这个金属楔块称为积屑瘤。在积屑瘤的生成过程中，它的高度不断增加，但由于切削过程中的冲击、振动、负荷不均匀及切削力的变化等原因，会

piece of metal adheres firmly to the tool face and can cut the workpiece instead of the cutting edge. This piece of metal is called built-up edge (BUE), as shown in Fig. 1-15. The built-up edge increases in size until it becomes so large that it may break off because of the shock, uneven load and the change of cutting force. Then it will grow again. So the built-up edge is unstable.

As shown in Fig. 1-15, the BUE adhered to the tool rake face near the tool tip increases the actual rake angle of the tool, which is beneficial to protect the cutting tool and reduce the cutting force. The increment Δh_D of cutting thickness would increase with the increase of BUE height. Once BUE is broken or removed from the rake face, Δh_D would decrease rapidly. The variation of cutting thickness is certain to cause cutting force fluctuation. The fluctuation of cutting force would cause vibration. Part of BUE adhered to the machined surface causes the finished surface to be rough. Though the BUE on rake face can act as cutting edge, its frequent breaking off from the tool face would speed up the tool wear and shorten tool life. Therefore, BUE should be avoided in finish machining.

4. Methods to restrain the formation of built-up edge

There are several methods to control or avoid built-up edge.

1) To avoid using the speeds at which built-up edge is easily formed. The condition for built-up edge formation mainly depends on the cutting speed when the workpiece material is given, as shown in Fig. 1-16.

2) To reduce the plasticity of material to be cut.

3) To select proper cutting fluid.

4) To increase the rake angle of cutting tool and to improve the sharpening quality of the cutting tool.

1.3.2 Cutting forces and cutting power

In the cutting process, the cutting force has direct influence on the generation of cutting heat, and further on tool wear, tool life and machined surface quality. In production, the cutting force is the important basis for the calculation of the cutting power, and the design and application of machine tools, cutting tools and fixture.

1. Sources of the cutting force, resultant force as well components of the cutting force

The cutting force comes from two aspects: one is the resistance of material to elastic and plastic deformation (F_{rN} and F_{aN}); the other is the frictional resistance (F_{rf} and F_{af}) exerted by the chip and the workpiece upon the tool point, as shown in Fig. 1-17. The resultant force F of these forces mentioned above is called the total cutting force which acts on the rake face near the cutting edge.

Take the cylindrical turning for example, as shown in Fig. 1-18, the resultant force F is decomposed into three components of forces.

Primary cutting force F_c—the component of the resultant force along the direction of the cutting speed v_c. It is also called the tangential cutting force. It is the principal basis used to calculate the power required in the design of the machine tool and to check the strength and stiffness of machine parts.

出现整个或部分积屑瘤破裂、脱落及再生成的现象。因此，积屑瘤的存在是不稳定的。

如图 1-15 所示，粘接在前面上的积屑瘤增大了刀具的实际前角，这有助于保护刀具和减小切削力。切削厚度的增量 Δh_D 会随着积屑瘤高度 H_b 的增加而增加。一旦积屑瘤折断或从前面上脱落，Δh_D 会迅速减小。切削厚度的变化必然引起切削力大小的波动，从而引起振动。部分积屑瘤粘接在已加工表面上，使加工过的表面变得粗糙。尽管前面上的积屑瘤可以充当切削刃，但它从前面上的频繁脱落反而会加快刀具磨损，缩短刀具寿命。因此，精加工时应避免产生积屑瘤，以确保加工质量。

4. 控制积屑瘤产生的措施

在生产实践中常采用以下措施来抑制或消除积屑瘤。

1）避开容易产生积屑瘤的切削速度范围。当工件材料一定时，切削速度是影响积屑瘤的主要因素，如图 1-16 所示。

2）降低被加工材料的塑性。

3）合理使用切削液。

4）增大刀具前角，提高刀具刃磨质量。

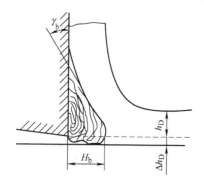

Fig. 1-15　Built-up edge 积屑瘤

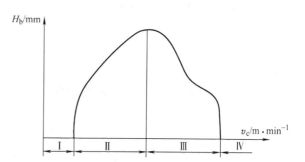

Fig. 1-16　Effect of cutting speed on BUE
切削速度对积屑瘤的影响

1.3.2　切削力与切削功率

在切削过程中，切削力决定着切削热的产生，并影响刀具磨损和已加工表面质量。在生产中，切削力又是计算切削功率，设计和使用机床、刀具、夹具的重要依据。

1. 切削力的来源、合力及其分力

切削力主要来源于两个方面：一是材料的弹、塑性变形所产生的抗力；二是刀具与切屑、工件表面间的摩擦阻力，如图 1-17 所示。上述这些力的合力 F 称为总切削力，它作用在切削刃附近的前面上。

以车削外圆为例（图 1-18），将合力 F 分解为三个互相垂直的分力。

主切削力 F_c——切削合力在主运动方向上的分力，与切削速度方向一致，又称切向力。该力是计算机床所需功率和校验工件刚度的主要依据。

背向力 F_p——切削合力在垂直于工作平面上的分力，又称径向力。作用在基面内，与进给方向垂直，其与主切削力的合力会使工件发生弯曲变形或引起振动，进而影响工件的加工精度和表面粗糙度。

Back force F_p—the component of the resultant force F perpendicular to the feed direction in the tool reference plane. It is also called radial cutting force. It will easily cause the deformation of the workpiece and bring about the vibration of the workpiece. Furthermore, it will also affect the machining accuracy and surface roughness.

Feed force F_f—the component of the resultant force F parallel to the feed direction in the tool reference plane. It is also called axial force. It may be used to check the strength and the stiffness of the feeding mechanism.

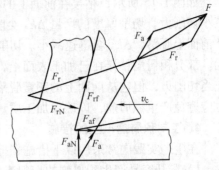

Fig. 1-17 Forces acting on the cutting tool
作用在刀具上的力

The relation between the resultant force and the components of the cutting forces is as follows:

$$F = \sqrt{F_c^2 + F_p^2 + F_f^2} \quad (1\text{-}19)$$

$$F_p = F_D \cos\kappa_r ; F_f = F_D \sin\kappa_r \quad (1\text{-}20)$$

Generally, F_c is the largest component of the resultant force, the next is F_p, the least is F_f.

$$F_p = (0.15 \sim 0.7)F_c$$

$$F_f = (0.1 \sim 0.6)F_c$$

2. Cutting power

Cutting power can be expressed as $P_c(\text{kW})$:

$$P_c = F_c v_c \times 10^{-3} \quad (1\text{-}21)$$

where, F_c—main cutting force(N);

v_c—cutting speed(m/s).

The power P_E required by the machine motor is:

$$P_E \geq \frac{P_c}{\eta_m} \quad (1\text{-}22)$$

where, η_m—machine transmission efficiency ($\eta_m = 0.75 \sim 0.85$).

3. Specific cutting force

Specific cutting force $K_C(\text{N/mm}^2)$ means the main cutting force in a certain removal unit area. It is calculated by the following equation:

$$K_C = F_c/A_D \quad (1\text{-}23)$$

where, A_D—nominal cross sectional area of cutting layer(mm^2), $A_D = a_p f$.

The main cutting force F_c can be calculated by the following equation:

$$F_c = K_C A_D = K_C a_p f \quad (1\text{-}24)$$

4. Factors of influencing on the cutting force

There are many factors which have an influence on cutting force, such as workpiece materials, cutting variables and geometric parameters of the cutting tool and so on.

(1) Workpiece material The effect of the work material on the cutting force depends upon the

进给力 F_f——切削合力在进给方向上的分力,又称轴向力。它是校验进给机构强度的主要依据。

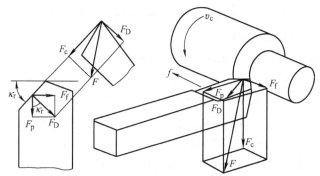

Fig. 1-18　Resultant forces and components of cutting forces　车削合力及分力

合力与分力之间的关系如下

$$F = \sqrt{F_c^2 + F_p^2 + F_f^2} \tag{1-19}$$

$$F_p = F_D \cos\kappa_r;\quad F_f = F_D \sin\kappa_r \tag{1-20}$$

一般情况下,F_c 最大,F_p 次之,F_f 最小。

$$F_p = (0.15 \sim 0.7)F_c$$

$$F_f = (0.1 \sim 0.6)F_c$$

2. 切削功率

切削功率 P_c(单位为 kW)可以表示为

$$P_c = F_c v_c \times 10^{-3} \tag{1-21}$$

式中　F_c——主切削力(N);
　　　v_c——切削速度(m/s)。

机床电动机所需功率 P_E 应满足

$$P_E \geqslant \frac{P_c}{\eta_m} \tag{1-22}$$

式中　η_m——机床传动效率,一般取 $\eta_m = 0.75 \sim 0.85$。

3. 单位切削力

单位切削力是指单位面积上的主切削力,用 K_C 表示(单位为 N/mm²)。

$$K_C = F_c / A_D \tag{1-23}$$

式中　A_D——切削层公称横截面积(mm²),$A_D = a_p f$。

主切削力 F_c 为

$$F_c = K_C A_D = K_C a_p f \tag{1-24}$$

4. 影响切削力的因素

影响切削力的因素很多,如工件材料、切削用量、刀具几何参数等。

(1) 工件材料　工件材料对切削力的影响取决于材料的剪切强度、材料的塑性变形程

shear strength of the materials, the degree of plastic deformation and the friction between the work and the cutting tool. The harder and the stronger of the materials, the higher the shear strength, the larger the cutting force. Comparing with steel 45, F_c increases by 4% when cutting steel 60, and decreases by 13% when cutting steel 35.

The higher the plasticity and toughness of material, the larger the cutting deformation, the larger the cutting force. For example, the elongation percentage of stainless steel 1Cr18Ni9Ti is 4 times higher than that of steel 45. Because of its large cutting deformation, difficulty in chip-breaking, and severe working hardening in machining, F_c increases by 25% comparing to machining steel 45. Again, the hardness of gray cast iron HT200 is close to that of steel 45, but F_c produced in cutting gray cast iron decreases by 40% comparing to cutting steel 45.

(2) Effect of cutting variables on cutting force

1) Effect of feed (f) and depth of cut (a_p). The cutting force F_c increases with the increase of f and a_p, but the influence degree of f and a_p on F_c is different. Experiments show that the influence index of a_p on F_c is 1, and that of f on F_c is 0.75. It means that the influence of depth of cut (a_p) on the cutting force is larger than that of feed (f).

2) Effect of cutting speed v_c. When v_c increases within the range of low cutting speed (5 ~ 20m/min), the height of BUE increases gradually, the cutting force F_c decreases. When v_c continues to increase within the range of 20 ~ 35m/min, BUE disappears gradually, the cutting force F_c increases. When v_c is over 35m/min, as the cutting temperature goes up, the frictional coefficient decreases, F_c decreases gradually and then tends to a stable value.

When cutting brittle metals, as the cutting deformation and frictional coefficient are less, the effect of the cutting speed on the cutting force is not clear.

(3) Effect of geometric parameters of the cutting tool on the cutting force

1) Effect of the rake angle γ_o. The rake angle γ_o has a great influence on cutting force. As γ_o increases, the cutting force decreases due to the decrease of the cutting deformation. The variation of rake angle has the largest influence on F_f, and the smallest influence on F_c. The influence of the rake angle on the cutting force when cutting the metals with higher plasticity is more significant than that generated in cutting the metals with lower plasticity.

2) Effect of the cutting edge angle κ_r. The variation of the cutting edge angle κ_r would lead to the change of cross-section area of undeformed chip A_D and the components of cutting force F_p and F_f. It can be seen from Fig. 1-19 that with the increase of κ_r within the range of 30° ~ 60°, as the undeformed chip thickness h_D increases and the cutting deformation decreases, the primary cutting force F_c decreases. When κ_r increases within the range of 60° ~ 90°, the influence of the round nose on the cutting process increases gradually. This increases the chip-tool friction and makes the chip flow more difficult, thus causing the cutting force F_c to increase gradually.

As the cutting edge angle κ_r within 60° ~ 75° can reduce the cutting forces F_c and F_p, the lathe cutters with $\kappa_r = 75°$ are widely used in production. When turning the long shafts, the lathe cutters with larger κ_r ($\kappa_r > 60°$) are often used in order to reduce F_p.

度以及刀具的几何参数。工件材料硬度、强度越高,其剪切强度就越高,那么切削力就越大。与切削 45 钢相比,切削 60 钢时 F_c 增加了 4%,而切削 35 钢时 F_c 减小了 13%。

工件材料的塑性、韧性越大,发生的变形也越大,所以切削力也越大。例如,1Cr18Ni9Ti 不锈钢的伸长率比 45 钢高 4 倍。由于这种钢的切削变形大、断屑难、加工硬化严重,所以 F_c 比切削 45 钢增大了 25%。同理,灰铸铁 HT200 的硬度与 45 钢不相上下,但其切削力 F_c 比切削 45 钢时下降了 40%。

(2) 切削用量对切削力的影响

1) 进给量 f 和背吃刀量 a_p 的影响。a_p 和 f 的增加均会使切削力增大,但两者的影响程度不同。实验研究表明,a_p 对 F_c 的影响指数为 1,而 f 对 F_c 的影响指数为 0.75。这就意味着背吃刀量 a_p 对切削力的影响比进给量 f 的影响大。

2) 切削速度 v_c 的影响。当切削速度 v_c 在低速范围(5~20m/min)内逐渐增加时,随着积屑瘤高度的渐渐增加,切削力 F_c 减小。当切削速度 v_c 在 20~35m/min 的范围内持续增加时,由于积屑瘤渐渐消失,切削力 F_c 又逐渐增大。当切削速度 v_c 大于 35m/min 时,随着切削温度升高,摩擦因数下降,切削力 F_c 逐渐减小,之后趋于某一稳定值。

当切削脆性材料时,由于切削变形和摩擦因数都很小,切削速度对切削力的影响不明显。

(3) 刀具几何参数对切削力的影响

1) 前角 γ_o 的影响。前角 γ_o 对切削力的影响很大。前角 γ_o 增加,由于切削变形减小,切削力减小。前角 γ_o 的变化对 F_f 的影响最大,而对 F_c 的影响最小。一般加工塑性较大的金属时,前角对切削力的影响比加工塑性较小的金属时更为显著。

2) 主偏角 κ_r 的影响。主偏角 κ_r 的变化会导致切削层面积 A_D 以及切削分力 F_p 和 F_f 的变化。由图 1-19 可知,当主偏角在 30°~60°范围内增大时,切削厚度增加,切削层变形减小,故主切削力 F_c 减小。当主偏角在 60°~90°范围内增大时,刀尖圆弧对切削过程的影响逐渐增大,使刀具与切屑之间的摩擦加剧,使得切屑流出更难,从而使主切削力 F_c 增大。

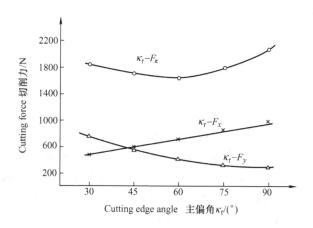

Fig. 1-19　Influence of the cutting edge angle κ_r on cutting force 主偏角 κ_r 对切削力的影响

由于在 60°~75°范围内的主偏角 κ_r 可以减小切削力 F_c 和 F_p,因此 $\kappa_r = 75°$ 的车刀在生产中得到广泛应用。车削轴类零件,尤其是细长轴时,为了减小径向分力 F_p 的作用,往往采用较大主偏角($\kappa_r > 60°$)的车刀切削。

1.3.3 Cutting heat and cutting temperature

1. Generation and transmission of cutting heat

Cutting heat comes from elastic deformation and plastic deformation resulting from metal cutting layers, as shown in Fig. 1-20. Suppose that the power consumed by primary motion is converted to heat, the cutting heat generated in unit time can be calculated as follows:

$$Q = F_c v_c \tag{1-25}$$

where, Q—cutting heat produced in unit time (J/s);

F_c—primary cutting force (N);

v_c—cutting speed (m/s).

The ways to conduct cutting heat are chip, workpiece, cutting tool and ambient medium (such as air, cutting fluid and so on). The main factors influencing on heat conduction are the thermal coefficient of both the workpiece and the cutting tool, as well as the ambient mediums.

2. Cutting temperature and its influencing factors

The average temperature in the contact region between the chips and the rake face is called cutting temperature. Cutting temperature depends mainly on the relationship between heat generation and heat dissipation in unit time. The main influence factors on the cutting temperature are as follows:

(1) Workpiece materials

The higher the hardness and strength of the workpiece material, the more power is consumed in the cutting process, the more heat is generated, and the higher is the cutting temperature. The larger the plasticity of the workpiece material, the higher the cutting temperature. The larger the thermal conductivity of the workpiece, the easier the heat dissipation, and the lower the cutting temperature.

(2) Cutting variables

The experimental formula of the cutting temperature θ obtained by the experiment is as follows:

$$\theta = C_\theta v_c^{x_\theta} f^{y_\theta} a_p^{z_\theta} \tag{1-26}$$

where, C_θ—coefficient related to experimental conditions (see Tab. 1-2);

$x_\theta, y_\theta, z_\theta$—influence indexes of cutting variables v_c, f, a_p on the cutting temperature (see Tab. 1-2).

It can be seen from Tab. 1-2 and Eq. (1-26) that v_c, f, a_p have different effect degree on the cutting temperature. Cutting speed v_c has the largest effect on cutting temperature, the next is feed f, the least is the depth of cut a_p. Cutting temperature has direct effect on the tool wear and tool life. According to the above theory, the use of larger a_p and f is more favorable than the use of higher cutting speed in order to improve the tool life and control the cutting temperature in a suitable range.

(3) Geometric parameters of the cutting tool

1) Influence of the rake angle γ_o on the cutting temperature. With the increase of the rake angle γ_o, the unit cutting force would decrease, causing the reduction of cutting heat generated. Therefore, the cutting temperature drops. But overlarge rake angle γ_o is unfavourable to reducing the cutting temperature, because large rake angle γ_o would lead to the decrease of wedge angle, i.e. the decrease of radiating volume.

1.3.3 切削热与切削温度

1. 切削热的产生与传出

切削热来源于切削层金属产生的弹性变形和塑性变形所做的功,如图 1-20 所示。假定主运动所消耗的功全部转化为热能,则单位时间内产生的切削热可由下式算出

$$Q = F_c v_c \tag{1-25}$$

式中 Q——单位时间内产生的切削热(J/s);
F_c——主切削力(N);
v_c——切削速度(m/s)。

切削热传散出去的途径主要是切屑、工件、刀具及周围的介质(如空气、切削液等)。影响热传导的主要因素是工件和刀具材料的导热系数以及周围介质的情况。

2. 切削温度及其影响因素

切削温度一般是指切屑与前面接触区域的平均温度。切削温度的高低受单位时间内产生的热量与传散的热量两方面的综合影响。影响切削温度的主要因素如下。

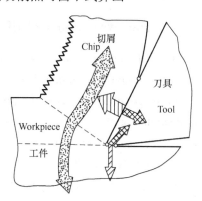

Fig. 1-20 Sources of cutting heat
切削热的来源

(1) 工件材料 工件材料的硬度和强度越高,切削时消耗的功越越多,产生的切削热越多,切削温度越高。工件材料的塑性越大,切削温度越高。工件材料的热导率越大,热量越易传出,切削温度就越低。

(2) 切削用量 通过实验获得切削温度 θ 的实验公式为

$$\theta = C_\theta v_c^{x_\theta} f^{y_\theta} a_p^{z_\theta} \tag{1-26}$$

式中 C_θ——与实验条件有关的系数,见表 1-2;
x_θ、y_θ、z_θ——切削用量 v_c、f、a_p 对切削温度的影响指数,见表 1-2。

由表 1-2 和式(1-26)可知,切削用量三要素 v_c、f、a_p 对切削温度的影响程度不一样,v_c 最大,f 次之,a_p 最小。切削温度对刀具磨损和刀具寿命有直接影响。由上述规律可知,为控制切削温度,提高刀具使用寿命,选用大的 a_p 和 f 比选用大的切削速度有利。

Table 1-2 Coefficients and indexes of cutting temperature 切削温度的系数和指数

Workpiece material 工件材料	Tool Material 刀具材料	Coefficient 系数	Index 指数		
		C_θ	x_θ	y_θ	z_θ
45 steel 45 钢	W18Cr4V	140~170	0.35~0.45	0.2~0.3	0.08~0.1
Gray Cast iron 灰铸铁	W18Cr4V	120	0.5	0.22	0.04
45 steel 45 钢	YT15	160~320	0.26~0.41	0.14	0.04
Titanium alloy 钛合金	YG8	429	0.25	0.1	0.019

(3) 刀具几何参数

1) 前角 γ_o 对切削温度的影响。前角增大时,单位切削力下降,使产生的切削热减少,切削温度下降。但是过大的前角会使刀具楔角减小,从而减小刀具的散热体积,对降低切削温度不利。

2) Influence of the cutting edge angle κ_r on the cutting temperature. The decrease of κ_r would lead to the increase of cutting width and tool nose angle. This situation can improve the heat dissipation condition of the cutting tool and reduce the cutting temperature.

Besides, the tool wear would cause the increase of the cutting temperature. Using cutting fluid can reduce the cutting temperature and increase the tool life.

1.3.4 Tool wear and tool life

When the cutting tool is used to cut the metal in cutting process, it wears gradually. When the cutting tool wears to a certain degree, it will become failure. Abnormal tool wear will increase the tool consumption and decrease the productivity. Therefore, the analysis of tool wear mechanism has an important significance for selecting cutting conditions reasonably, using the cutting tool correctly and determining the tool life.

1. Tool wear forms

Tool wear can be divided into the normal wear and abnormal wear. The normal wear refers to the phenomenon that the cutter material particles on the contact surface between the cutter and the workpiece or chips are taken away by chips or workpiece. The abnormal wear, also named as breakage, refers to the tool tipping, cutting edge curling, cracking, flaking caused by impact, or vibration, or thermal effect.

The normal wear includes the following forms:

(1) Rake face wear (Crater wear) As shown in Fig. 1-21b, c, the crater wear is denoted by KT.

(2) Flank wear As shown in Fig. 1-21a, the wear is denoted by VN. In the middle part of the cutting edge (B), the wear is homogeneous and the average value is denoted by VB.

(3) Wear on both rake face and flank The tool wear on both rake face and flank occurs simultaneously. It usually appears when the plastic metal is machined and the thickness of undeformed chip is moderate ($h_D = 0.1$-0.5mm).

2. Causes of tool wear

Cutting tool wear is caused by mechanical and thermochemistry effects under the conditions of high temperature and high pressure. The main causes of tool wear are as follows:

(1) Abrasive wear Abrasive wear occurs in metal cutting when some small hard particles on the friction surface of the chips or workpiece slide over the surface of the tool to lead to grooves. Actually, abrasive wear is the main cause of tool wear on low cutting speed.

(2) Adhesive wear Adhesive wear is a phenomenon of cold welding caused by adsorption affinity between molecules on the new surface of a friction pair between the tool and the workpiece. If the two surfaces undergo relative sliding motion, the welded points at the interface will be broken and adhesive wear is produced. Adhesive wear depends on the temperature and chemical components of the workpiece and tool material (affinity function of elements). When cutting at low and medium cutting speed, adhesive wear is the main cause of tool wear.

(3) Diffusion wear The phenomenon that chemical elements of tool materials and workpiece

2) 主偏角对切削温度的影响。主偏角减小时,致使切削宽度增大,刀尖角增大,刀具散热条件改善,有利于降低切削温度。

另外,刀具磨损会使切削温度升高。使用切削液有助于降低切削温度,提高刀具寿命。

1.3.4 刀具磨损与刀具寿命

金属切削过程中,刀具在切除金属的同时,其本身也逐渐被磨损。当磨损到一定程度时,刀具便失去切削能力。刀具磨损增加刀具消耗、降低生产率。分析刀具磨损机理对合理选择切削条件,正确使用刀具及确定刀具使用寿命具有重要意义。

1. 刀具的磨损形式

刀具的磨损形式可分为正常磨损和非正常磨损两大类。刀具正常磨损是指在刀具与工件或切屑的接触面上,刀具材料的微粒被工件或切屑带走的现象。若由于冲击、振动、热效应等原因致使刀具崩刃、卷刃、断裂、表层剥落而损坏称为非正常磨损或刀具的破损。

刀具的正常磨损形式一般有以下几种:

(1) 前面磨损(月牙洼磨损) 如图 1-21b、c 所示,月牙洼磨损量以其深度 KT 表示。

(2) 后面磨损 如图 1-21a 所示,该区的磨损量用 VN 表示。在切削刃的中部(B 区),其磨损均匀,并以平均磨损值 VB 表示。

(3) 前、后面同时磨损 当切削塑性金属时,如果切削层公称厚度适中($h_D = 0.1 \sim 0.5$mm),则经常发生这种磨损。

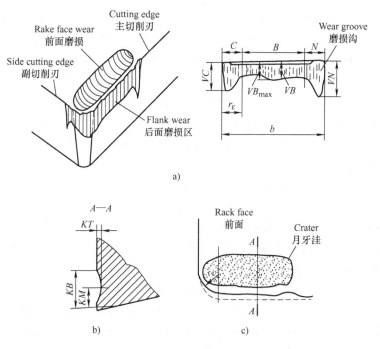

Fig. 1-21 Schematic diagram of tool wear 刀具磨损形式示意图

2. 刀具的磨损原因

刀具是在高温和高压下受到机械的和热化学的作用而发生磨损的,主要原因如下:

materials occur mutual diffusion at high temperature is called diffusion wear. Diffusion wear depends mainly on the temperature in tool-work interface.

(4) Oxidizing wear When the cutting temperature reaches to 700-800℃, Co, WC and TiC of the carbide alloy combine with O_2 in the air, generating the oxidizing films (such as CoO, WO_2, TiO_2) with low hardness and strength. These oxides are loosely joined with the tool material and are easily rubbed away by chip or workpiece, causing the oxidation wear of tool.

To sum up, tool wear is determined by the cutting temperature. Diffusion wear and oxidizing wear are mainly the major causes of tool wear at high temperature. Within low-medium cutting temperature regions adhesive wear is sometimes the major causes, but abrasive wear exists in all kinds of cutting temperature.

3. Tool wear process and wear criteria

When the tool wear is up to a certain volume—wear criteria, the tool should be replaced or resharpened.

(1) Tool wear process The changing process of flank wear VB with time is called tool wear process. Usually, it is divided into three stages, as shown in Fig. 1-22.

1) Initial wear stage. After sharpening, as the tool surface roughness is quite large, the surface is easy to wear, so the wear land is formed quickly in this stage. The degree of the initial wear depends on the tool sharpening quality.

2) Normal wear stage. Due to the rugged parts on the tool surface have been ground, so the amount of the tool wear steadily increases as time goes by. In the normal wear stage, the wear curve is basically a straight line inclined upwards, and its slope denotes the intensity of tool wear. Wear intensity is one of the important indexes in evaluating the tool cutting performance. Normal wear stage is the effective work stage of the tool.

3) Severe wear stage. When tool wear VB developed to a certain degree, the tool wear is aggravated with the cutting force and cutting temperature increasing respectively. In order to guarantee the machining quality and reduce the tool consumption, the cutting tool should be resharpened or replaced before this wear stage.

(2) Tool wear criteria The wear limit which determines the cutting tool should be re-sharpened is called wear criterion. Based on ISO standards, "wear criterion" means the allowable maximum value of the average wear in the middle part of the wear zone and is denoted by VB. The tool wear criterion depends on the tool materials, workpiece materials and cutting conditions.

4. Empirical formula of tool life

(1) Tool service life (tool life) In practice, in order to judge tool wear situation easily, quickly and accurately, tool life is usually used to reflect the wear criterion indirectly. Tool life is the time in which a new sharpened cutting tool is used actually for cutting from the start of cut to the wear limits, denoted by T, and its unit is s(or min.).

(2) Relationship between cutting variables and tool life

1) v_c—T relationship. Based on his experimental work, Taylor proposed the formula for tool life in 1906.

(1) 磨料磨损　磨料磨损也称机械磨损，由于切屑或工件的摩擦面上有一些微小的硬质点，能在刀具表面刻划出沟纹，这就是磨料磨损。低速切削时磨料磨损是主要原因。

(2) 粘结磨损　粘结磨损是由工件与刀具组成摩擦副的新鲜表面上分子间的吸附力引起的冷焊现象。由于有相对运动，刀具上的微粒被对方带走而造成磨损。粘结磨损与切削温度有关，也与刀具及工件两者的化学成分有关（元素的亲和作用）。在中、低切削速度下，粘结磨损往往是刀具磨损的主要原因。

(3) 扩散磨损　扩散磨损是刀具材料和工件材料在高温下化学元素相互扩散而造成的磨损。扩散磨损主要决定于接触面之间的温度。

(4) 氧化磨损　当切削温度达 700~800℃ 时，空气中的氧与硬质合金中的钴及碳化钨、碳化钛等发生氧化作用，产生较软的氧化物（如 CoO、WO_2、TiO_2）被切屑或工件摩擦掉而形成的磨损称为氧化磨损。

综上所述，切削温度对刀具磨损起决定性的影响。高温时主要出现扩散和氧化磨损，中低温时，粘结磨损占主导地位，磨料磨损则在不同的切削温度下都存在。

3. 刀具的磨损过程及磨钝标准

刀具磨损到一定程度（磨钝标准）就需要换刀或重磨。

(1) 刀具的磨损过程　磨损过程就是后面磨损量 VB 随时间 t 增长的变化过程，如图 1-22 所示。

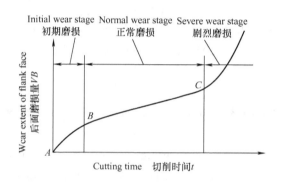

Fig. 1-22　Typical tool wear curve 典型的刀具磨损曲线

1) 初期磨损阶段。这一阶段磨损较快。这是因为刀具在刃磨后，刀具表面粗糙度值较大，表层组织不耐磨所致。初期磨损量的大小与刀具刃磨质量有很大关系。

2) 正常磨损阶段。由于刀具表面高低不平之处已被磨去，压强减小，磨损缓慢，这一阶段磨损曲线基本上是一条直线，其斜率代表刀具正常工作时的磨损强度。磨损强度是比较刀具切削性能的重要指标之一。正常磨损阶段也是刀具的有效工作阶段。

3) 剧烈磨损阶段。刀具磨损量 VB 增长到一定程度时，切削力增大，切削温度升高，刀具磨损加剧。生产中为保证质量，减少刀具消耗，应在这阶段之前及时重磨或更换刀具。

(2) 刀具的磨钝标准　刀具磨损到一定限度就不能再继续使用，这个磨损限度称为磨钝标准。国际标准 ISO 统一规定，刀具磨钝标准是指后面磨损带中间部分平均磨损量允许达到的最大值，以 VB 表示。制订磨钝标准时主要根据刀具材料、工件材料和加工条件的具体情况而定。

$$v_c T^m = A \tag{1-27}$$

where, A is the coefficient related to experiment condition; m is an exponent depending on the characteristics of the workpiece and tool material. Its magnitude reflects the influence degree of v_c on T. The larger the value of m, the lower the influence of v_c on T.

2) f—T, a_p—T relationship. In the same way, the relationship between f and T, a_p and T can be found by cutting tests.

$$f T^n = B \tag{1-28}$$

$$a_p T^p = C \tag{1-29}$$

Test shows that the larger the feed or depth of cut, the lower the tool life. But the effect of feed on T is larger than that of depth of cut.

3) Equation of tool life. Synthesizing the relationship of v_c, f, a_p to tool life T, the equation of tool life can be written as follows:

$$T = \frac{C_T}{v_c^{1/m} f^{1/n} a_p^{1/p}} \tag{1-30}$$

or

$$v_c = \frac{C_v}{T^m f^{y_v} a_p^{x_v}} \tag{1-31}$$

where, C_T and C_v represent the coefficients related to workpiece material, tool material and other cutting conditions; x_v and y_v represent indexes, $x_v = m/p$, $y_v = m/n$.

When turning the carbon steels with $R_m = 0.637$ GPa by means of carbide tool (YT5), $1/m = 5$, $1/n = 2.25$, $1/p = 0.75$, Eq. (1-30) can be expressed as follows:

$$T = \frac{C_T}{v_c^5 f^{2.25} a_p^{0.75}} \tag{1-32}$$

It can be seen from Eq. (1-32) that the effect of v_c on T is the largest, the next is f, and a_p is the least.

(3) Determination of tool life When using cutting tool in production, a rational tool life T should be determined firstly. Based on the predetermined T, the cutting speed v_c is calculated or selected, and then the cutting efficiency and cost are calculated. At present, there are two methods used to determine the tool life.

1) Tool life T_p with the maximum productivity. The tool life determined based on the shortest operation time is called the tool life with the maximum productivity, expressed in T_p.

$$T_p = \left(\frac{1-m}{m}\right) t_c \tag{1-33}$$

where, t_c — time required by tool changing once.

The cutting speed v_{cp} with the maximum productivity corresponding to T_p is:

$$v_{cp} T_p^m = A \tag{1-34}$$

【Discussion】 It can be known from Eq. (1-33) that if the tool life selected is larger than T_p, it means the cutting variables are too smaller. This is unfavourable for increasing the productivity. On the other hand, if the tool life selected is smaller than T_p, it means the cutting variables are too larger. This would increase the time for the tool resharpening and tool setting, and the productivity will

4. 刀具寿命经验公式

（1）刀具使用寿命　在生产实际中，为了更方便、更快速、更准确地判断刀具的磨损情况，一般以刀具的使用寿命来间接地反映刀具的磨钝标准。刀具的使用寿命是指新刃磨的刀具从开始切削一直到磨损量达到磨钝标准时的切削时间，用符号 T 表示，单位为 s（或 min）。

（2）切削用量与刀具寿命的关系

1）v_c 与 T 的关系。泰勒根据自己所做的切削实验，于1906年提出了刀具寿命的计算公式，即

$$v_c T^m = A \tag{1-27}$$

式中　A——与实验条件有关的系数；

m——取决于刀具和工件材料的指数，其大小反映切削速度对刀具寿命的影响程度。m 值大，表明切削速度对刀具寿命的影响小，即刀具的切削性能较好。

2）f 与 T，a_p 与 T 的关系。同理，通过切削实验可以得出进给量 f 与刀具寿命 T，切削深度 a_p 与刀具寿命 T 的关系。即

$$f T^n = B \tag{1-28}$$

$$a_p T^p = C \tag{1-29}$$

实验表明，进给量与切削深度越大，刀具寿命越短。但进给量对刀具寿命的影响比切削深度对刀具寿命的影响大。

3）刀具寿命方程。综合切削用量三要素与刀具寿命的关系，刀具寿命方程可以写成如下形式

$$T = \frac{C_T}{v_c^{1/m} f^{1/n} a_p^{1/p}} \tag{1-30}$$

或

$$v_c = \frac{C_v}{T^m f^{y_v} a_p^{x_v}} \tag{1-31}$$

式中　C_T 和 C_v——与工件材料、刀具材料和其他切削条件有关的系数；

x_v 和 y_v——指数，$x_v = m/p$，$y_v = m/n$。

当用 YT5 硬质合金车刀切削 $R_m = 0.637\text{GPa}$ 的碳素钢时，$1/m = 5$，$1/n = 2.25$，$1/p = 0.75$，则式（1-30）可写成

$$T = \frac{C_T}{v_c^5 f^{2.25} a_p^{0.75}} \tag{1-32}$$

可见，切削速度 v_c 对刀具寿命的影响最大，进给量 f 次之，背吃刀量 a_p 最小。

（3）刀具寿命的确定　当在实际生产中使用切削刀具时，应当首先确定一个合理的刀具寿命。然后根据预先确定的刀具寿命，就可以计算或选择切削速度，进而计算切削效率和成本。目前，确定刀具寿命的方法有两种。

1）最高生产率刀具寿命。根据单件工序工时最短的原则确定刀具寿命，称为最高生产率刀具寿命，用 T_p 表示。

be decreased too.

2) Tool life T_c with the minimum production cost. The tool life determined based on the lowest operation cost is called the tool life with the minimum production cost, expressed in T_c. T_c is also named as economic tool life.

$$T_c = \frac{1-m}{m}\left(t_c + \frac{C_t}{M}\right) \qquad (1\text{-}35)$$

where, C_t — the required cost due to sharpening tool once (including the expenses of tool, grinding wheel and labor cost);

M — the machining cost allocating the factory expenses per minute to this operation.

The cutting speed v_{cc} corresponding to T_c is:

$$v_{cc} T_c^m = A \qquad (1\text{-}36)$$

【Discussion】 It can be known from Eq. (1-35) that if the tool life selected is larger than T_c, that means the cutting variables are relative small, cutting efficiency is low, and the economic effects would not be high. Besides, as the machining time is overlong, the machine expense would increase.

It can be seen by comparing Eq. (1-33) with Eq. (1-35) that T_c is larger than T_p, that is, $v_{cc} < v_{cp}$. The lower the tool cost C_t, the nearer T_c to T_p.

Generally, T_c is usually used as the tool life standard in production. But in the emergency period or under special conditions, the tool life can also be determined in light of the maximum productivity. It can be seen from Eq. (1-35) that the cutting tools with higher manufacturing cost should have longer tool life than other cutting tools. For example, the tool life of gear cutters and broaching tools is longer than that of ordinary lathe cutters and drills. In addition, the cutting tools used on the transfer line, aggregate machine tools and CNC machine tools should have higher tool life; Those forming tools, or the tools for finishing large-size workpieces should also have higher tool life.

(4) Other factors affecting tool life As stated above, the cutting variables have an influence on the tool life. Of the three cutting variables, the cutting speed has the largest influence on the tool life. Besides, there are many other factors which affect the tool life. Generally speaking, all the factors which affect the cutting temperature would have a influence on tool life in the same way. It is necessary to point out that the performance of the cutter material has very large effect on the tool life. The higher the hot hardness of tool material is, the better the wear resistance of it, so the longer the tool life. But under impact cutting or heavy cutting conditions, or when cutting the hard-to-machine materials, the main influencing factors on tool life are the strength and impact toughness of the tool material. The higher the strength and toughness of tool material are, the longer the tool life is.

1.4 Applications of basic theory in the cutting process

The application of the basic laws in the metal cutting will be introduced in this section. Main issues include how to control the chip, how to improve the machinability of the work materials, how

$$T_{\text{p}} = \left(\frac{1-m}{m}\right)t_{\text{c}} \tag{1-33}$$

式中 t_{c}——换刀一次所需的时间。

与 T_{p} 相对应的最大生产率的切削速度 v_{cp} 为

$$v_{\text{cp}} T_{\text{p}}^m = A \tag{1-34}$$

【讨论】由式（1-33）可知，如果所选择的刀具寿命大于 T_{p}，那就意味着切削用量过小，不利于提高生产率；另一方面，如果所选择的刀具寿命小于 T_{p}，那就意味着切削用量过大，这会增加刀具的刃磨和调整时间，从而也会引起生产率的降低。

2）最低生产成本刀具寿命。根据单件工序成本最低的原则确定的刀具寿命，称为最低生产成本刀具寿命，又称经济刀具寿命，用 T_{c} 表示

$$T_{\text{c}} = \frac{1-m}{m}\left(t_{\text{c}} + \frac{C_{\text{t}}}{M}\right) \tag{1-35}$$

式中 C_{t}——换刀一次所需的费用，包括刀具、砂轮消耗和工人工资等；

　　　M——该工序单位时间内所分担的全厂开支。

与 T_{c} 相对应的经济切削速度 v_{cc} 为

$$v_{\text{cc}} T_{\text{c}}^m = A \tag{1-36}$$

【讨论】由式（1-35）可知，如果所选择的刀具寿命大于 T_{c}，那就意味着切削用量较小，切削效率低，经济效益也就不会好。此外，由于加工时间过长，加工成本也就随之升高。

对比式（1-33）和式（1-35），可知 $T_{\text{c}} > T_{\text{p}}$，即 $v_{\text{cc}} < v_{\text{cp}}$。刀具成本 C_{t} 越低，则 T_{c} 越接近 T_{p}。

一般情况下应采用最低生产成本刀具寿命 T_{c}。当遇到紧急任务或特殊情况时，则可采用最高生产率刀具寿命。从最低生产成本刀具寿命的公式看出，制造成本高的刀具应有较长的刀具寿命。例如，齿轮刀具、拉刀的寿命应比普通车刀、钻头的寿命长。此外，在自动线、组合机床和数控机床上使用的刀具应有较长的刀具寿命；成形刀具或用于精加工大型工件的刀具也应具有较长的刀具寿命。

(4) 影响刀具寿命的其他因素　如上所述，切削用量影响刀具寿命。在切削用量三要素中，对刀具寿命影响最大的是切削速度。此外，还有许多其他因素对刀具寿命也有影响。一般说来，凡是影响切削温度的因素都会对刀具寿命造成影响。有必要指出的是，刀具材料的性能对刀具寿命有很大的影响。刀具材料的高温硬度越高，耐磨性越好，刀具寿命也越高。但在有冲击切削、重型切削和难加工材料切削时，影响刀具寿命的主要因素是冲击韧度和抗弯强度。韧性越好，抗弯强度越高，刀具寿命越长。

1.4　金属切削基本规律的应用

本节将要介绍金属切削基本规律的应用。讨论的主要问题包括如何控制切屑，如何改进工件材料的切削加工性，如何选择切削液，如何确定刀具的几何参数，以及如何选择切削用量等。

to select the cutting fluid, how to determine the tool geometric parameters, how to select the machining variables and so on.

1.4.1 Control of chips

In high speed turning, the control and disposal of chips are important to the operator and cutting tools. Long curling chips snarl about the workpiece and machine. Their sharp edges and high tensile strength make their removal from the cutting area difficult and hazardous (dangerous), particularly when the machine is in operation.

After the chip is deformed severely in the deformation zones I and II, its hardness increases and its plasticity decreases. When it encounters some barriers such as the flank, transiting surface or surface to be cut and so on in flowing away from parent material, if the strain of a part of chip exceeds the rupture strain value of the chip, the chip would be broken. Fig. 1-23 indicates the fracture situation when the chip collides with the workpiece or tool flank.

Researches showed that the larger the brittleness of the material, and the larger the chip thickness, the smaller the chip curl radius, the easier the chip breaking off. The following measures can be taken in practice to control chip.

(1) Chip breaker The chip breaker means the groove, step machined, or block clamped on the rake face, as shown in Fig. 1-24. It is provided to control the ribbon-like chips that are formed at high cutting speeds. The chip breaker curls and highly stresses the chips and causes it to break into small pieces so that they are easily removed from the machine.

There are three types of chip breakers on the rake face of inserts, as shown in Fig. 1-25: Fig. 1-25a is parallel type, Fig. 1-25b is inclined type to outside, and Fig. 1-25c is inclined type to inside. The second type can break the chip in the shape of "C" or "6" in the wide cutting variable range. It has the broadest chip breaking range. The next is the parallel type. The third type often breaks the chip in the form of long and tight spiral roll and has narrow chip breaking range, which is suitable for semi-finishing and finishing with small depth of cut.

(2) Proper selection of tool angles The cutting edge angle κ_r is the main influencing factor on chip breaking. The increase of the cutting edge angle κ_r can make the cutting thickness increase, which makes for breaking chip. The lathe cutters with good chip breaking effect in practice often have larger κ_r (60°-90°). To decrease the rake angle γ_o can enhance chip deformation and makes it easy to break the chip. To change the inclination angle λ_s can control the direction of the curled chip.

(3) Change of cutting variables Of the three cutting variables, feed f has the largest influence on chip breaking, the next is the depth of cut a_p, the smallest one is the cutting speed v_c. The increase of f can increase the chip thickness and is favorable to chip breaking, but it would increase the surface roughness. The decrease of v_c can increase cutting deformation and is also favorable to chip breaking, but it would decrease the material removal efficiency. Therefore, cutting variables should be selected properly according to the practical condition.

1.4.1 切屑的控制

在高速车削时,切屑的控制与排除对于操作工和刀具都是很重要的。长长卷曲的切屑会缠绕工件和机床。它们锋利的刃口和高的抗拉强度增加了从切削区切离的难度和危险,当机床运转时尤其如此。

当切屑在第Ⅰ和第Ⅱ变形区出现严重变形时,其硬度升高而塑性下降。它在离开母材向外流出的过程中碰到一些如后面、过渡表面或待加工表面等障碍后,如果部分切屑的应变超过切屑的断裂应变值,切屑就会折断。图1-23所示为当切屑碰到工件或刀具后面后的折断情况。

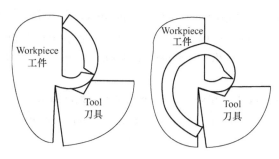

Fig. 1-23 Chip fracture after colliding with the workpiece or tool flank
切屑碰到工件或刀具后面后而折断

研究表明,材料的脆性越大,切屑的厚度越大,切屑的卷曲半径越小,切屑就越容易折断。在生产实践中可采取以下措施来控制切屑。

(1) 断屑器 断屑器是指在刀具前面上做出的槽、台阶或固定的一个附加挡块,如图1-24所示。它是用来控制高速切削形成的带状切屑。断屑器使切屑卷曲并产生很大的应力,从而使它断成碎片以便于排屑。

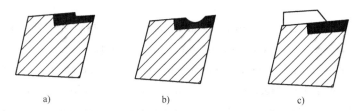

Fig. 1-24 Chip breakers used on single-point tools 单刃刀具上使用的断屑器
a) Step type 台阶式   b) Groove type 槽式   c) Clamp type 挡块式

图1-25所示为刀片前面上三种类型的断屑器:平行式(图1-25a);向外倾斜式(图1-25b);向内倾斜式(图1-25c)。其中第二种类型的断屑器可以在很大的切削用量范围内使切屑断成"C"形屑或"6"形屑。它的断屑范围最宽,其次是平行式。第三种类型的断屑器常把切屑断成既长又紧的螺旋卷状,但其断屑范围较窄,适合于切削深度小的半精和精加工。

(2) 正确选择刀具角度 主偏角 κ_r 是影响断屑的主要因素。增加主偏角 κ_r 可以增加切屑厚度,从而有利于断屑。在生产实际中具有良好断屑效果的车刀常具有较大的主偏角 κ_r

1.4.2 Machinability

Chips may be cut from some materials with relative ease and from others with the greater difficulty. This difference may be attributed to the machinability of the respective materials. Machinability is defined as the degree of difficulty with which the material is cut satisfactorily for the purpose intended. In general, good machinability is associated with the removal of material with moderate cutting force, the formation of rather small chips, longer tool life and good surface finish. It is commonly observed that the materials with high hardness have poor machinability because of high temperature, large power consumption and rapid tool wear. However, the hardness alone would not be able to specify the machinability, since it also depends on the other factors such as ①chemical composition of material; ② micro-structure of material; ③ physical properties such as tensile strength, ductility and hardness of the workpiece material; ④rigidity of tool and work holding devices; ⑤cutting variables, etc.

1. Primary indexes used for assessing material machinability

1) The tool life or cutting speed permitted under a certain tool life and relative machinability.

$$K_r = v_i / v_s \tag{1-37}$$

where, K_r—relative machinability index;

v_i—cutting speed of material to be investigated when tool life is T. $T = 60$ min for plain metallic materials, or $T = 20$ min for hard-to-machine materials;

v_s—cutting speed of material standard steel for 60 min(for plain metallic materials) tool life, or 20 min(for hard-to-machine materials) tool life.

For example, taking the magnitude of v_{60} of steel 45(170~225HBW, $R_m = 0.637$GPa) as the reference and designating it v_{060}, then the ratio of v_{60} of other materials to be machined to the v_{060} is named as relative machinability.

$$K_r = v_{60} / v_{060}$$

If $K_r > 1$, it means the work material investigated is easier to cut than steel 45; If $K_r < 1$, it means the work material investigated is more difficult to cut than steel 45.

2) Surface roughness. Under the same machining conditions, the grades of surface roughness can be compared. The smaller the surface roughness, the better the machinability of the work material.

3) Cutting force. The larger the cutting force generated in cutting, the more the power consumption, the worse the machinability of the work material.

4) Ease in chip control. The easier to control chip flow direction and break chip, the better the machinability of the work material.

2. Measures to improve the machinability of work materials

1) To regulate the chemical composition of materials. The machinability of plain carbon steels falls steadily as its carbon content increases. Steel up to the hardness with 300HBW do not present great machining difficulty unless large amounts of alloy elements are added.

The addition of small amount of certain elements (for example up to 0.1% S or up to 0.2% Pb) can improve the machinability of steels without appreciable changing mechanical properties.

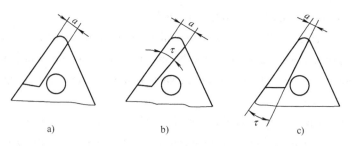

Fig. 1-25 Three types of chip breakers 三种类型的断屑器
a) Parallel type 平行式   b) Inclined type to outside 向外倾斜式   c) Inclined type to inside 向内倾斜式

（60°～90°）。减小刀具前角可以增大切屑变形，从而使得断屑容易。改变刃倾角的大小可以控制切屑的流出方向。

（3）改变切削用量 在切削用量三要素中，进给量 f 对断屑影响最大，其次是切削深度 a_p，而切削速度 v_c 的影响最小。进给量 f 增大可使切削厚度增大，容易断屑，但它会增大表面粗糙度值。切削速度 v_c 降低使切削变形增大，也有利于断屑，但会降低金属切除率。因此切削用量应根据实际情况进行正确选择。

1.4.2 切削加工性

在切削加工中，有些材料容易切削，有些材料却很难切削。这种差异是由材料各自的切削加工性决定的。材料的切削加工性是指材料加工满足所要求指标的难易程度。一般来讲，良好的切削加工性意味着切除材料时的切削力适中，切屑很小，刀具寿命长和良好的表面粗糙度。观察表明，高硬度材料的切削加工性差，因为在加工时易产生高温、消耗功率大和刀具磨损快。然而，仅凭材料硬度是不能确定其切削加工性好坏的，还取决于其他一些因素，如：①材料的化学成分；②材料的金相组织结构；③材料的抗拉强度、塑性、硬度等物理力学性能；④刀具与工件夹持装置的刚性；⑤切削用量等。

1. 评定材料切削加工性的主要指标

1）刀具寿命或一定刀具寿命下允许的切削速度指标和相对加工性。生产中通常采用相对加工性来衡量工件材料的切削加工性。

$$K_r = v_i / v_s \tag{1-37}$$

式中 K_r——相对切削加工性指标；

v_i——刀具寿命为 T 时所研究材料的切削速度。对于一般金属材料，$T=60\mathrm{min}$；对于难加工材料，$T=20\mathrm{min}$；

v_s——刀具寿命为 T 时基准钢材的切削速度。对于一般金属材料，$T=60\mathrm{min}$；对于难加工材料，$T=20\mathrm{min}$。

例如，以 45 钢（170～225HBW，$R_m=0.637\mathrm{GPa}$）的 v_{60} 为基准，写作 v_{060}，其他被切削的工件材料的 v_{60} 与之比 K_r 称为相对加工性，即

$$K_r = v_{60} / v_{060}$$

如果 $K_r > 1$，说明所考察的材料比 45 钢易切削；若 $K_r < 1$，则比 45 钢难切削。

2）表面粗糙度。在相同的加工条件下可以比较表面粗糙度等级。表面粗糙度值越小，

These kinds of steels are labeled "free cutting steels".

2) Proper heat treatment. To transform the microstructure of metallic materials by means of heat treatment is a primary method to improve the machinability. For instance, the machinability of low carbon steel can be improved by normalizing, whereas the machinability of high carbon steel can be improved by spheroidizing.

3) To select the structure state of the material. The machinability of low carbon steel in the cold drawing state is better than that in the natural state. The machinability of hot rolled blanks is better than that of forged blanks.

1.4.3 Cutting fluids

1. Introduction to cutting fluids and their functions

The medium in which the cutting process takes place is called the cutting fluid. A cutting fluid may be defined as any liquid, solid or gas which is applied to the workpiece material or to the cutting tool, to assist in the cutting operation and which facilitates easier machining.

When cutting fluids were first introduced they were to assist in preventing plain carbon steel tools from overheating, which caused the cutting edge to soften and wear quickly. Keeping the cutting tool in low temperature resulted in a much longer tool life. As there had to be considerable cooling, water was the traditional main ingredient in the cutting fluid, with soda added to prevent machine corrosion. Since then there has been considerable progress in both tool technology and the development of cutting fluids.

2. Functions of cutting fluids

Cutting fluids have to perform a variety of tasks in several different ways. Their duties may be divided into four areas dealing with cooling, lubricating, washing and surface finishing.

(1) Cooling Overheating of the cutting area may produce the following difficulties: ①Expansion of the workpiece leading to measuring difficulties and possible errors in sizes of the workpieces; ②Warping caused if the heating is in one place along the work, producing local distortions; ③Welding between the work and the center in the lathe or grinding machine as the job expands and frictional welding occurs; ④Reduction of tool life.

The cooling action of the cutting fluid consists in abstracting the heat generated during the cutting process. The cooling properties of the cutting fluid depend to a large extent on the cooling method. The cooling efficiency increases with heat conductivity and specific heat of the cutting fluid. Cooling effect increases, too, with the flow rate of the fluid, the area of the application and with the decreasing temperature of the cutting fluid.

(2) Lubrication As the tool is forced through the workpiece there are areas of high friction between the swarf and the tool face. Lubrication helps to reduce this friction, and has the following advantages: ①Reducing the frictional heating effect; ②Lowering the chip's resistance to sliding, which means less force on the tool; ③Lessening the power used in cutting; ④Reducing the tool wear, as metal-to-metal contact is avoided.

(3) Washing Cutting fluids help to wash away swarf from the cutting area, which has the fol-

工件材料的切削加工性越好。

3）切削力。切削中产生的切削力越大，消耗的功率就越大，工件材料的切削加工性越差。

4）切屑控制的难易程度。控制切屑的流向和断屑越容易，工件材料的切削加工性越好。

2. 改善工件材料切削加工性的途径

1）调整材料的化学成分。普通碳素钢的切削加工性随着其含碳量的增加而下降。硬度在 300HBW 以下的钢加工难度不大，除非添加大量的合金元素。

添加一些少量的化学元素（如 0.1% S 或 0.2% Pb）可以改善钢的切削加工性，而其力学性能不会发生明显变化。这种钢材被冠名为"易切钢"。

2）适当的热处理。通过热处理改变金属材料的金相组织结构是改善其切削加工性的主要方法。例如，低碳钢可以通过正火、而高碳钢可以通过球化处理改善各自的切削加工性。

3）选择材料的结构状态。冷拔低碳钢的切削加工性比正火状态的要好。热轧毛坯的切削加工性比锻件的好。

1.4.3 切削液

1. 切削液的概念

切削过程发生的介质称为切削液。凡是作用于工件材料或切削刀具以辅助切削和使切削更容易的任何液体、固体或气体都可以称为切削液。

首次在切削中采用切削液旨在防止普通碳素钢刀具产生过热，过热会引起切削刃软化，从而加速刀具磨损。始终保持刀具处于低的切削温度可延长刀具寿命。由于水有相当好的冷却效果，所以水是切削液的主要成分，其中添加苏打是为了防止机床腐蚀。自那以后，刀具技术和切削液的研制都取得了相当大的进步。

2. 切削液的功用

切削液必须以几种不同的方式执行各种各样的任务。其职责可以分为四个方面，即冷却、润滑、清洗和使表面光洁。

（1）冷却　过热的切削区会引起一系列不良影响：①工件热膨胀会导致测量困难和工件尺寸误差；②工件受热不均匀，使局部出现变形，引起工件翘曲；③当工件膨胀和出现摩擦焊接现象时，工件与车床或磨床顶尖之间产生粘接；④缩短刀具寿命。

切削液的冷却作用在于其吸收切削过程中产生的切削热。切削液的冷却性能很大程度上取决于冷却方法。冷却效率随着切削液导热性和比热容的增加而提高，冷却效果也随着切削液流动速度、施加面积的增加以及切削液温度的降低而增强。

（2）润滑　由于刀具强行切入工件，在刀具前面和切屑之间有很大的剧烈摩擦区。润滑有助于减少摩擦，并且还具有以下优点：①减小摩擦热效应；②降低切屑滑动阻力，即意味着可减小切削力；③减小切削消耗的功率；④由于避免了金属与金属的直接接触，从而减小刀具磨损。

（3）清洗　切削液有助于冲洗切削区的切屑，从而具有以下好处：①易于监视切削过程，从而易于注意到切削过程是否正常；②由于在钻削、铣削、锯削或磨削中由切屑堵塞引起刀具破裂的机率小，故切削安全。

lowing advantages: ①It is easier to observe the cutting action, so that incorrect cutting can be easily noticed; ②It is safer since there is less chance of tooling breakage caused by the swarf clogging during drilling, milling, sawing or grinding.

(4) Surface finishing Because of the lubricating action, there is a smoother cutting action resulting in: ①A better surface finish on the work, since there is less chance of tearing occurring; ②Less chance of the forming of BUE which may lead to eventual tool failure; ③More accurate workpieces, especially when cutting screw threads, as the surfaces are much smoother.

3. Types of cutting fluids

There are three basic types of cutting fluids used in metal cutting. They are water-based emulsions, straight mineral oils and mineral oils with additives.

(1) Water-based emulsions Pure water is by far the best cutting fluid available because of its highest heat-carrying (high specific heat) capacity. Besides, it is cheap and easily available. Its low viscosity makes it flow at high speed through the cutting fluid system and also penetrates the cutting zone. However, water corrodes the work material very quickly, particularly at high temperature prevalent in the cutting zone; the machine tool parts on which water splashes would be corroded.

Therefore, other materials should be added to water to improve its wetting characteristics, rust inhibitors and other additives to improve lubrication characteristics. These are also called water soluble oils. The concentrated oil is normally diluted in water to any desired concentration.

(2) Straight mineral oils The term straight when applied to lubricants and coolants means undiluted. These are pure mineral oils without any additives. Their main function is lubrication and rust prevention. They are chemically stable and lower in cost. However their effectiveness as cutting fluids is limited and therefore they are normally used for light machining operations such as lathes and single spindle automatics where free cutting brasses and steels are being machined.

(3) Mineral oils with additives (neat oils) This is by far the largest variety of cutting fluids available. A number of additives have been developed which when added to the mineral oils would produce the desirable characteristics for different machining situations. Many difficult-to-machine situations would be eased by using these cutting fluids. These are generally named as neat oils.

The additives generally improve the loading capacity as well as chemical activity. The blended oils are mixtures of mineral and fatty oils. They are suitable for heavier duty operations such as threading. Extreme pressure (EP) cutting oils are mineral oils containing extreme pressure additives like sulphur, chlorine or a combination of both. When both chlorine (up to 3%) and sulphur (up to 5%) are presented in the mineral oil, they give the oil extreme pressure and are suitable for severe cutting operations on strong and tough materials such as stainless steels and nickel alloys. In broaching operation also these oils are quite commonly used.

4. Selection of cutting fluids

The cutting fluid should be carefully selected. It is observed that each metal being machined and even each type of machining has its optimum cutting fluid. The selection of a specific cutting fluid depends on the following factors: ①cutting variables; ②workpiece material; ③tool material; ④expected tool life; ⑤cost of the cutting fluid and so on.

（4）表面光洁　由于润滑作用，切削过程平稳，从而导致：①由于工件表面撕裂的机率很小，表面光洁；②不易形成导致刀具失效的积屑瘤；③由于表面非常光洁，加工的工件更精确，切削螺纹时尤其如此。

3. 切削液的种类

在金属切削中有三种基本类型的切削液。它们分别是水基乳化液、纯矿物油和具有添加剂的矿物油。

（1）水基乳化液　纯水是目前可以得到的最好的切削液，因为其具有最强的散热能力。此外，它价格便宜，随时可用。其低浓度使其能快速流过切削液系统，且能穿透切削区域。然而，水腐蚀零件很快，尤其是在切削区高温情况下；机床上水能溅到的零部件也会被腐蚀。

因此，常在水中添加一些其他材料去提高其湿润性、抗锈蚀能力，以及别的任何添加材料以提高其润滑性。这种切削液也称水溶油。这种浓缩油可以用水稀释到所要求的浓度。

（2）直接矿物油　当"直接"这个词用于冷却剂和润滑剂时指的是未稀释过的。这是没有任何添加成分的纯矿物油。其主要功能是润滑和防锈。化学性能稳定，成本低。但是作为切削液，其效力有限，因此它们通常用于轻载切削，如在车床和单轴自动机床上加工易切削黄铜和钢。

（3）具有添加剂的矿物油（净油）　这是目前可利用的品种最多的切削液。已研制出了名目繁多的添加材料。当把它们添加到矿物油中后，可以在不同的加工情况下获得理想的性能。使用这些切削液可使许多难加工材料变得易加工。这些切削液通常称为净油。

添加材料通常可以提高切削液的承载能力和化学稳定性。调和油是矿物油与动物油的混合物。它们适用于像车螺纹那样的较重负荷的切削。极压切削油是含有像硫、氯或其混合物等极压添加剂的矿物油。如果矿物油中氯的含量达3%、硫含量达5%，可以给油赋予极压，因而适用于加工如不锈钢和镍合金等强度高、韧性大的材料。在拉削工序中，这种油也很常用。

4. 切削液的选择

要慎重选择切削液。研究发现，每一种被加工的金属，甚至每一种类型的加工都有其最佳的切削液。切削液的具体选择取决于下列因素：①切削用量；②工件材料；③刀具材料；④期望的刀具寿命；⑤切削液的成本等。

（1）加工性质　由于粗加工会产生大量的切削热，应当优先选择冷却性能好的切削液。在精加工中应选择润滑性能好的切削液，以提高表面质量并延长刀具寿命。

（2）工件材料　尽管在加工铸铁之类的脆性材料时有时会使用可溶于水的油性乳化液，但常常是干切削。铝合金和铜合金具有较好的切削加工性，其中一些可以进行干切削。通常采用极压乳化液或极压切削油，但不应用含硫的切削液。对于软钢和低中碳钢，可采用乳白色乳化液或极压切削油。极压切削油或甚至某些乳化液可以用来加工高碳钢和合金钢。当加工不锈钢或耐热合金时，应采用含有高浓度添加剂氯的高性能净油，但应避免使用硫添加剂。

（3）刀具材料　高速钢刀具通常都有较好的热硬性。对于一般加工可采用水基切削液。对于切削力大的场合，最好使用极压净油。大部分用于螺纹切削、铰削、拉削和剃削的刀具都是用高速钢制造的，而且切削速度低；刀具引导部分与已加工表面产生剧烈摩擦。因此，

(1) Machining nature As a large amount of cutting heat would be generated in rough machining, the cutting fluids with good cooling property should be selected as a priority. In finishing the cutting fluids with good lubricating property should be selected so as to improve the surface quality and prolong the tool life.

(2) Workpiece material Brittle materials like grey cast iron are often machined dry although water-soluble oily emulsions are sometimes used. Aluminum alloys and copper alloys have better machinability, and some of them could be machined dry. Generally the soluble oils or oils with inactive EP additive may be used. But the sulphurized oils should not be applied because of the corrosion to them. For the mild steel and low or medium carbon steels, milky type soluble oil or EP cutting oil could be used. EP cutting oil or even some milky soluble oils can be used for machining high carbon steels and alloy steels. When machining stainless steels or heat resistant alloys, high performance neat oils with high concentration of chlorinated additives should be used, but sulphur additives should be avoided.

(3) Cutting tool material HSS tools generally have better hot hardness. For general machining water based cutting fluids can be used. For heavy cutting EP neat oils are preferable. For screw cutting, reaming, broaching and shaving, most of the cutting tools are made of HSS and always work at a low cutting speed; the friction between the leading part and the machined surfaces is very severe. Therefore, EP cutting oils with good lubricity or EP emulsions with high concentration of oils should be used.

Carbide tools have a better heat resistance and can be used dry at a low cutting speed. But an emulsion or water solution with low concentration may be used if necessary, and it must be poured continuously and sufficiently, otherwise the carbide blade will be cracked because of sudden cooling.

1.4.4　Proper choice of tool geometric parameters

Tool geometric parameters include the tool angle, shape of cutting edge, types of rake face and flank, etc. Whether the choice of tool geometric parameters is rational has an important influence on the tool service life, machining quality, productivity and machining cost.

1. The choice of the rake angle and types of rake faces

The rake angle is one of the most important parameters. It has a direct influence on the cutting force, cutting temperature, cutting power, and further on the tool life and even surface quality.

(1) The choice of the rake angle γ_o To increase the rake angle can decrease the cutting deformation, thus decrease the cutting force, cutting heat and cutting power, and improve the tool service life. It can also restraint the build up edge (BUE) and improve the surface quality. But the overlarge rake angle would weaken the strength of the cutting edge and heat dissipation, and even cause tipping. Therefore, the increase of the rake angle has both advantages and disadvantages. The rake angle must have a rational value under a specific cutting condition.

The general principle in the selection of the rake angle is to choose larger rake angle as far as possible on condition that the tool life can be ensured. The factors should be considered as follows:

应当采用具有良好润滑性能的极压切削油或具有高浓度油的极压乳化液作为切削液。

硬质合金刀具有较好的耐热性，在低速加工时可以不使用切削液。必要时可以采用低浓度的乳化液或水溶液，并且必须连续和充分地浇注，否则，硬质合金刀片就会因急冷而出现裂纹。

1.4.4 刀具几何参数的合理选择

刀具几何参数包括刀具角度、切削刃的形状、前面和后面的形式等。刀具几何参数的选择是否合理对刀具的使用寿命、加工质量、生产率以及加工成本等都有重大影响。

1. 前角与前面的选择

前角是最重要的参数之一。它对切削力、切削温度、切削功率、进而对刀具寿命甚至表面质量都有直接影响。

（1）前角 γ_o 的选择 增大前角 γ_o 可以降低切削变形，从而减小切削力、切削热和切削功率，并提高刀具寿命。它也可以抑制积屑瘤的形成并提高表面质量。但是过大的前角会削弱切削刃的强度并减少散热，甚至会引起崩刃。因此，增大前角既有优点也有缺点。前角在具体的切削条件下必须有一个合理值。

选择刀具前角 γ_o 的一般原则是在保证刀具寿命的前提下，尽可能选择较大的前角。选择时应考虑以下因素：

1）工件材料。加工塑性材料时应选择较大的 γ_o，加工脆性材料时应选择较小的 γ_o。工件材料的强度和硬度越高，γ_o 应越小。工件材料的塑性越好，γ_o 应越大。

2）刀具材料。高速钢刀具可选择较大的 γ_o，而硬质合金和陶瓷刀具则选择较小的 γ_o。

3）加工精度要求。粗加工时应选择较小的 γ_o，而精加工时应选择较大的 γ_o。成形刀具的前角较小，其对减小切削刃的形状误差有利，从而可提高工件的加工精度。加工不同材料时硬质合金刀具前角的参考值见表1-3。

Table 1-3 Rake angles of carbide tool 硬质合金刀具的前角

Workpiece material 工件材料	Carbon steel 碳素钢 R_m/GPa				40Cr	Stainless steel 不锈钢	ZGMn 铸钢	Ti-alloy Steel 钛合金钢
	≤0.445	≤0.558	≤0.784	≤0.98				
Rake angle 前角	25°~35°	15°~20°	12°~15°	10°	10°~18°	15°~30°	3°~-3°	5°~10°

Workpiece material 工件材料	Hardened steel 淬火钢（HRC）					Cast iron 铸铁（HBW）		Copper 铜		Al-alloy 铝合金
	38~41	44~47	50~52	54~58	60~65	≤220	>220	Brass 黄铜	Bronze 青铜	
Rake angle 前角	0°	-3°	-5°	-7°	-10°	12°	8°	15°~25°	5°~15°	25°~30°

（2）前面的种类及其选择 图1-26 所示为几种类型的前面，其中 $b_{\gamma 1}$ 代表第一前面倒棱的宽度。γ_{o1} 代表第一前面倒棱的前角。

图1-26a 所示类型用于单刃或多刃精密切削刀具，如成形刀具、铣刀、螺纹刀具和齿轮刀具等。这是前面的基本形式；图1-26b 所示类型用于制造切削具有高强度和高硬度材料的硬质合金刀具；图1-26c 所示类型适合于前、后面都出现磨损的刀具。它可以减少刀具重磨的面积，增加刀片的刃磨次数。图1-26d 所示类型刀具主要用于粗加工塑性金属和镗孔。图1-26e 所示类型具有带负前角的窄平面，其切削刃强度得到加强。由于棱带宽度 $b_{\gamma 1}$ 非常小，

1) Workpiece material. It is rational to choose larger γ_o for ductile metals and smaller γ_o for brittle metals. The higher the strength and hardness of work material, the smaller the γ_o. The better the plasticity of work material, the larger the γ_o.

2) Tool point material. It is rational to choose larger γ_o for HSS and smaller γ_o for carbides and ceramics.

3) Machining requirements. It is rational to choose smaller γ_o for roughing and larger γ_o for finishing. The smaller γ_o for the forming tool has the advantage of reducing the form error of the cutting edge, thus improving the machining accuracy of the workpiece. The rake angles of the carbide tool for machining various materials are listed in Tab. 1-3.

(2) Types of rake faces and its choice There are several types of rake faces, as shown in Fig. 1-26, where $b_{\gamma 1}$ represents the land width of the first face, and γ_{o1} represents the rake angle on the first face land.

The type shown in Fig. 1-26a is used for single-point or multi-point precision cutting tools, such as the forming tool, milling cutter, screw cutting tool and gear cutting tool, etc. This is a basic type of the rake face. The type shown in Fig. 1-26b is used for making the carbide cutting tools, which are used to cut the materials with higher strength and hardness. The type shown in Fig. 1-26c is suitable for the tools with wear arising on both rake face and flank simultaneously. It can reduce the resharpening area of the rake face and increase the resharpening times of the tool blade. The type shown in Fig. 1-26d is mainly used for roughing plastic metals or boring. The type shown in Fig. 1-26e has a very narrow flat surface with $-\gamma_{o1}$. The strength of cutting edge gets increased. Because b_{r1} is very small, the land has nearly no influence on the cutting action.

In general, $b_{r1} = (0.3$-$0.8)f$, $\gamma_{o1} = -5° \sim -10°$ when the carbide tools are used to cut the steels; $b_{r1} = (0.3$-$0.8)f$, $\gamma_{o1} = -25°$ for roughing or cutting with impact and vibration.

2. The choice of the clearance angle and flank

The clearance angle is also one of the important geometric parameters. To increase the clearance angle can decrease the friction between the flank and the machined surface, and improve the quality of the machined surface. But with the increase of the clearance angle, the wedge angle and the cutting edge strength decrease. At the same time, the volume of the dispersing heat becomes smaller, and causes the cutting temperature rise and speed up the tool wear.

For the same *VB* value, when the clearance angle increases from α_{o1} to α_{o2}, as shown in Fig. 1-27, the material volume consumed by resharpening tool increases. Besides, the radial wear value *NB* becomes large. The radial position of the tool point is changed clearly, thus affecting the machining accuracy.

(1) Choice of the clearance angle (α_o) The principle in determining α_o is to choose smaller α_o as far as possible under the condition of not producing friction. The factors should be considered as follows:

1) Machining accuracy. Because of the small cutting force in finishing, larger clearance angle ($\alpha_o = 8°$-$12°$) can be selected to prevent friction. In rough machining, smaller α_o ($6°$-$8°$) should be selected to increase the cutting tool strength.

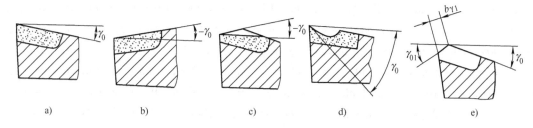

Fig. 1-26　Types of rake faces 前面的类型

故该棱带几乎不妨碍切削效果。

通常，当用硬质合金刀具加工钢材时，$b_{\gamma 1} = (0.3 \sim 0.8)f$，$\gamma_{o1} = -5° \sim -10°$；当粗加工或在具有冲击或振动情况下切削时，$b_{\gamma 1} = (0.3 \sim 0.8)f$，$\gamma_{o1} = -25°$。

2. 后角与后面的选择

后角也是重要的几何参数之一。增大后角可以减小后面与已加工表面的摩擦，提高已加工表面的质量。但是随着后角的增加，楔角和切削刃的强度减小。与此同时，刀具的散热体积减小，从而引起切削温度升高，加速刀具磨损。

对于相同的 VB 值，当后角由 α_{o1} 增大到 α_{o2} 时，如图 1-27 所示，由磨刀消耗的刀具材料体积增大了。此外，刀具的径向磨损值 NB 也增大了。刀具径向位置变化明显，从而影响加工精度。

（1）后角 α_o 的选择　确定 α_o 的原则是在不至于引起磨损的条件下尽可能选择较小的后角。具体来说应考虑以下因素：

1）加工精度。由于精加工时切削力较小，故可以选择较大的后角（$\alpha_o = 8° \sim 12°$）以防摩擦。在粗加工时，应选择较小的后角（$\alpha_o = 6° \sim 8°$）以增加刀具的强度。

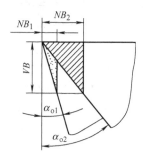

Fig. 1-27　Effect of the clearance angle on tool wear volume 后角对刀具磨损量的影响

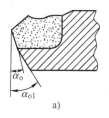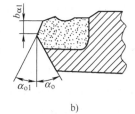

Fig. 1-28　Types of flanks 后面的类型

2）工件材料。切削塑性材料时，应选择较大的 α_o 以减小磨损；切削脆性材料或切削强度、硬度高的材料时，应选择较小的后角以增加切削刃的强度。

（2）后面的种类及其选择　后面是具有 $+\alpha_o$ 的平面。但硬质合金和陶瓷刀具还有另外两种形式的后面。

1）图 1-28a 所示为带有两个正后角的双后面。其优点是能够减轻刀具刃磨的工作量，减小刀具磨损量。

2) Workpiece material. The larger α_o should be selected for cutting ductile metals so as to reduce friction; the smaller α_o should be selected for cutting brittle metals or cutting steels with high strength and high hardness so as to increase the tool point strength.

(2) Types and choice of flanks and its choice Often the flank has a plane surface with $+\alpha_o$. But there are other two types of flanks which are used on the carbide alloy tools and ceramics tools.

1) Double flanks with two positive clearance angles, as shown in Fig. 1-28a. Its advantages are to relieve the tool resharpening workload and decrease the tool wear amount.

2) Land of the first flank (see Fig. 1-28b). Its parameters are $b_{\alpha 1} = 0.1$-0.3mm, $\alpha_{o1} = -5°$-$-10°$, where $b_{\alpha 1}$ represents the land width of the first flank, and α_{o1} represents the clearance angle on the first flank.

The function of the land of the first flank is to increase friction between the land and the cutting surface, forming damp effectiveness to enhance the cutting stability and avoid the vibration. Therefore the land of the first flank is called the damping land.

3. Choice of the leading and minor cutting edge angles

(1) Choice of the leading cutting edge angle κ_r To decrease the cutting edge angle κ_r can prolong the tool service life. When the depth of cut a_p and the feed f are kept constant, the decrease of κ_r would reduce the undeformed chip thickness h_D and increase the undeformed chip width b_D, thus lightening the load born by the cutting edge per unit length. To decrease κ_r is good for reducing the machined surface roughness. But the decrease of κ_r would cause the radial-thrust force to increase and increase the deformation of the technological system. Besides, to increase κ_r is helpful for chip breaking and chip removal.

The specific principles in the selection of the leading cutting edge angle κ_r are: ① to choose the larger κ_r if the technological system rigidity is insufficient. Generally, $\kappa_r = 60°$-$75°$. ② to choose the smaller κ_r to cut metal materials with high strength and hardness; ③ to choose the turning tool with $\kappa_r = 90°$ to machine the shaft with shoulder.

(2) Choice of the minor cutting edge angle κ_r' The minor cutting edge angle κ_r' is the primary angle affecting the surface roughness. The smaller the κ_r', the smaller the surface roughness. The magnitude of κ_r' also affects the strength of the tool point. But too small κ_r' would cause severe friction between the trailing flank and the machined surface and vibration, which is harmful for improving the surface quality.

The principle in the selection of the minor cutting edge angle κ_r' is to choose the smaller κ_r' under the premise of not producing friction and vibration.

The values of κ_r and κ_r' under different cutting conditions are listed in Tab. 1-4.

4. Choice of the cutting edge inclination angle λ_s

(1) Functions of λ_s

1) Effect on the surface quality. To change λ_s can control the chip flowing direction to prevent the chip from scratching the machined surface. The influence of λ_s on the chip flowing direction is illustrated in Fig. 1-29. It can be seen that the chip flows to the machined surface when $\lambda_s < 0$ and to the surface to be machined when $\lambda_s > 0$.

2)第一后面的棱带(图 1-28b)。其参数是 $b_{\alpha 1} = 0.1 \sim 0.3$ mm, $\alpha_{o1} = -10° \sim -5°$,其中,$b_{\alpha 1}$ 代表第一后面的棱带宽度,α_{o1} 代表第一后面的后角。

第一后面棱带的作用是增加后面与切削表面的摩擦,从而产生阻尼效应以提高切削稳定性和避免振动。因此该棱带也称消振棱。

3. 主、副偏角的选择

(1) 主偏角 κ_r 的选择 减小主偏角 κ_r 可以延长刀具寿命。在切削深度 a_p 和进给量 f 不变时,主偏角 κ_r 的减小能使切削层厚度 h_D 减小、切削层宽度 b_D 增加,从而减轻切削刃单位长度上的负荷。减小 κ_r 对降低已加工表面粗糙度有利。但是减小 κ_r 会使径向力增大,从而增大工艺系统的变形。此外,增大 κ_r 有助于断屑和排屑。

选择主偏角 κ_r 的具体原则是:①如果工艺系统刚度不足,应选择较大的 κ_r。通常,$\kappa_r = 60° \sim 75°$;②加工高强度、高硬度的金属材料时,应选择较小的 κ_r;③在车削台阶轴时,应选择 $\kappa_r = 90°$ 的车刀。

(2) 副偏角 κ_r' 的选择 副偏角 κ_r' 是影响表面粗糙度的主要角度。副偏角 κ_r' 越小,表面粗糙度越小。κ_r' 的大小也影响切削刃的强度。但是过小的 κ_r' 会引起副后面与已加工表面的剧烈摩擦和振动,这对提高表面质量不利。

选择副偏角 κ_r' 的原则是在不至于引起摩擦和振动的前提下选择较小的 κ_r'。

在不同的切削条件下 κ_r 和 κ_r' 的取值见表 1-4。

Table1-4　Reference values of κ_r and κ_r'　κ_r 和 κ_r' 的参考值

Cutting Conditions 切削条件	High rigidity, hardened steel 高刚度、淬硬钢	Better rigidity, Cylindrical turning, facing, chamfering 刚性好、车外圆、车端面、车倒角	Poor rigidity, rough turning, power turning 刚性差、粗车、机动车削	Poor rigidity, multi-diameter shaft, slender shaft 刚性差、车台阶轴和细长轴	Parting-off, slotting 切断、切槽
κ_r	10°~30°	45°	60°~70°	75°~93°	≥90°
κ_r'	10°~5°	45°	15°~10°	10°~6°	1°~2°

4. 刃倾角 λ_s 的选择

(1) 刃倾角 λ_s 的功用

1) 影响表面质量。改变 λ_s 可以控制切屑流出方向以防止切屑划伤已加工表面。λ_s 对切屑流出方向的影响如图 1-29 所示。由图看出,当 $\lambda_s < 0$ 时,切屑流向已加工表面;当 $\lambda_s > 0$ 时,切屑流向待加工表面。

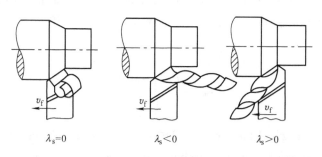

Fig. 1-29　Influence of λ_s on chip flowing direction　λ_s 对切屑流出方向的影响

2) Effect on the sharpness of the cutting edge. When $\lambda_s \leqslant 45°$, the working rake angle and clearance angle will increase with the increase of λ_s, thus increasing the sharpness of the cutting edge.

3) Effect on the tool bit strength and heat dissipation condition. To choose $-\lambda_s$ can increase the tool bit volume, and further increase tool strength and improve the heat dissipation condition. If $-\lambda_s$ is used for intermittent cutting, the workpiece will first contact with the cutting edge at a point far from the tool nose, thus the impact on the tool nose can be avoided. Many turning tools with good cutting performance often have a larger rake angle with $-\lambda_s$, thus, solving the contradiction between sharp and strong.

4) Effect on the magnitude and direction of the cutting force. In general, the cutting force of the cutting tool with $\lambda_s > 0$ is smaller than that of the cutting tool with $\lambda_s < 0$. With the increase of $-\lambda_s$, the radial-thrust force would increase significantly, thus leading to the deformation of the workpiece and the vibration of the technological system.

(2) Choice of λ_s To sum up, λ_s should be selected according to the tool bit strength, chip flow direction and specific machining condition. Take $\lambda_s = 0°\text{-}5°$ for finishing and $\lambda_s = 0°\text{-}-5°$ for rough turning. For the situations such as discontinuous surface, ununiform surplus material and so on, the cutting tool with $-\lambda_s$ should be selected. If the rigidity of the technological system is insufficient, $-\lambda_s$ should be avoided as far as possible.

1.4.5 Proper choice of cutting variables

The magnitude of cutting variables has a critical influence on the machining quality, productivity and machining cost, especially for batch production and the processing on the automatic machines, transfer lines and CNC machine tools. At present, cutting variables are mainly determined by means of cutting handbooks, practical experience or technological tests in many factories. The principles and methods to select cutting variables rationally are mainly introduced in this section.

1. Principles to select cutting variables

The following factors should be considered when selecting cutting variables.

(1) Productivity The productivity means the numbers of qualified products produced in unit time. It is related to the machining time and metal removal rate.

1) Machining time (t_m). It is an index to measure whether the productivity is high or low. Take the cylindrical turning for example, as shown in Fig. 1-30, t_m can be calculated by Eq. (1-38).

$$t_m = \frac{(l + l_1 + l_2)h}{nfa_p} \tag{1-38}$$

where, l represents the length of the surface to be cut; l_1 represents the cut-in length; l_2 represents the cut-out length; h represents the total machining allowance.

Substitute $v_c = n\pi D/1000$ into Eq. (1-38), then

2)影响切削刃的锋利程度。当 $\lambda_s \leq 45°$ 时,随着 λ_s 的增加,刀具的工作前角和后角也随之增大,从而使切削刃更锋利。

3)影响刀头强度和散热性。选择 $-\lambda_s$ 可以增加刀头体积,进而提高刀具强度并改善散热条件。在断续切削时采用 $-\lambda_s$,工件就会先和远离刀尖的切削刃接触,从而可以避免刀尖与工件的冲击。许多切削性能好的车刀常具有较大的前角和 $-\lambda_s$,从而解决了刀具强固与锋利的矛盾。

4)影响切削力的大小和方向。通常,$\lambda_s > 0$ 刀具的切削力比 $\lambda_s < 0$ 刀具的小。随着 $-\lambda_s$ 的增加,径向力就会大大地增加,从而导致工件变形和工艺系统的振动。

(2)λ_s 的选择 总之,λ_s 应根据刀头强度、切屑流出方向以及具体的切削条件来确定。精车时取 $\lambda_s = 0° \sim 5°$,粗车时取 $\lambda_s = 0° \sim -5°$。对于如断续表面和加工余量不均匀等情况,应选择具有 $-\lambda_s$ 的刀具。如果工艺系统刚度不足,则应尽量避免选择具有 $-\lambda_s$ 的刀具。

1.4.5 切削用量的合理选择

切削用量的大小对加工质量、生产率和加工成本有着重要影响,对批量生产以及在自动机床、自动线以及数控机床上的加工来说尤其如此。目前在许多工厂,切削用量主要是根据切削加工手册、实践经验或工艺实验来确定的。本节主要介绍合理选择切削用量的原则和方法。

1. 切削用量的选择原则

在选择切削用量时应考虑以下因素。

(1)生产率 生产率指的是单位时间内生产合格产品的数量。它与加工时间和金属切除率有关。

1)加工时间(t_m)。它是衡量生产率高低的指标。以外圆车削为例,如图 1-30 所示,t_m 可以用式(1-38)来计算。

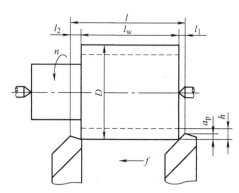

Fig. 1-30 Calculation for machining time in cylindrical turning 车削外圆时加工时间的计算

$$t_m = \frac{(l + l_1 + l_2)h}{nfa_p} \tag{1-38}$$

式中 l——待切削表面的长度;

l_1——切入长度;

$$t_m = \frac{(l + l_1 + l_2)\pi Dh}{1000 v_c f a_p} \tag{1-39}$$

2) Metal removal rate (Q). Q stands for the volume of the metal cut per minute. It is the other index to measure the cutting efficiency. Q can be determined by Eq. (1-40)

$$Q = 1000\, v_c f a_p \tag{1-40}$$

where, Q represents the metal removal rate; a_p represents the depth of cut; f represents the feed; v_c represents the cutting speed.

It can be seen from Eq. (1-39) and Eq. (1-40) the increase of any of variables (v_c, f, a_p) can shorten the machining time, increase the metal removal rate and raise the cutting efficiency in the same way.

(2) Machine power   It is known from the Eq. (1-21) that the larger F_c and v_c, the larger P_c. As for the influence of f and a_p on F_c, the test has shown that the influence of a_p on F_c is larger than that of feed f. So we'd rather choose larger f than larger a_p within the allowable machine power.

(3) Tool life (T)   According to the tool life Eq. (1-32), we should give priority to the choice of larger a_p, and then f. At last we should determine the cutting speed according to the tool life.

(4) Surface roughness   The machining variables should be determined according to the requirements for the surface roughness in finishing and semi-finishing. It is well known that BUE has a harmful effect on the surface roughness. Because BUE is formed easily at a middle cutting speed (15-50 m/min), it is rational to choose either lower cutting speed (<5 m/min) or higher cutting speed (>70 m/min) in finishing to avoid forming BUE so as to decrease the surface roughness.

The effect of the feed on the roughness can be expressed by the following:

For the cutting tools with straight cutting edge, i.e. the nose radius $r_\varepsilon = 0$, as shown in Fig. 1-31(a), the maximum surface roughness R_{max} is given by

$$R_{max} = f/(\cot\kappa_r + \cot\kappa_r') \tag{1-41}$$

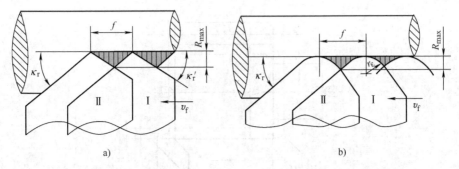

Fig. 1-31   Surface geometry model 表面几何模型

For the cutting tools with nose radius $r_\varepsilon > 0$, as shown in Fig. 1-31b, the maximum surface roughness R_{max} is given by

$$R_{max} = r_\varepsilon - [r_\varepsilon^2 - (f/2)^2]^{1/2} \approx f^2/8r_\varepsilon \tag{1-42}$$

It can be seen from above that the feed has direct effect on the surface roughness. Therefore, the increase of the feed is chiefly limited by the surface roughness.

l_2——切出长度；

h——待加工表面的总余量。

把 $v_c = n\pi D/1000$ 带入式(1-38)，那么

$$t_m = \frac{(l + l_1 + l_2)\pi Dh}{1000 v_c f a_p} \tag{1-39}$$

2）金属切除率(Q)。Q 代表单位时间内切除的金属体积。它是衡量加工效率高低的另一个指标。Q 的大小可用式(1-40)来计算。

$$Q = 1000 v_c f a_p \tag{1-40}$$

式中 Q——金属切除率；

a_p——切削深度；

f——进给量；

v_c——切削速度。

由式（1-39）和式（1-40）可知，切削用量三要素中任何一个增大都能缩短加工时间，提高金属切除率和切削效率。

（2）机床功率 由式（1-21）可知，主切削力 F_c 和切削速度 v_c 越大，切削功率 P_c 也越大。至于 f 和 a_p 对 F_c 的影响，实验表明，a_p 对 F_c 的影响大于 f 的影响。因此在机床功率允许的范围内，最好选择较大的 f 而不是 a_p。

（3）刀具寿命（T） 根据刀具寿命计算公式（1-32）可知，应当优先选择较大的 a_p，然后是 f，最后根据规定的刀具寿命确定切削速度。

（4）表面粗糙度 切削用量应当根据半精和精加工的表面粗糙度要求来确定。众所周知，积屑瘤对表面粗糙度有负面影响。由于积屑瘤易在中速切削（15～50m/min）时形成，因此在精加工时为避免积屑瘤，最好采用低速（<5m/min）或采用高速（>70m/min）以降低粗糙度。

进给量对表面粗糙度的影响可以分为两种情况来讨论。

对于直刃刀具，如图 1-31a 所示，其刀尖圆弧半径 $r_\varepsilon = 0$，最大表面粗糙度 R_{max} 用式（1-41）表示。

$$R_{max} = f/(\cot\kappa_r + \cot\kappa_r') \tag{1-41}$$

对于刀尖圆弧半径 $r_\varepsilon > 0$ 的刀具，如图 1-31b 所示，最大表面粗糙度 R_{max} 用式（1-42）表示。

$$R_{max} = r_\varepsilon - [r_\varepsilon^2 - (f/2)^2]^{1/2} \approx f^2/8r_\varepsilon \tag{1-42}$$

由此可知，进给量对表面粗糙度有直接影响。因此进给量的增大与否主要受表面粗糙度的限制。

对于 a_p，它对表面粗糙度没有明显影响。但是随着 a_p 的增加，切削过程中会出现颤振。因此在精加工时应选择较小的 a_p。

由此可知，选择切削用量的规则是首先选择尽可能大的切削深度 a_p，然后选择较大的进给量 f，最后根据允许的刀具寿命、机床功率和其他具体情况来正确确定切削速度 v_c。

2. 切削用量的选择方法

（1）切削深度的选择 切削深度 a_p 主要根据工件的加工余量来确定。通常，粗加工余量应尽可能在一次进给中切除。以外圆车削为例，有

As for a_p, it has not clearly effect on the roughness. But as a_p increases, the chatter would appear in the cutting process. Therefore, the small a_p should be selected in finish machining.

As seen from above, the rules of selecting machining variables are to choose a_p as large as possible firstly, and then a larger feed, lastly, a proper cutting speed v_c should be determined according to the allowable tool life, machine power and other specific conditions.

2. Methods of selecting machining variables

(1) Selection of the depth of cut   The depth of cut a_p is determined mainly according to the excess metal on the workpiece surface. In general, the roughing allowance should be removed in one pass as far as possible. Take the cylindrical turning for instance, there is:

$$a_p = Z = \frac{d_w - d_m}{2} \tag{1-43}$$

where, Z is the machining allowance in radial direction, d_w is the diameter before machining, d_m is the diameter after machining.

When machining a workpiece, it is advisable to take one roughing and one finishing if possible. The roughing cut should be as deep as possible. In rough turning, the depth of cut will depend on the following factors: ①condition of the machine tool; ②type and shape of the cutting tool used; ③rigidity of the workpiece; ④feed rate.

When Z is too large or when the rigidity of the technological system is insufficient, Z can be removed in two or more passes. Suppose Z is removed by two cuts, then

$$a_{p1} = (2/3 \sim 3/4)Z$$
$$a_{p2} = (1/3 \sim 1/4)Z$$

Generally, the depth of cut should not be larger than 2/3 length of the cutting edge of the cutter blade.

Whereas in finish turning, the depth of cut will depend on the type of the workpiece and the surface finish required. Because the machining allowance is small in finishing, it is usual to remove all surplus materials. But when the machining allowance is larger, it should be removed by two or more cuts. In any case, the depth of cut should not be less than 0.10 mm.

(2) Selection of feed   After a_p is determined, f should be determined according to the cutting force permitted by machining system. Besides, other factors should also be considered as follows: ①strength of the feed mechanism; ②rigidity of the workpiece; ③rigidity of the tool shank; ④strength of the blade and so on.

The allowable f can be calculated by means of the strength or rigidity calculation formula related to above factors. In workshop practice, f is determined by looking up the handbooks on cutting variables.

Some reference values of the feed in roughing with carbide and HSS turning tools are listed in Tab. 1-5. In finish machining, f is usually determined according to the requirement for the surface roughness.

(3) Selection of the cutting speed   After a_p and f are determined, according to the stipulated tool life T, the allowable cutting speed v_T(m/min) can be calculated by means of the Eq. (1-44):

$$a_p = Z = \frac{d_w - d_m}{2} \tag{1-43}$$

式中 Z——径向加工余量；

d_w——加工前的直径；

d_m——加工后的直径。

当加工某工件时，可能的话，建议采用粗精加工各一次。粗切深度应尽可能大。在粗车时，切削深度取决于以下因素：①机床情况；②所用刀具的种类和形状；③工件的刚性；④进给速度。

当余量 Z 过大或当工艺系统刚度不足时，Z 可以在两次或两次以上的进给中去除。假设 Z 由两次进给去除，那么

$$a_{p1} = (2/3 \sim 3/4)Z$$
$$a_{p2} = (1/3 \sim 1/4)Z$$

通常，切削深度不应大于刀片切削刃长度的 2/3。

而在精车时，切削深度取决于工件的类型和要求的表面粗糙度。由于精加工余量小，通常一次切除全部余量。但当加工余量较大时应当分两次进给切除。在任何情况下，切削深度不应小于 0.10 mm。

（2）进给量的选择　在 a_p 确定后，进给量 f 应根据加工系统允许的切削力来确定。此外，另外一些因素也应考虑：①进给机构的强度；②工件的刚性；③刀柄的刚度；④刀片的强度等。

允许的 f 应当利用与上述因素有关的强度或刚度计算公式来计算。在实际生产中，进给量 f 是通过查阅切削用量手册确定的。

在使用硬质合金和高速钢车刀粗车时进给量的一些参考值见表 1-5。精加工时，进给量 f 通常依据表面粗糙度要求来确定。

Table 1-5　Feeds for rough cylindrical turning or facing with carbide and HSS cutters

硬质合金和高速钢车刀粗车外圆和端面时的进给量

Work material 工件材料	Tool bar size B×H 刀杆尺寸/mm	Workpart diameter 工件直径 d_w/mm	Depth of cut 切削深度 a_p/mm				
			≤3	>3~5	>5~8	>8~12	>12
			Feed 进给量 f/mm				
Carbon structural steels, alloy structural steels 碳素结构钢，合金结构钢	16×25	20	0.3~0.4	—	—	—	—
		40	0.4~0.5	0.4~0.5	—	—	—
		60	0.5~0.6	0.5~0.7	0.3~0.5	—	—
		100	0.6~0.9	0.6~0.9	0.5~0.6	0.4~0.5	—
		400	0.8~1.2	0.8~1.2	0.6~0.8	0.5~0.6	—
	20×30 25×25	20	0.3~0.4	—	—	—	—
		40	0.4~0.5	0.3~0.4	—	—	—
		60	0.6~0.7	0.5~0.7	0.4~0.6	—	—
		100	0.8~1.0	0.7~0.9	0.5~0.7	0.4~0.7	—
		600	1.2~1.4	1.0~1.2	0.8~1.0	0.6~0.9	0.4~0.6

（3）切削速度的选择　在 a_p 和 f 确定后，可根据规定的刀具寿命，利用式（1-44）计算切削速度（m/min）

$$v_T = \frac{C_v}{T^m a_p^{x_v} f^{y_v}} K_v \tag{1-44}$$

The coefficients and exponents in Eq. (1-44) can be obtained by looking up the machining process handbooks.

In practical production, the general principles of choosing the cutting speed are as follows:

1) Lower cutting speeds should be selected in rough machining, and higher cutting speeds can be used in finish machining.

2) Lower cutting speeds should be selected when cutting the materials with higher strength and hardness. The poorer the machinability of material to be cut is, the lower cutting speed is used. The better the cutting performance of tool material is, the higher the cutting speed.

3) In intermittent cutting or when cutting casting and forging blanks with a hard surface, the cutting speed should be reduced appropriately in order to decrease the impact and thermal stress.

4) Lower cutting speeds should be selected when cutting large workpieces, long shafts and thin-wall workpieces. If the process system has lower rigidity, the cutting speeds which would cause self-exciting vibration should be avoided.

$$v_T = \frac{C_v}{T^m a_p^{x_v} f^{y_v}} K_v \tag{1-44}$$

式中的系数和指数可在加工工艺手册中查得。

在实际生产中,选择切削速度的一般原则是:

1)粗加工时应选择较低的切削速度;精加工时可选择较高的切削速度,同时应避免积屑瘤的产生。

2)加工材料的强度及硬度较高时,应选较低的切削速度。材料的加工性越差,切削速度应选得越低。刀具材料的切削性能越好时,切削速度应选得越高。

3)在断续切削或是加工锻、铸件等带有硬皮的工件时,为了减小冲击和热应力,要适当降低切削速度。

4)加工大件、细长轴和薄壁工件时,要选用较低的切削速度;如果工艺系统的刚度较差,切削速度就应避开产生自激振动的临界速度。

Chapter 2
Machine Tools and Cutting Tools

第2章
机床与刀具

2.1 Introduction

A machine tool is the one that holds the cutting tools to remove the surplus material from a workpiece in order to generate the requisite part of the given size, configuration and surface roughness. It is different from the ordinary machine which is essentially a means of converting the source of power from one form to the other. Machine tools are the machines used to make other machines. Therefore, machine tools are named as mother machines, since without them no component can be finished.

In the last two centuries, new inventions have led to the development of a number of machine tools. Machine tool versatility has grown to meet the various needs of new inventors. For example, James Watt's steam engine, which was designed in England around 1763, could become reality only after a satisfactory method was found by John Wilkinson around 1775 to bore the engine cylinder with a cutting tool mounted on a boring bar. The boring bar could be rotated and fed through the cylinder, thus generating an internal cylindrical surface. Most of conventional machine tools such as lathes, milling machines, shapers and planers, boring machines, drill presses used today have the same design as the early versions developed during the last two centuries. With the advent of numerical control technology, all types of machine tools are equipped with their own computer numerical controls and almost every machining process can now be efficiently automated with an exceptional degree of accuracy, reliability and repeatability. At present, the CNC machining center appears to be the most capable and versatile automatic machine tool that can perform drilling, milling, boring, reaming and tapping operations.

2.1.1 Classification and model of machine tools

1. Classification of machine tools

There are a variety of machine tools. Machine tools have to be classified for the convenience of distinguishing, application and management. There are many ways in which machine tools can be classified as follows.

(1) Classification based on the cutting tools used on machine tools

①Lathe; ② Milling machine; ③ Drilling machine; ④Planer and shaper; ⑤Grinding machine; ⑥Boring machine; ⑦Broaching machine; ⑧Threading machine; ⑨Gear cutting machine; ⑩Sawing machine; ⑪Others. Every class of machine tools can be divided into several groups based on their technological range, layout and configuration, and every group can be divided further into several series.

(2) Classification based on the degree of specialization

1) General purpose machine tools. They are designed for performing a great variety of machining operations on a wide range of workpieces. They are employed chiefly in job and small lot production. Examples are center lathe, column and knee type milling machine and universal cylindrical grinding machine.

2.1 概述

机床是指夹持刀具从工件表面上切除多余材料以获得所要求零件的尺寸、形状和表面粗糙度的机器。它不同于一般的机器。一般的机器实质上是一种把动力源从一种形式转换成另一种形式的装置，而机床是用于制造机器的机器。因此机床又称为工作母机，因为没有机床，任何零件都难以加工得很精确。

在过去的200年间，新的发明创造致使大量的机床研制成功。机床的多用途可以满足各种新发明的需求。例如，1763年左右在英格兰设计的瓦特蒸汽机，只有在John Wilkinson于1775年找到了用镗刀加工发动机缸体的方法后才得以问世。镗刀旋转并沿着缸筒进给，从而加工出内圆柱面。大多数如今使用的车床、铣床、刨床、镗床、钻床等通用机床仍与过去200年间研制的早期机床有着相同的结构形式。随着数控技术的问世，各种类型的机床都配备了各自的计算机数控装置，几乎每一种加工方法现在都实现了高效自动化，而且加工精度高、可靠性好、重复制造精度高。目前，CNC加工中心似乎是可以进行钻削、铣削、镗削、铰削和攻螺纹等操作的、用途最广的自动化机床。

2.1.1 机床的分类及型号

1. 机床的分类

机床的品种和规格繁多。为了便于区别、使用和管理，应对机床加以分类。机床的分类方法很多，常用的分类方法如下。

（1）根据机床上所用的刀具分类

①车床；②铣床；③钻床；④刨床；⑤磨床；⑥镗床；⑦拉床；⑧螺纹加工机床；⑨齿轮加工机床；⑩锯床；⑪其他机床。每一类机床都可以根据各自的工艺范围、布局以及结构分成若干组，而每组又可以细分为若干系。

（2）根据机床的专业化程度分类

1）通用机床。通用机床是为对多种零件进行各种各样的加工而设计的。它们主要适用于单件小批生产。卧式车床、升降台铣床以及万能外圆磨床等都属于通用机床。

2）专门化机床。专门化机床是指专门用于加工某一类或几类零件到一定尺寸和通过改变零件的结构性状完成几道特定工序的机床。曲轴车床、凸轮轴车床、螺纹车床以及花键铣床等都属于专门化机床。

3）专用机床。专用机床的工艺范围最窄，专门用于加工某一种零件的某一道特定工序。它们在大批大量生产中已得到广泛应用。如加工机床主轴箱的专用镗床、加工车床导轨的专用磨床等。各种组合机床也属于专用机床。

（3）根据机床的质量和规格分类

1）小型机床。其质量小于1t。

2）中型机床。其质量为1~10t。

3）大型机床。其质量为10~30t。

4）重型机床。其质量为30~100t。

5）超重型机床。其质量大于100t。

2) Specialized machine tools. These machine tools are only used to machine a certain class or several classes of workpieces in a certain dimension limit and to complete some specific operations by making the necessary changes in their construction. Examples are crankshaft turning lathe, camshaft grinding machine, threading lathe and spline milling machine.

3) Special purpose machine tools. These machine tools are those which are designed specifically for doing a single operation on a class of workpieces or on a single workpiece. They have found application in large lot and mass production. Special purpose boring machines for boring holes on the headstock of lathe and grinding machine for grinding the guideways of lathe are included in this group. Aggregate machine tools also belong to the special purpose machine tools.

(3) Classification based on the weight and size of machine tools

1) Small-size machine tools. They weigh up to one ton.

2) Medium-size machine tools. Their weight varies from 1 to 10 tons.

3) Large-size machine tools. They weigh varies from 10 to 30 tons.

4) Heavy-duty machine tools. They weigh varies from 30 to 100 tons.

5) Super-heavy-duty machine tools. They weigh over 100 tons.

Besides, the same type of machine tools can be classified based on the accuracy, such as machine tools with standard accuracy, precision machine tools and high precision machine tools. There are other kinds of classifications, such as automatic machine tool and semi-automatic machine tool, multi-spindle machine tool, multi-cutter machine tool and so on. In general, machine tools are classified based on the machining feature firstly, and then described further based on some characteristics, such as semi-automatic & multi-cutter machine tool, universal cylindrical grinding machine with high precision.

The method to classify machine tools will develop with the development of the machine tool technology. Modern machine tools are developing in the direction of the CNC machine tool. The functions of the CNC machine tool are increasingly diversifying. With high operational concentration, a CNC machine tool can complete many tasks which could be finished by means of several traditional machine tools.

2. Technical parameters and dimension series of machine tools

Technical parameters of a machine tool are various technical data to express the size and machining capacity of it. They include the primary parameter, the second primary parameter, structural dimension of major assembly, traveling distance of major assembly, various moving speeds and speeds series, power of motors, overall dimension of the machine tool and so on. All these parameters are listed in the instruction book of the machine tool. They are the important technical data for users to select, check and use the machine tool.

The primary parameter is the one which reflects the maximum working capacity. It affects directly the other parameters and structure size of the machine tool. The primary parameter is generally expressed by means of the maximum workpiece size, which can be machined on the machine tool, or the assembly size related to it. For example, the primary parameter of the engine lathe is the maximum swing diameter of the workpiece over the bed. The primary parameter of the drilling machine is

此外，同类型机床还可按工作精度分为普通精度机床、精密机床和高精度机床。机床也有其他分类方法，如自动机床、半自动机床、多轴机床和多刀机床等。通常，机床首先按照加工特征划分，然后进一步根据某些特点来描述，如半自动多刀机床、高精度万能外圆磨床等。

机床的分类方法将随着机床技术的发展而发展。当代机床正朝着 CNC 机床的方向发展。CNC 机床的功能日益扩大，CNC 机床的工序高度集中，可以完成通常由多台通用机床完成的多项任务。

2. 机床的技术参数和尺寸系列

机床的技术参数是表示其规格和加工能力的各种技术数据。它们包括主参数、第二主参数、主要部件的结构尺寸及其运动距离、各种运动速度以及速度系列、电动机的功率、机床的总体尺寸等。所有这些技术参数都列入机床的说明书内。它们是供用户选择、审查和使用机床的重要技术参数。

主参数是反映机床最大加工能力的参数。它直接影响机床的其他参数和结构尺寸。主参数通常用机床上能够加工的最大工件尺寸，或与之有关的部件尺寸来表示。例如，卧式车床的主参数是床身上工件的最大回转直径，即最大加工直径。钻床的主参数是最大钻孔直径。类似地，外圆磨床的主参数是最大磨削直径，卧式镗床的主参数是主轴直径，升降台铣床和龙门铣床的主参数都是其工作台的宽度。但也有一些机床，如拉床，其主参数是用力而不是用尺寸表示的。对于某些通用机床，当仅用主参数仍不能清楚表示其加工能力和规格时，就要使用第二主参数。第二主参数通常指的是主轴数目、最大跨距、工件的最大长度等。

3. 机床型号

机床型号是赋予每种机床的一个代号，用以简明地表示机床的类型、通用特性和结构特性以及主要技术参数等。GB/T 15375—2008《金属切削机床型号编制方法》规定，我国的机床型号由汉语拼音字母和阿拉伯数字按一定的规律组合而成。通用机床的型号如图 2-1 所示。其中△用阿拉伯数字表示，○用大写的汉语拼音字母表示。括号"（ ）"中表示可选项，无内容时不表示，有内容时不带括号，◎用大写的汉语拼音字母，或阿拉伯数字，或两者兼有之表示。

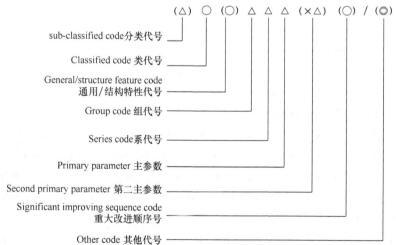

Fig. 2-1 Expression of the general-purpose machine tool model 通用机床型号的表示

the maximum hole-drilling diameter, and similarly, the maximum grinding diameter for cylindrical grinding machine, the spindle diameter for horizontal boring machine, the width of the working table for column and knee type and planer type milling machines. But the primary parameter of some machine tools, such as broaching machine, is expressed by means of the force rather than the size. For some general purpose machine tools, when their working capacity and size can not be expressed clearly by means of a primary parameter, the second primary parameter should be used. The second primary parameter generally means the numbers of the spindle, the maximum. span, the maximum. length of the work, etc.

3. Model of the machine tool

The model of the machine tool is expressed with the code of the machine tool, which is used to show the type, general feature, structural feature, and primary technical parameters. etc. It is specified in "Planning of Machine Tool Model" (GB/T 15375—2008) that the model of domestic machine tools is formed by combining Chinese phonetic alphabet with Arabic numerals in light of certain regulations. The expression of the general purpose machine tool model is shown in Fig. 2-1. △ is expressed with numeral; ○ is expressed with the capital Chinese phonetic letter; the content in bracket "(   )" is a selectable item, and it is not presented without any content; ◎ is expressed with the capital Chinese phonetic letter, or Arabic numeral, or both of them.

Take the model CA6140 for example, where "C" means the lathe, "A" stands for structural feature of the lathe, "6" means that the lathe falls into the group of floor-type and horizontal lathe, and "1" means that the lathe belongs to the horizontal lathe series, "40" is a primary parameter, which means the max-turning diameter is 400mm.

For the model MG1432A, M—grinding machine; G—high precision; 1—cylindrical grinding machine group; 4—universal cylindrical grinding machine series; 32—the max-grinding diameter is 320 mm; A—significant improvement for the first time.

2.1.2 Surface generating methods and kinematic analysis of machine tools

A large number of surfaces could be generated or formed as the case may be with the help of the motions given to the tool and workpiece. The shape of the tool also makes a very important contribution to the final surface obtained.

The components produced by machine tools, as shown in Fig. 2-2, are essentially combination of the following geometric forms: cylinder; cone; sphere; plane surface; helical surface.

1. Methods to produce surfaces

A geometrical surface is usually defined as the trace obtained in the motion of one geometrical generating line called the generatrix 1 along other geometrical generating line called the directrix 2, as shown in Fig. 2-3. A cone is formed by rotating the generatrix around a directrix with one of its end fixed at a point called vertex. The relative motion of the geometrical lines in producing surfaces is called formative motion.

The required workpiece surface can be produced by forming or by generating or by a combination of these two fundamental methods. Generating is where the required profile is obtained by ma-

以 CA6140 型车床为例,"C"指的是车床,"A"代表车床的结构特性,"6"表示落地车床和卧式车床组,"1"表示卧式车床系,"40"是车床的主参数,表示其最大车削直径为 400mm。又如 MG1432A 型机床,其中 M—磨床;G—高精度;1—外圆磨床组;4—万能外圆磨床系;32—最大磨削直径为 320 mm;A—第一次重大改进。

2.1.2 表面的形成方法与机床运动分析

利用刀具和工件的相对运动可以产生大量各种各样的表面。刀具的形状也对工件最终表面的获得起到非常重要的作用。

机床加工出的零件,如图 2-2 所示,基本上是由圆柱面、圆锥面、球面、平面、螺旋面等几何形状及其组合构成的。

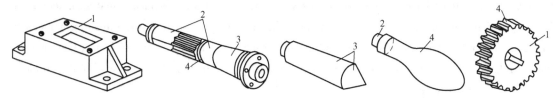

Fig. 2-2　Components with various geometric forms 具有各种几何形状的零件
1—Plane 平面　2—Cylindrical surface 圆柱面　3—Conical surface 圆锥面　4—Forming surface 成形表面

1. 各种表面的形成方法

一种几何表面通常认为是母线 1 沿着导线 2 移动形成的轨迹,如图 2-3 所示。圆锥就是母线 1 绕着一条一端固定在一点(顶点)的导线 2 旋转形成的。在形成表面的过程中几何线之间的相对运动称为成形运动。

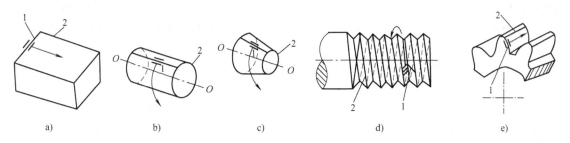

Fig. 2-3　Forming of workpiece surfaces 各种工件表面的形成
1—Generatrix 母线　2—Directrix 导线

所要求的工件表面可以用成形法,也可以用展成法,还可以用这两种基本方法的组合来获得。在展成法中,所要求的表面是通过控制工件和切削刃的相对运动获得的。在成形法中,所形成的表面形状就是切削刃的直接复制。因此所获得的形状精度取决于成形刀具的精度。

事实上,形成几何表面的方法不外乎有以下四种及其组合,即轨迹法、成形法、相切轨迹法和展成法,如图 2-4 所示。

2. 机床运动分析

(1) 机床的运动

nipulating the relative motion of the workpiece and cutting edge. In forming, the shape of the work produced is direct replica of the cutting tool. Thus the accuracy of the shape obtained is dependent on the accuracy of the form cutting tool.

In fact, various methods of producing geometrical surfaces are forming, generating, tracing, tangent tracing, as shown in Fig. 2-4.

2. Analysis of machine tool motions

(1) Motions of machine tool

1) Surface forming motion. There are simple forming motion and compound forming motion. The former means that an independent forming motion includes a rotation or a linear motion only. For example, the rotation of the workpiece and the linear motion of the lathe cutter in cylindrical turning are two simple forming motions, as shown in Fig. 2-5. The latter means that an independent forming motion consists of two or more rotations or linear motions, such as the threading motions in Fig. 2-6. The rotation of the workpiece and the linear motion of the threading tool must keep a strict motion relation, i.e. when the workpiece turns a circle, the cutter must move a lead of thread along the axis of the workpiece.

2) Auxiliary motion. Its function is to realize some auxiliary actions of machine tools in machining process. There are many types of auxiliary motions, such as engaging and disengaging motions, indexing motion, speed changing, clamping and loosening of the workpiece, etc.

(2) Motion connection of the machine tool   Machine tools must be of three basic parts to realize various motions required in machining process. They are:

1) Executive element. It is the executive part whose function is to carry the work and cutter to complete a certain form of movement in light of correct trace, such as spindle, tool rest, working table, etc.

2) Power source. It provides power for the moving parts of the machine tool, such as AC motor, servo motor, etc.

3) Transmitting device. It is used to transmit the power and motion from power source to the executive part, and to transform the movement form, moving direction and speed. The transmitting device can link power source to executive part or executive part to executive part and make them keep a specific motion connection. There are many types of transmitting devices, such as mechanical transmission, electrical transmission, hydraulic transmission, pneumatic transmission and so on. Belt drive, gear drive, worm drive, chain drive and lead screw drive are mechanical transmission devices. There are two types of transmission mechanisms in mechanical transmission devices. One is named as the transmission mechanism with fixed transmission ratio as its transmission ratio and direction are unchangeable, such as gear drive with fixed transmission ratio, lead screw and nut, worm and worm gear; the other is named as the transmission mechanism with position-changeable elements, that is, the position of the transmission member can be changed to change its transmission ratio and direction, such as interpolating gear, sliding gear and clutch, etc.

The transmission connection, which is constituted by transmitting devices from power sources to executive member or from an executive member to the other executive members, is called transmission chain. Transmission principle of thread turning is shown in Fig. 2-7. It can be seen there are

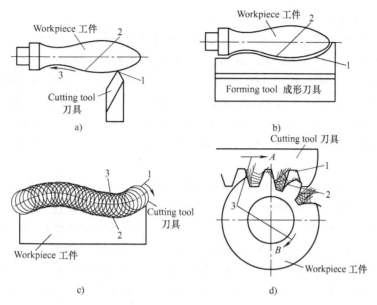

Fig. 2-4　Methods of producing geometrical surfaces 产生几何表面的方法
1—Cutting edge 切削刃　2—Generating line 发生线　3—Trace motion 轨迹运动

1) 表面成形运动。表面成形运动有简单表面成形运动和复合表面成形运动之分。前者是指一个独立的成形运动，由单独的旋转运动或直线运动构成。例如，用普通车刀车削外圆柱面时，工件的旋转运动和刀具的直线移动就是两个简单运动，如图 2-5 所示。后者是由两个或两个以上的旋转运动或直线运动，按照某种确定的运动关系组合而成的，如图 2-6 所示，车削螺纹时，工件旋转运动和刀具的直线移动之间必须保证严格的运动关系，即工件每转 1 转，刀具就必须沿工件轴线移动一个导程。

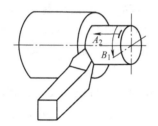

Fig. 2-5　Simple forming motion
简单成形运动

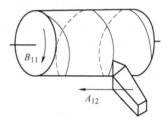

Fig. 2-6　Compound forming motion
复合成形运动

2) 辅助运动。其作用是实现机床加工过程中所需的各种辅助动作。它的种类很多，如切入切出运动、分度运动、变速、夹紧和松开工件所需要的运动等都属于辅助运动。

(2) 机床的运动联系　为了实现加工过程中所需的各种运动，机床必须具备以下三个基本部分：

1) 执行件。执行件是机床运动的执行部件，其作用是带动工件和刀具按照正确的轨迹完成某种形式的运动。机床主轴、刀架、工作台等都是执行件。

2) 动力源。动力源是向执行件提供运动和动力的装置。如交流电动机、伺服电动机等。

two kinds of transmission chains in Fig. 2-7.

1) External-connection transmission chain. This transmission chain connects the executive members with power source (motor), having the executive member get a certain speed and moving direction. In Fig. 2-7, the transmission chain from the motor to the spindle (motor—1—2—u_v—3—4—spindle) is external-connection transmission chain. The change of the transmission radio affects only the productivity or surface finish, and it does not affect the component shape. Therefore, there can be some friction transmission with inaccurate transmission ratio in the external-connection transmission chain.

2) Internal-connection transmission chain. This chain connects one executive member with another to compose a compound forming motion. In Fig. 2-7, the transmission chain from the spindle to the toolrest (spindle—4—5—u_f—6—7—toolrest) is an internal-connection transmission chain. Since it determines the trace of the compound motion (the shape of generating line), there are serious requirements for the relative velocity or relative displacement between the connected executive members in the chain. Therefore, the transmission ratio between the transmission members must be accurate. There should not be any friction transmission member or other transmission pairs with the variation of instantaneous transmission ratio, such as belt drive and chain drive in the internal-connection transmission chain.

2.2 Lathes and lathe cutters

Lathes have found very widely applications in machine manufacturing industry. They take the largest proportional part of machine tools in an ordinary manufacturing factory with about 20% – 35% of the total machine tools. Their common characteristic is using turning tools as major cutting tools to do external and internal cylindrical turning, facing and threading. Generally, the primary motion is the rotary motion of the workpiece, and the feed motion is the linear motion of the cutting tool in lathe operations.

Based on the differences in the usage, performance and structure, lathes are classified into ordinary lathe and floor-type lathe, turret lathe, vertical lathe, automatic lathe, multi-spindle automatic and semi-automatic lathe, copying and multi-cutter lathe, instrument lathe and so on. Besides, there are many other specialized and special purpose lathes, such as threading lathe with high precision, tooth-relieved lathe, crankshaft lathe, camshaft lathe and so on.

2.2.1 Basic contents of turning

Most lathes are capable of performing both external and internal cylindrical machining operations. External cylindrical machining operations include straight turning, taper turning, turning grooves, facing, threading, knurling and cutting off (parting off) stock. Internal cylindrical machining operations include all common hole-machining operations such as drilling, boring, reaming, counterboring, countersinking, center drilling, tapping and threading with a single-point cutting tool. Various lathe operations are illustrated in Fig. 2-8.

3）传动装置。传动装置是传递运动和动力的装置，它把动力源的运动和动力传给执行件，并完成运动形式、方向、速度的转换等工作。传动装置可以连接动力源和执行件，也可以连接执行件和执行件，从而在动力源和执行件之间建立起运动联系，使执行件获得一定的运动。传动装置有机械、电气、液压、气动等多种形式。机械传动装置有带传动、齿轮传动、齿轮齿条传动、链传动、蜗轮蜗杆传动、丝杠螺母传动等多种传动形式。有两种类型的机械传动机构。一种称为定比传动机构，因为其传动比和传动方向是不变的，如定比齿轮传动、丝杠螺母传动、蜗轮蜗杆传动等。另一种称为换置机构，即通过改变传动件的位置，可以改变传动机构的传动比和传动方向，如交换齿轮、滑移齿轮和离合器等。

由动力源到执行件，或执行件到执行件之间通过传动装置构成的传动联系称为传动链。图 2-7 所示为车螺纹传动原理图。由图可知有两种类型的传动链。

1）外联系传动链。该传动链把执行件与动力源（电动机）连接起来，使执行件获得一定的速度和运动方向。在图 2-7 中，从电动机到主轴（电动机—1—2—u_v—3—4—主轴）的传动链即为外联系传动链。该传动链中传动比的变化只会影响生产率或表面粗糙度，不会影响零件的形状。因此，该传动链可以采用传动比不准确的摩擦传动。

2）内联系传动链。该传动链是联系执行件和执行件之间的传动链，以形成复合运动。在图 2-7 中，从主轴到刀架（主轴—4—5—u_f—6—7—刀架）的传动链即为内联系传动链。由于它决定着复合运动的轨迹（即表面形成线的形状），所以对传动链中执行件之间的相对速度和相对位移都有严格要求。因此，内联系传动链的传动比必须准确，不应有摩擦传动或瞬时传动比变化的传动副，如带传动和链传动等。

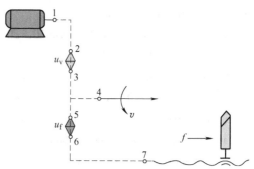

Fig. 2-7 Transmission principle of thread turning
车螺纹传动原理

2.2 车床与车刀

车床在机械制造业中已得到广泛应用。一般机械工厂中，车床在金属切削机床中所占的比例最大，约占金属切削机床总台数的 20%~35%。车床的共同特征是以车刀为主要切削工具，车削各种零件的外圆、内孔、端面及螺纹等。对车削加工来说，通常工件的旋转运动为主运动，车刀的直线移动为进给运动。

根据车床的用途、性能和结构的不同，车床分为卧式车床和落地车床、转塔车床、立式车床、自动车床、多轴自动和半自动车床、仿形和多刀车床、仪表车床等。此外，还有许多专门化车床和专用车床，如高精度螺纹车床、铲齿车床、曲轴车床、凸轮轴车床等。

2.2.1 车削的基本内容

大多数车床可以进行内、外圆表面的加工。外圆表面的加工包括车外圆柱面、车锥面、车沟槽、车端面、车螺纹、滚花和切断。内圆表面的加工包括所有常见孔的加工，如钻孔、镗孔、铰孔、锪沉头孔、钻中心孔、攻丝和车螺纹等。各种车削工艺内容如图 2-8 所示。

2.2.2 Constructional features of a center lathe (CA6140)

1. Components of the lathe CA6140

The layout of the lathe with model CA6140 is shown in Fig. 2-9. The major components of the lathe are as follows.

(1) Headstock 1 It is installed on the left end of the lathe bed 6. There is a spindle in the headstock, which is used to rotate the workpiece through the chuck or other fixtures. Its main functions are to carry the spindle and realize its start, stop, reversing, brake and changing speed, and to transmit the drive motion to the spindle and the feed motion from the spindle to the feed system.

(2) Compound rest 3 It locates on the middle of the lathe bed 6 and move longitudinally along the guideways on the lathe bed. The compound rest is composed of several tool rests. Its function is to hold the lathe cutter and make the cutter move longitudinally and transversely.

(3) Tailstock 5 It is mounted on the inner guideways on lathe bed 6. Its position can be set along the inner guideways. Its function is to support the workpiece by means of back center. The tailstock can also be used to hold the drill and other hole-making cutting tools to machine holes.

(4) Bed 6 It is fixed on the base 10. The bed is a fundamental support part. The primary assemblies are mounted on the bed.

(5) Apron 9 It is fixed on the bottom of the compound rest 3, and carries the compound rest to move longitudinally. Its function is to transmit the movement from the feed box to the compound rest, and to bring along the compound rest to realize longitudinal feed and cross feed, quick travel, or threading. Various manipulating handles and buttons are mounted on the apron for operators to operate the machine tool conveniently.

(6) Feed box 11 It is fixed on the left front side of the bed 6. The feed box is a primary speed-change mechanism. Its function is to change the pitch of the thread to be cut or the longitudinal feed rate.

2. Major specification of the lathe CA6140

(1) Swing over the bed　　　　　　400mm

(2) Distance between centers　　　750;1000;1500;2000mm

(3) Swing over the cross slide　　　210mm

(4) Horse power/rotation speed of the main motor 7.5kW/1450 r/min

(5) RPM series and rotation speed of the spindle.

Counter-clockwise rotation speed from 10 r/min to 1400 r/min with 24 series in total

Clockwise rotation speed from 14 r/min to 1580 r/min with 12 series in total

(6) Feed range

Longitudinal feeds from 0.028mm/r to 6.33mm/r with 64 series in total

Transverse feeds from 0.014 mm/r to 3.16 mm/r with 64 series in total

(7) Screw cutting capacity

Metric system thread with 44 kinds of lead: $S = 1 - 192$mm.

British system thread with 20 kinds of a: $a = 2 - 24$ tooth number/in.

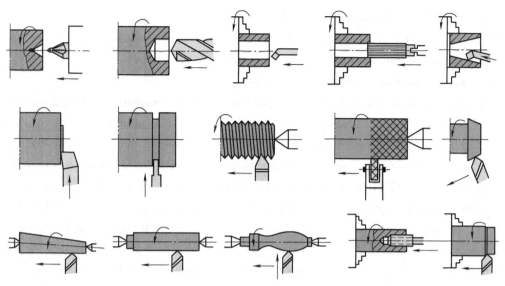

Fig. 2-8　Lathe operations　车床工艺内容

2.2.2　CA6140 卧式车床的结构特征

1. CA6140 卧式车床的组成

图 2-9 所示为 CA6140 型卧式车床的外形图，其主要组成部件如下：

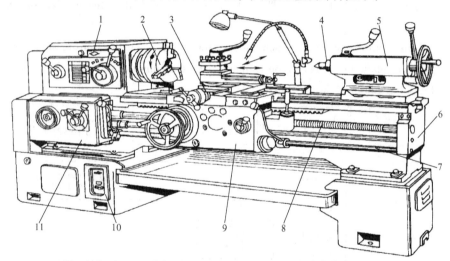

Fig. 2-9　Layout of the CA6140 lathe　CA6140 型卧式车床的布局
1—Headstock 主轴箱　2—Chuck 卡盘　3—Compound rest 刀架部件　4—Back center 尾座顶尖
5—Tailstock 尾座　6—Lathe bed 床身　7—Feed rod 光杠　8—Lead screw 丝杠
9—Apron 溜板箱　10—Base 底座　11—Feed box 进给箱

（1）主轴箱 1　主轴箱固定在床身 6 的左端。装在主轴箱中的主轴，通过卡盘等夹具，装夹工件。主轴箱的主要功用是支承并实现主轴的起动、停止、换向、制动和变速，并把驱动运动传给主轴，把进给运动从主轴传到进给系统。

（2）刀架部件 3　它位于床身 6 的中部，并可沿床身上的刀架导轨作纵向移动。刀架部件由几层刀架组成，它的功用是装夹车刀，并使车刀作纵向、横向或斜向运动。

Module thread with 39 kinds of m: $m = 0.25 - 48$.

Diametral pitch thread with 37 kinds of DP: $DP = 1 - 96$ tooth number/in.

(8) Others: accuracy available, spindle nose diameter and hole size.

3. Transmission system of the lathe CA6140

(1) Transmission system diagram of a machine tool   A transmission system diagram of a machine tool is generally used to analyze the transmission situation of a machine tool. The transmission system diagram is a sketch map used for indicating the transmission relations of all motions in a machine tool. In the drawing, a specific drive mechanism of a transmission chain is expressed by a simple prescribed symbol. Tooth number of gear and worm gear, thread number of worm, lead of screw, pulley diameter, power and revolution of electric motor, and sequence number of shaft should also be marked on a specific drive member. The drive mechanisms in transmission chain are arranged in turn in the light of motion transmitting sequence. The transmission system diagram is made in the form of development drawing in the outline contour of the machine tool. In order to spread a three-dimensional transmission structure and plot it in an ichnography, sometimes, some transmission pairs which have lost transmission connection in development drawing have to be connected by means of brace or dashed line. A transmission system diagram can only express the transmission relation, and doesn't reflect the practical dimension and special position of each element. Starting from power source (such as electric motor), the serial numbers of transmission shaft are expressed with Roman numerals (Ⅰ, Ⅱ, Ⅲ...) in turn in the light of the motion transmitting sequence.

(2) Block diagram of the lathe transmission system   Based on the transmission sequence and the relative position relation of transmission elements, the transmission system of the lathe can be expressed in block diagram as shown in Fig. 2-10.

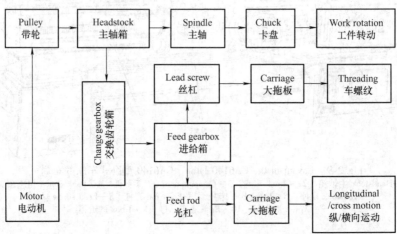

Fig. 2-10   Block diagram of the lathe transmission system 车床传动系统框图

(3) Transmission system diagram of the lathe   Transmission system diagram of the lathe CA6140 is shown in Fig. 2-11.

The lathe must have two motions: one is the rotary motion of the workpiece—primary motion;

(3) 尾座 5　它装在床身 6 的尾座导轨上,并可沿此导轨纵向调整位置。尾座的功用是用后顶尖支承工件。在尾座上还可以安装钻头等孔加工刀具,以进行孔加工。

(4) 床身 6　床身固定在底座 10 上。床身是车床的基本支承件。在床身上安装着车床的各个主要部件。

(5) 溜板箱 9　它固定在刀架部件 3 的底部,可带动刀架一起作纵向运动。溜板箱的功用是把进给箱传来的运动传递给刀架,使刀架实现纵向进给、横向进给、快速移动或车螺纹。在溜板箱上装有各种操纵手柄及按钮,工作时工人可以方便地操作机床。

(6) 进给箱 11　它固定在床身 6 的左前侧。进给箱是进给运动传动链中主要的传动比变换装置(变速装置,变速机构),它的功用是改变被加工螺纹的螺距或纵向进给的进给量。

2. CA6140 车床的主要技术规格

(1) 床身上最大工件回转直径　　　　　　　　　　　　　　　　400mm
(2) 两顶尖之间的距离　　　　　　　　　　　　　　750；1000；1500；2000mm
(3) 滑鞍上最大工件回转直径　　　　　　　　　　　　　　　　210mm
(4) 主电动机的功率/转速　　　　　　　　　　　　　　7.5kW/1450r/min
(5) 转速级数和主轴转速
　　正转 24 级　　　　　　　　　　　　　　　　　　　10 ~ 1400r/min
　　反转 12 级　　　　　　　　　　　　　　　　　　　14 ~ 1580r/min
(6) 进给量范围
　　纵向进给量 64 级　　　　　　　　　　　　　　　0.028 ~ 6.33mm/r
　　横向进给量 64 级　　　　　　　　　　　　　　　0.014 ~ 3.16mm/r
(7) 螺纹加工能力
　　米制螺纹　　　　　44 种　　　　　　　　　　$S = 1 ~ 192$mm
　　英制螺纹　　　　　20 种　　　　　　　　　　$a = 2 ~ 24$ 扣/in
　　模数螺纹　　　　　39 种　　　　　　　　　　$m = 0.25 ~ 48$mm
　　径节螺纹　　　　　37 种　　　　　　　　　　$DP = 1 ~ 96$ 牙/in

(8) 其他,如可达到的精度,主轴端直径和主轴孔的尺寸等。

3. CA6140 车床的传动系统

(1) 机床的传动系统图　机床的传动系统图通常用来分析机床的传动情况。机床的传动系统图是表示机床全部运动传动关系的示意图,在图中用简单的规定符号代表各种传动元件。齿轮和蜗轮的齿数、蜗杆头数、螺纹导程、带轮直径、电动机的功率和转速,以及各轴的序号也要标在具体的传动件上。机床的传动系统图画在一个能反映机床外形和各主要部件相互位置的投影面上,并尽可能绘制在机床外形的轮廓线内。在图中,各传动元件是按照运动传递的先后顺序,以展开图的形式画出来的。要把一个立体的传动结构展开并绘制在一个平面图中,有时,对于展开后失去联系的传动副,要用大括号或虚线连接起来以表示它们的传动联系。传动系统图只能表示传动关系,并不代表各元件的实际尺寸和空间位置。传动轴的编号,通常从动力源(如电动机等)开始,按运动传递顺序,顺次地用罗马数字Ⅰ、Ⅱ、Ⅲ……表示。

(2) 车床传动系统框图　根据传动顺序以及各传动件的相互位置关系,车床的传动系统可以用图 2-10 所示的框图来表示。

(3) 车床的传动系统图　CA6140 车床的传动系统如图 2-11 所示。

the other is the linear motion of the cutter—feed motion.

In order to realize these motions, the transmission system of the lathe should have two transmission chains. One is the transmission chain of primary motion, the other is the transmission chain of longitudinal and transverse feed motions. Let's take the lathe CA6140 for example to analyze its transmission system.

4. To analyze the transmission chain of the lathe CA6140

(1) Transmission chain of the primary motion   The functions of the transmission chain of the primary motion are to transmit the motion and power from power source to the spindle, and make the spindle rotate at proper RPM. Both the rotating speed and the direction of the spindle can be changed.

To understand the transmission route of a machine tool is the foundation of recognizing and analyzing a machine tool. Whenever a transmission chain is analyzed, the general way is to " hold two ends and connect the middle". That is to say, when analyzing the transmission route of a transmission chain, it should be clear firstly that what the first part is and what the last part is in the transmission chain. Then the transmission connection between two end parts is to be found. As the transmission chain of the primary motion is an external-connection transmission chain, the first part is the electric motor, and the last part is the spindle, i.e. electric motor—spindle. Then starting from the electric motor, you try to find out the transmission connection between the motor and the spindle in the light of the motion transmitting sequence.

1) Transmission route. It can be seen from Fig. 2-11 that the motion of the electric motor is transmitted via V-belt to shaft I. A bidirectional multi-piece friction clutch M_1 is assembled on shaft I to make spindle VI rotate counter-clockwise, clockwise, and stop. The spindle rotates counter-clockwise if the left part of M_1 is engaged, or clockwise if the right part of M_1 is engaged. When M_1 is placed in the middle position, the spindle stops turning. The motion of shaft I is transmitted to shaft III firstly via M_1 and shaft II, and then to the spindle VI via two transmission routes: when the sliding gear Z_{50} on the spindle VI is shifted to the left, the motion from shaft III is directly transmitted to the spindle VI via gear pairs 63/50, and the spindle VI can obtain higher RPMs; when the sliding gear Z_{50} is shifted to the right and engaged with M_2, the motion from shaft III is transmitted to spindle VI via shafts III-IV-V and make the spindle obtain the middle and lower rotation speeds.

Transmission route expression of the primary motion:

$$\text{Motor}(电动机) \atop (7.5\text{kW},1450\text{r/min}) - \frac{\varphi130}{\varphi230} - \text{I} - \left\{ \begin{array}{l} M_1 \rightarrow \text{Left}(左) - \left\{ \begin{array}{c} \frac{56}{38} \\ \frac{51}{43} \end{array} \right\} \\ M_1 \rightarrow \text{Right}(右) - \frac{50}{34} - \text{VII} - \frac{34}{30} \end{array} \right\} - \text{II} - \left\{ \begin{array}{c} \frac{39}{41} \\ \frac{30}{50} \\ \frac{22}{58} \end{array} \right\} - \text{III} -$$

$$\left\{ \begin{array}{l} \left\{ \begin{array}{c} \frac{20}{80} \\ \frac{50}{5} \end{array} \right\} - \text{IV} - \left\{ \begin{array}{c} \frac{20}{80} \\ \frac{50}{51} \end{array} \right\} - \text{V} - \frac{26}{58} - M_2(\text{engaged } 合) \\ \frac{63}{50} \end{array} \right\} - \text{Spindle VI}(主轴)$$

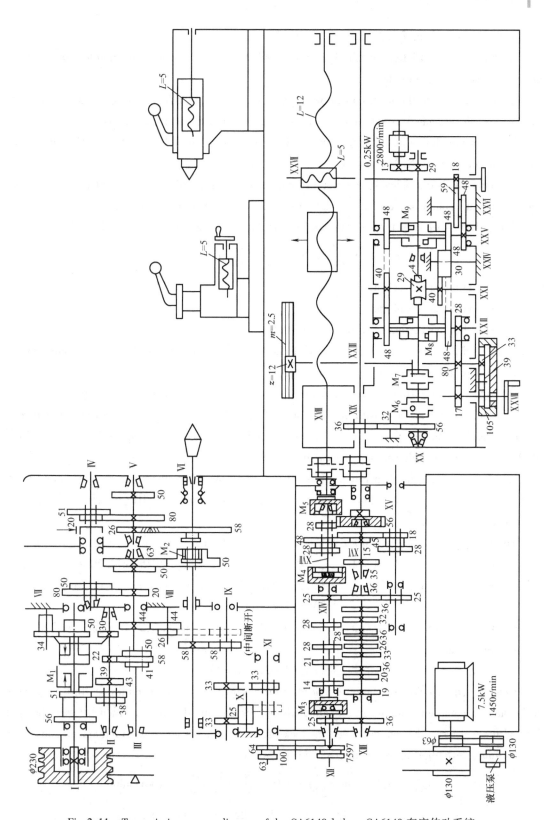

Fig. 2-11 Transmission system diagram of the CA6140 lathe CA6140 车床传动系统

It can be seen from Fig. 2-11 that:

① A multi-piece friction clutch M_1 is assembled on shaft Ⅰ to make the spindle rotate counter-clockwise, clockwise, and stop.

② Join of gears to the shaft can be with single key, spline, sliding bearing or radial ball bearing, which can be named as stationary gear, sliding gear and free gear or idle gear.

③ When the sliding gear Z_{50} on spindle Ⅵ is shifted to right end to mesh with M_2, the spindle can obtain low 18-level RPMs with 10r/min – 500r/min;

④ When M_2 is disengaged, the motion from shaft Ⅲ can be directly transmitted to spindle Ⅵ. The spindle can obtain 6 – level high rotation speeds with 450r/min – 1400r/min.

The rotation speeds of the spindle can be calculated by the motion balance Eq. (2-1).

$$n_s = n_m \times \frac{130}{230} \times (1-\varepsilon) u_{Ⅰ-Ⅱ} \times u_{Ⅱ-Ⅲ} \times u_{Ⅲ-Ⅳ} \qquad (2-1)$$

where, n_s —rotation speed of spindle, r/min;

n_m —rotation speed of motor, r/min;

ε —sliding coefficient of V pulley, $\varepsilon = 0.02$;

$u_{Ⅰ-Ⅱ}$ —variable drive ratio between shaft Ⅰ and shaft Ⅱ. The rest can be deduced similarly.

(2) Rotation speed of the spindle and rotation speed series.

① High speed transmission route. When the sliding gear Z_{50} on spindle Ⅵ is shifted to the left (M_2 is disengaged) and mesh with gear Z_{63} on shaft Ⅲ, it can be seen from the transmission route that the spindle can get 6-level high rotation speeds (2 × 3 = 6) ranging from 450r/min to 1400r/min when it rotates counter clockwise.

② Low speed transmission route. When the sliding gear Z_{50} on the spindle Ⅵ is shifted to the right to mesh with clutch M_2, then the motion is transmitted from shaft Ⅲ, via shaft Ⅳ and shaft Ⅴ, finally to the spindle.

Theoretically, the spindle can get 24 – level (2 × 3 × 2 × 2 = 24) low speeds. In fact, $u_1 = 20/80 \times 20/80 = 1/16$, $u_2 = 50/50 \times 20/80 = 1/4$, $u_3 = 20/80 \times 51/50 \approx 1/4$, $u_4 = 50/50 \times 51/50 \approx 1$. Because $u_2 \approx u_3$, there are only 3 kinds of transmission ratios. Therefore, the real rotation speed levels are

$$2 \times 3 \times (2 \times 2 - 1) = 18$$

③ Rotation speed levels in the clockwise rotation. It is clear that the rotation speed levels in the clockwise rotation are

$$3 \times (1 + 2 \times 2 - 1) = 12$$

Generally, the clockwise rotation is not used for cutting. It is used to return the threading cutter to the starting point along thread in threading without disengaging the drive relation between the spindle and the tool rest. The rotation speed in clockwise rotation is higher so as to reduce auxiliary time.

④ Limit revolution in the counter-clockwise rotation

$$n_{max} = 1450 \text{r/min} \times \frac{130}{230} \times 0.98 \times \frac{56}{38} \times \frac{39}{41} \times \frac{63}{50} \approx 1400 \text{r/min}$$

$$n_{min} = 1450 \text{r/min} \times \frac{130}{230} \times 0.98 \times \frac{51}{43} \times \frac{22}{58} \times \frac{20}{80} \times \frac{20}{80} \times \frac{26}{58} \approx 10 \text{r/min}$$

根据车床的功用，车床必须具有两个运动：一是工件的回转运动，即主运动；二是刀具的直线运动，即进给运动。

为了实现这些运动，车床的传动系统必须具有两个传动链，一个是主运动传动链，另一个是纵向/横向进给运动传动链。下面就以 CA6140 车床为例来分析其传动系统。

4. CA6140 车床的传动链分析

（1）主运动传动链　主运动传动链的功用是将电动机的旋转运动及动力传递给主轴，使主轴以合适的转速带动工件旋转。车床的主轴应能变速与换向。

看懂机床的传动路线是认识和分析机床的基础。每当分析机床的传动链时，通常的方法是"抓两端，连中间"。也就是说，当分析某一条传动链的传动路线时，首先应搞清楚该传动链中哪是首端件，哪是末端件，然后再找它们之间的传动联系，这样就可以很容易地找出传动路线。由于车床主运动传动链是外联系传动链，因此其首端件是电动机，而末端件是主轴，即电动机—主轴；然后从电动机开始，按照运动传递顺序，设法找出电动机和主轴之间的传动联系。

1）传动路线。由图 2-11 看出，运动由电动机经 V 型带传至主轴箱中的轴 I。在轴 I 上装有双向多片式摩擦离合器 M_1，其作用是使主轴（轴Ⅵ）正转、反转或停止。离合器 M_1 左半部分接合时，主轴正转；右半部分接合时，主轴反转；左右都不接合时，轴 I 空转，主轴停止转动。轴 I 的运动经 M_1—轴Ⅱ—轴Ⅲ，然后分成两条路线传给主轴：当轴Ⅵ上的滑移齿轮 Z_{50} 移至左边位置时，运动从轴Ⅲ经齿轮副 63/50 直接传给主轴Ⅵ，使主轴得到高转速；当滑移齿轮 Z_{50} 向右移，使齿轮式离合器 M_2 接合时，则运动经轴Ⅲ—轴Ⅳ—轴Ⅴ传给主轴Ⅵ，使主轴获得中、低转速。

主运动传动路线表达式为

$$\text{Motor（电动机）} \atop (7.5\text{kW},\,1450\text{r/min}) - \frac{\varphi 130}{\varphi 230} - \text{I} - \begin{Bmatrix} M_1 \rightarrow \text{Left（左）} - \begin{Bmatrix} \frac{56}{38} \\ \frac{51}{43} \end{Bmatrix} \\ M_1 \rightarrow \text{Right（右）} - \frac{50}{34} - \text{Ⅶ} - \frac{34}{30} \end{Bmatrix} - \text{Ⅱ} - \begin{Bmatrix} \frac{39}{41} \\ \frac{30}{50} \\ \frac{22}{58} \end{Bmatrix} - \text{Ⅲ} -$$

$$\begin{Bmatrix} \begin{Bmatrix} \frac{20}{80} \\ \frac{50}{50} \end{Bmatrix} - \text{Ⅳ} - \begin{Bmatrix} \frac{20}{80} \\ \frac{50}{51} \end{Bmatrix} - \text{Ⅴ} - \frac{26}{58} - M_2\,(\text{engaged 合}) \\ \frac{63}{50} \end{Bmatrix} - \text{Spindle Ⅵ（主轴）}$$

从主传动系统图 2-11 可以看出：

① 在轴 I 上安装的双向多片式摩擦离合器 M_1 可以使主轴正转、反转和停止。

② 齿轮可以用单键、花键、滑动轴承或深沟球轴承与轴连接，从而分别称为固定齿轮、滑移齿轮和空套齿轮（惰轮）。

③ 当主轴Ⅵ上的滑移齿轮 Z_{50} 移到右端与离合器 M_2 啮合时，主轴可以获得从 10~500r/min 共 18 种低转速。

④ 当离合器 M_2 脱开时，来自第Ⅲ轴的运动可以传到主轴Ⅵ上，从而使主轴获得从 450~1400r/min 共 6 种高转速。

主轴的转速可用运动平衡式（2-1）进行计算。

$$n_s = n_m \times \frac{130}{230} \times (1-\varepsilon) u_{\text{I-Ⅱ}} \times u_{\text{Ⅱ-Ⅲ}} \times u_{\text{Ⅲ-Ⅳ}} \tag{2-1}$$

(2) Transmission chain of the feed motion Its functions is to make the tool rest move longitudinally or transversely at the required speed. It consists of threading transmission chain and power-driven feed transmission chain.

To find the first and last part: Spindle—Toolrest

1) Threading transmission chain. CA6140 lathe can be used to machine 4 kinds of standard threads: metric system, British system, module thread and diametral pitch thread, no matter whether it is right-hand or left-hand. It can also be used to machine large lead threads, non-standard threads and precision threads.

To cut an accurate thread, it is essential to maintain the correct relationship between the movement of the carriage and the rotation speed of the work. This can be achieved by means of the lead screw, which is driven by a train of gears from the spindle. In thread turning, the feed of the tool per revolution is equal to the lead of the thread to be cut.

Calculating displacement:

$$1\text{r} \mid \text{work} = L \mid \text{tool}(\text{mm})$$

where, L—lead of the thread. This displacement expression means that while the workpiece being cut turns one circle, the cutting tool must move the distance which is equal to the lead of the thread to be cut.

Motion balance equation:

$$1_{\text{spindle}} \times u \times L_{\text{thread}} = L_{\text{work}} \qquad (2\text{-}2)$$

where, u—total drive ratio from the spindle to the lead screw;

L_{thread}—pitch of lead screw (12mm);

L_{work}—lead of the screw to be machined.

The transmission route expression of the threading transmission chain is shown in Fig. 2-12.

Based on the transmission route expression mentioned above, a motion balance equation for each thread can be listed so as to make the necessary analysis and calculation. In the transmission route expression, u_j represents 8 kinds of transmission ratios $\left(\dfrac{26}{28}, \dfrac{28}{28}, \dfrac{32}{28}, \dfrac{36}{28}, \dfrac{19}{14}, \dfrac{20}{14}, \dfrac{33}{21}, \dfrac{36}{21}\right)$ between shaft XIII and shaft XIV, and is named as basic shifting group in the feed gearbox. u_b represents 4 kinds of transmission ratios $\left(\dfrac{28}{35} \times \dfrac{35}{28}, \dfrac{18}{45} \times \dfrac{35}{28}, \dfrac{28}{35} \times \dfrac{15}{48}, \dfrac{18}{45} \times \dfrac{15}{48}\right)$ between shaft XV and shaft XVII, and is named as doubled shifting group.

① To cut metric system thread. Transmission route expression:

$$\text{Spindle VI} - \dfrac{58}{58} - \text{IX} - \begin{cases} \dfrac{33}{33}(\text{right} - \text{hand}) \\ \dfrac{33}{25} - \text{X} - \dfrac{25}{33}(\text{left} - \text{hand}) \end{cases} - \text{XI} - \dfrac{63}{100} \times \dfrac{100}{75} - \text{XII} - \dfrac{25}{36} - \text{XIII}$$

$$- u_j - \text{XIV} - \dfrac{25}{36} \times \dfrac{36}{25} - \text{XV} - u_b - \text{XVII} - \overrightarrow{\text{M}_5} - \text{XVIII}(\text{Lead screw})\text{-Toolrest}$$

The motion balance equation is expressed by Eq. (2-3)

式中 n_s——主轴的转速（r/min）；

n_m——电动机的转速（r/min）；

ε——V 带传动的滑动系数，$\varepsilon = 0.02$；

u_{I-II}——轴 I 与轴 II 之间的可变传动比，其余类推。

2）主轴的转速以及转速级数。

① 高速传动路线。当主轴 VI 上的滑移齿轮 Z_{50} 移到左端（离合器 M_2 脱开）并与轴 III 上的齿轮 Z_{63} 啮合时，由传动路线图可以看出，当主轴逆时针旋转（正转）时，主轴可以获得从 450 ~ 1400r/min 共 6 种高转速（$2 \times 3 = 6$）。

② 低速传动路线。当主轴 VI 上的滑移齿轮 Z_{50} 移到右端与离合器 M_2 啮合时，来自轴 III 的运动相继经过轴 IV 和轴 V 最后传递到主轴上。理论上，主轴应该获得 24 级（$2 \times 3 \times 2 \times 2 = 24$）低转速。但事实上，

$$u_1 = \frac{20}{80} \times \frac{20}{80} = \frac{1}{16}; \quad u_2 = \frac{20}{80} \times \frac{51}{50} \approx \frac{1}{4}; \quad u_3 = \frac{50}{50} \times \frac{20}{80} = \frac{1}{4}; \quad u_4 = \frac{50}{50} \times \frac{51}{50} \approx 1$$

因为 $u_2 \approx u_3$，所以实际上只有 3 种不同的传动比。因此，由低速路线传动时，使主轴获得的有效的转速级数不是 $2 \times 3 \times 4 = 24$ 级，而是从 10 ~ 500r/min 共有 $2 \times 3 \times (4-1) = 18$ 级低转速。

③ 主轴顺时针旋转（反转）时的转速级数。显然，主轴反转时的转速级数为

$$3 \times (1 + 2 \times 2 - 1) = 12$$

主轴反转通常不是用于切削，而是为了车螺纹时退刀。这样，就可以在不断开主轴和刀架之间传动链的情况下退刀。为了节省辅助时间，主轴反转的转速较高。

④ 主轴正转时的极限转速。

$$n_{max} = 1450 \text{r/min} \times \frac{130}{230} \times 0.98 \times \frac{56}{38} \times \frac{39}{41} \times \frac{63}{50} \approx 1400 \text{r/min}$$

$$n_{min} = 1450 \text{r/min} \times \frac{130}{230} \times 0.98 \times \frac{51}{43} \times \frac{22}{58} \times \frac{20}{80} \times \frac{20}{80} \times \frac{26}{58} \approx 10 \text{r/min}$$

（2）进给运动传动链　进给运动传动链的功用是使刀架以要求的速度作纵向或横向移动。它由车螺纹进给传动链和纵、横向机动进给传动链构成。

查找首、末两端件：主轴和刀架。

1）车螺纹进给传动链。CA6140 型卧式车床可以车削米制、英制、模数和径节四种标准螺纹。除此之外，还可以车削大导程螺纹、非标准螺纹和精密螺纹。

要切削精确的螺纹，必须保证溜板移动与工件转动之间的关系准确。这可以通过丝杠传动来保证，而丝杠是由主轴通过一系列齿轮传动的。在车螺纹时，车刀的每转进给量大小等于被加工螺纹的导程。

计算位移：

$$1\text{r} | \text{工件} = L | \text{刀具 （mm）}$$

其中，L 为丝杠的导程。这个位移表达式意思是每当加工的工件转 1 转，刀具就必须沿工件走过一个被加工螺纹的导程。

运动平衡方程：

$$1_{\text{主轴}} \times u \times L_{\text{丝杠}} = L_{\text{工件}} \tag{2-2}$$

$$L = 1 \times \frac{58}{58} \times \frac{33}{33} \times \frac{63}{100} \times \frac{100}{75} \times \frac{25}{36} \times u_j \times \frac{25}{36} \times \frac{36}{25} \times u_b \times 12 \tag{2-3}$$

After simplifying, Eq. (2-3) is changed into Eq. (2-4).

$$L = 7 u_j \times u_b \tag{2-4}$$

where, u_j, u_b—drive ratio of the basic shifting group and doubled shifting group. If u_j, u_b are chosen properly, all lead values (1 – 12mm) in the metric system can be obtained.

32 kinds of metric system leads listed in Tab. 2-1 can be machined by changing u_j and u_b, where there are 20 kinds of standard leads.

It can be seen from Tab. 2-1 that CA6140 lathe has the ability to cut the metric system threads whose maximum lead is only 12 mm. When the leads larger than 12 mm are required to cut on this lathe, the lead-enlarging transmission mechanism must be used. Here the sliding gear Z_{58} on shaft Ⅸ should be shifted to the right (the position of the dashed line shown in Fig. 2-11) to engage with the gear Z_{26} on shaft Ⅷ. And then the connection between shaft Ⅵ and shaft Ⅸ is realized by means of a train of gears via shafts Ⅴ, Ⅳ, Ⅲ and Ⅷ, instead of the gear drive pairs 58/58. Therefore, the lead-enlarging transmission route is expressed as follows:

Table 2-1 Metric system leads machined by lathe CA6140 CA6140 车床加工的米制螺纹

u_b \ L \ u_j	$u_{b1} = \frac{18}{45} \times \frac{15}{48} = \frac{1}{8}$	$u_{b2} = \frac{28}{35} \times \frac{15}{48} = \frac{1}{4}$	$u_{b3} = \frac{18}{45} \times \frac{35}{28} = \frac{1}{2}$	$u_{b4} = \frac{28}{35} \times \frac{35}{28} = 1$
$u_{j1} = \frac{26}{28} = \frac{6.5}{7}$	—	—	—	—
$u_{j2} = \frac{28}{28} = \frac{7}{7}$	—	1.75	3.5	7
$u_{j3} = \frac{32}{28} = \frac{8}{7}$	1	2	4	8
$u_{j4} = \frac{36}{28} = \frac{9}{7}$	—	2.25	4.5	9
$u_{j5} = \frac{19}{14} = \frac{9.5}{7}$	—	—	—	—
$u_{j6} = \frac{20}{14} = \frac{10}{7}$	1.25	2.5	5	10
$u_{j7} = \frac{33}{21} = \frac{11}{7}$	—	—	5.5	11
$u_{j8} = \frac{36}{21} = \frac{12}{7}$	1.5	3	6	12

$$\text{Spindle Ⅵ} - \begin{cases} (\text{Enlarging lead}) \dfrac{58}{26} - \text{Ⅴ} - \dfrac{80}{20} - \text{Ⅳ} - \left(\dfrac{\frac{50}{50}}{\frac{80}{20}}\right) - \text{Ⅲ} - \dfrac{44}{44} \times \dfrac{26}{58} \\ (\text{Normal lead}) \dfrac{58}{58} \end{cases} - \text{Ⅸ} -$$

It can be seen from the lead-enlarging transmission route that the drive ratio from spindle Ⅵ to

式中 u——从主轴到丝杠之间的总传动比；

$L_{丝杠}$——丝杠的螺距（12mm）；

$L_{工件}$——被加工螺纹的导程。

车螺纹传动链的传动路线表达式如图 2-12 所示。

根据上述传动路线表达式，可以列出每种螺纹的运动平衡式，进行分析和计算。表达式中 u_j 代表轴 XIII 至轴 XIV 间的 8 种可供选择的传动比 $\left(\dfrac{26}{28}, \dfrac{28}{28}, \dfrac{32}{28}, \dfrac{36}{28}, \dfrac{19}{14}, \dfrac{20}{14}, \dfrac{33}{21}, \dfrac{36}{21}\right)$，称进给箱的基本变速组，简称基本组；$u_b$ 代表轴 XV 至轴 XVII 间的 4 种传动比 $\left(\dfrac{28}{35}\times\dfrac{35}{28}, \dfrac{18}{45}\times\dfrac{35}{28}, \dfrac{28}{35}\times\dfrac{15}{48}, \dfrac{18}{45}\times\dfrac{15}{48}\right)$，称增倍组。

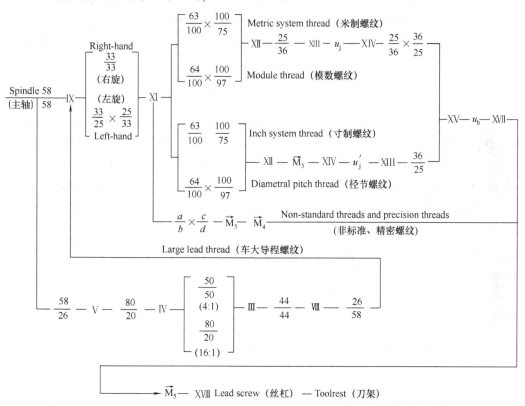

Fig. 2-12 Transmission route of threading transmission chain 车螺纹传动链的传动路线

① 切削米制螺纹。传动路线表达式为

$$主轴 VI - \dfrac{58}{58} - IX - \left\{\begin{array}{l}\dfrac{33}{33}(右旋螺纹)\\ \dfrac{33}{25} - X - \dfrac{25}{33}(左旋螺纹)\end{array}\right\} - XI - \dfrac{63}{100}\times\dfrac{100}{75} - XII - \dfrac{25}{36} - XIII - u_j - XIV - \dfrac{25}{36}\times\dfrac{36}{25} - XV - u_b - XVII - \vec{M}_5 - XVIII (丝杠) - 刀架$$

车削米制螺纹时的运动平衡式为

$$L = 1\times\dfrac{58}{58}\times\dfrac{33}{33}\times\dfrac{63}{100}\times\dfrac{100}{75}\times\dfrac{25}{36}\times u_j\times\dfrac{25}{36}\times\dfrac{36}{25}\times u_b\times 12 \tag{2-3}$$

shaft Ⅸ can be expressed as Eq. (2-5) when the rotation speed of the spindle is 40 – 125r/min,

$$u_{k1} = \frac{58}{26} \times \frac{80}{20} \times \frac{50}{50} \times \frac{44}{44} \times \frac{26}{58} = 4 \qquad (2\text{-}5)$$

When the rotation speed of the spindle is 10 – 32r/min, the drive ratio from spindle Ⅵ to shaft Ⅸ can be expressed as Eq. (2-6).

$$u_{k2} = \frac{58}{26} \times \frac{80}{20} \times \frac{80}{20} \times \frac{44}{44} \times \frac{26}{58} = 16 \qquad (2\text{-}6)$$

Therefore, the function of the lead-enlarging transmission mechanism is to enlarge the lead of the screw up to either 4 times or 16 times so as to cut the screws with the large lead.

② To cut other threads. As the lead of the lead screw in CA6140 lathe is expressed in mm, the pitch and lead of all screws should be expressed in mm in threading.

a. For the British system thread, as it is expressed by tooth number per inch α, its lead L_α should be expressed in mm, as shown in Eq. (2-7).

$$L_\alpha = \frac{1}{\alpha}\text{in} = \frac{25.4}{\alpha}\text{mm} \qquad (2\text{-}7)$$

This means that there should be a special factor 25.4 presented in the British system transmission route. Compared with the metric system transmission route, the 25.4 can be produced by changing the drive ratio of a part of transmission route, i.e., exchanging the initiative and passive relation between two shafts in basic shifting group so that the denominator is arithmetic progression. The rest is the same as the metric system transmission route. The motion balance equation is expressed as Eq. (2-8).

$$L_\alpha = 1_{(\text{spindle}\,主轴)} \times \frac{58}{58} \times \frac{33}{33} \times \frac{63}{100} \times \frac{100}{75} \times \frac{1}{u_j} \times \frac{36}{25} \times u_b \times 12 = \frac{4}{7} \times 25.4 \times \frac{1}{u_j} \times u_b \qquad (2\text{-}8)$$

After simplifying, Eq. (2-8) is changed into Eq. (2-9).

$$\alpha = \frac{7}{4} \times \frac{u_j}{u_b}(\text{tooth number/in.}) \qquad (2\text{-}9)$$

b. For the module thread (metric worm), the relation between the pitch and the module is expressed as Eq. (2-10).

$$p_m = \pi m (\text{mm}) \qquad (2\text{-}10)$$

The lead of the module thread is expressed as Eq. (2-11), where k is the number of threads.

$$L_m = k\pi m (\text{mm}) \qquad (2\text{-}11)$$

The transmission route for the cutting module thread is nearly the same as that in the cutting metric thread except for interpolating gears. As there is a factor of π in the pitch of the module thread, the interpolating gears $\frac{64}{100} \times \frac{100}{97}$ should be used. Here the motion balance equation can be expressed as Eq. (2-12).

$$L_m = 1_{(\text{spindle}\,主轴)} \times \frac{58}{58} \times \frac{33}{33} \times \frac{64}{100} \times \frac{100}{97} \times \frac{25}{36} \times u_j \times \frac{25}{36} \times \frac{36}{25} \times u_b \times 12 \qquad (2\text{-}12)$$

Here, $\frac{64}{100} \times \frac{100}{97} \times \frac{25}{36} \approx \frac{7\pi}{48}$. After simplifying, the relation of m to u_j and u_b can be expressed

化简后得
$$L = 7u_j \times u_b \tag{2-4}$$

式中 u_j、u_b——分别代表基本组、增倍组的传动比。如果选择合适的 u_j、u_b，就可以得到米制中从 1~12mm 的全部导程。

通过改变 u_j、u_b 的大小可以加工表 2-1 列出的 32 种导程的米制螺纹，其中 20 种为标准导程的螺纹。

从表 2-1 可以看出，CA6140 车床能够车削米制螺纹的最大导程为 12mm。当要加工导程大于 12mm 的螺纹时，如车削多线螺纹和油槽时，就得使用扩大导程机构。此时应将轴Ⅸ上的滑移齿轮 Z_{58} 移至右端（图 2-11 的虚线）位置，与轴Ⅷ上的齿轮 Z_{26} 相啮合。于是主轴Ⅵ与轴Ⅸ之间不再是通过齿轮副 58/58 直接联系，而是经轴Ⅴ、Ⅳ、Ⅲ以及Ⅷ之间的齿轮副实现运动联系。所以，扩大导程的传动路线表达式为

$$\text{主轴Ⅵ} \begin{cases} (\text{扩大导程}) \dfrac{58}{26} - \text{Ⅴ} - \dfrac{80}{20} - \text{Ⅳ} \begin{cases} \dfrac{50}{50} \\ \dfrac{80}{20} \end{cases} - \text{Ⅲ} - \dfrac{44}{44} \times \dfrac{26}{58} \\ (\text{正常导程}) \dfrac{58}{58} \end{cases} \text{Ⅸ} - (\text{接正常导程传动路线})$$

从扩大导程传动路线可以看出，从主轴Ⅵ到轴Ⅸ之间的传动比 u_k 有以下两种表示。当主轴转速为 40~125r/min 时，传动比为

$$u_{k1} = \frac{58}{26} \times \frac{80}{20} \times \frac{50}{50} \times \frac{44}{44} \times \frac{26}{58} = 4 \tag{2-5}$$

当主轴转速为 10~32r/min 时，传动比则为

$$u_{k2} = \frac{58}{26} \times \frac{80}{20} \times \frac{80}{20} \times \frac{44}{44} \times \frac{26}{58} = 16 \tag{2-6}$$

所以，扩大导程机构的功用是将导程扩大至 4 倍或 16 倍，以便车削大导程螺纹。

② 其他类型螺纹的加工。由于 CA6140 车床的丝杠是米制螺纹，即导程用 mm 表示，因此车螺纹时，其他螺纹的螺距和导程均要换算成以 mm 为单位的米制。

a. 对于英制螺纹，由于它以每英寸长度上的螺纹扣（牙）数 α 表示（牙/吋），它的导程 L_α 也要转换成以 mm 为单位来表示，即

$$L_\alpha = \frac{1}{\alpha}\text{in} = \frac{25.4}{\alpha}\text{mm} \tag{2-7}$$

这就意味着在英制螺纹传动路线表达式中应该出现一个特殊的因子 25.4。对照米制螺纹的传动路线，25.4 可以通过改变传动链中部分传动副的传动比来获得，即把基本组中主动轴和被动轴交换，可使各传动比的分母呈等差数列。其余传动路线与米制螺纹的传动路线相同。车削英制螺纹的运动平衡式为

$$L_\alpha = 1_{(\text{Spindle 主轴})} \times \frac{58}{58} \times \frac{33}{33} \times \frac{63}{100} \times \frac{100}{75} \times \frac{1}{u_j} \times \frac{36}{25} \times u_b \times 12 = \frac{4}{7} \times 25.4 \times \frac{1}{u_j} \times u_b \tag{2-8}$$

化简得

$$\alpha = \frac{7}{4} \times \frac{u_j}{u_b} \text{（牙/in）} \tag{2-9}$$

as Eq. (2-13).

$$m = \frac{7}{4k} u_j u_b \quad (2\text{-}13)$$

It can be seen from Eq. (2-13) that the threads with different m can be cut by changing u_j and u_b.

c. For the diametral pitch thread (British worm), its standard parameter is the diametral pitch, expressed with DP. The axial tooth spacing of the British worm is expressed as Eq. (2-14).

$$P_{DP} = \frac{\pi}{DP}\text{in} = \frac{25.4k\pi}{DP}\text{mm} \quad (2\text{-}14)$$

When cutting the diametral pitch thread, the transmission route for cutting the British thread can be used, but the interpolating gears should be changed into $\frac{64}{100} \times \frac{100}{97}$. The motion balance equation is expressed with Eq. (2-15).

$$L_{DP} = 1_{(\text{Spindle 主轴})} \times \frac{58}{58} \times \frac{33}{33} \times \frac{64}{100} \times \frac{100}{97} \times \frac{1}{u_j} \times \frac{36}{25} \times u_b \times 12 \quad (2\text{-}15)$$

After simplifying, the relation of the diametral pitch (DP) to u_j and u_b can be expressed as Eq. (2-16).

$$DP = 7k \frac{u_j}{u_b} \quad (2\text{-}16)$$

24 kinds of commonly used diametral pitch threads can be gained by changing u_j and u_b.

2) Power feed transmission chains. They are used primarily for cylindrical turning and facing. In order to reduce the wear of the lead screw and ensure the transmission accuracy of the thread turning, the power feed is driven by means of a feed rod instead of a lead screw.

To find the first part and the last part

spindle—toolrest

Displacement calculation

1r | work = f | tool(mm)

In the power feed transmission chains of CA6140 lathe, the transmission route from the headstock to the shaft XVII of the feedbox is the same as the transmission route of the threading. Then the motion is transmitted to the feed rod XIX via gears 28/56, and the rotation of the feed rod is transmitted via the transmission elements in apron to the gear and rack, transverse lead screw XXVII respectively, thus making the tool rest feed longitudinally and transversely. The transmission route expression of the power feed transmission chain is shown in Fig. 2-13.

There are 64 kinds of longitudinal feeds and transverse feeds respectively in CA6140 lathe, where there are 32 kinds of normal feeds, 8 kinds of large feeds, 16 kinds of larger feeds and 8 kinds of fine feeds.

① Normal feeds. It is known from Fig. 2-11 that the transmission route of the normal feed is from the spindle via the normal pitch and metric thread transmission route to the tool rest. 32 kinds of normal feeds ranging from 0.08mm/r to 1.22 mm/r can be obtained.

② Large feeds. It is known from Fig. 2-11 that the transmission route of the large feed is from

b. 对于模数螺纹（米制蜗杆），螺距与模数之间的换算关系为

$$p_m = \pi m \tag{2-10}$$

模数螺纹的导程为

$$L_m = k\,\pi m \tag{2-11}$$

式中 k——螺纹的线数。

除了使用的交换齿轮不同外，切削模数螺纹的传动路线与切削米制螺纹几乎一样。由于在模数螺纹的螺距中有 π 因子，就要选用交换齿轮 $\frac{64}{100} \times \frac{100}{97}$。于是，车削模数螺纹的运动平衡式为

$$L_m = 1_{(\text{Spindle 主轴})} \times \frac{58}{58} \times \frac{33}{33} \times \frac{64}{100} \times \frac{100}{97} \times \frac{25}{36} \times u_j \times \frac{25}{36} \times \frac{36}{25} \times u_b \times 12 \tag{2-12}$$

而 $\frac{64}{100} \times \frac{100}{97} \times \frac{25}{36} \approx \frac{7\pi}{48}$。化简后，模数 m 与 u_j、u_b 的关系可以表示为

$$m = \frac{7}{4k} u_j u_b \tag{2-13}$$

由式（2-13）可知，通过改变 u_j 和 u_b，可以加工不同模数的螺纹。

c. 对于径节螺纹（英制蜗杆），其标准参数是径节，用 DP 表示。径节螺纹的轴向齿间距为

$$P_{DP} = \frac{\pi}{DP}\text{in} = \frac{25.4k\pi}{DP}\text{mm} \tag{2-14}$$

切削径节螺纹时，可以使用英制螺纹的传动路线，但必须使用 $\frac{64}{100} \times \frac{100}{97}$ 的交换齿轮。其运动平衡式为

$$L_{DP} = 1_{(\text{Spindle 主轴})} \times \frac{58}{58} \times \frac{33}{33} \times \frac{64}{100} \times \frac{100}{97} \times \frac{1}{u_j} \times \frac{36}{25} \times u_b \times 12 \tag{2-15}$$

化简后，径节 DP 与 u_j、u_b 的关系可以表示为

$$DP = 7k\frac{u_j}{u_b} \tag{2-16}$$

通过改变 u_j 和 u_b，可以加工 24 种常用的径节螺纹。

2) 机动进给传动链。机动进给传动链主要用于车削内、外圆柱面和端面。为了减少丝杠的磨损，保证螺纹的加工精度，机动进给使用光杠驱动而不是丝杠驱动。

查找首、末两端件：主轴和刀架。

计算位移：

$$1\text{ 转}\mid\text{工件} = f\mid\text{刀具（mm）}$$

CA6140 车床的纵向和横向机动进给传动链，从主轴至进给箱 XVII 的传动路线与加工螺纹的传动路线相同。其后经过齿轮副 28/56 传至光杠 XIX，再由光杠经溜板箱中的传动元件，分别传至齿轮齿条机构和横向进给丝杠 XXVII，使刀架实现纵向或横向进给。其传动路线表达式如图 2-13 所示。

CA6140 车床的纵向和横向进给量各有 64 种，其中有 32 种正常进给量，8 种大进给量，16 种较大进给量和 8 种细进给量。

① 正常进给量。运动经由正常螺距的米制螺纹传动路线传动时，可得到 0.08 ~ 1.22mm/r 的 32 种正常进给量。

the spindle via the normal pitch and inch thread transmission route to tool rest. Set $u_b = \dfrac{28}{35} \times \dfrac{35}{28} = 1$, then 8 kinds of large feeds ranging from 0.86mm/r to 1.59 mm/r can be obtained.

③ Larger feeds. When the spindle rotates at low 12-level speeds (10 ~ 125r/min), the movements from the spindle to the tool rest is carried out by the pitch-enlarging mechanism and inch thread transmission route. The feeds can be enlarged to 4 times or 8 times, and 16 kinds of larger feeds ranging from 1.71mm/r to 6.33 mm/r can be obtained.

④ Fine feeds. When the spindle rotates at high speeds (450 ~ 1400r/min), the movements from the spindle to the tool rest is carried out by the pitch-enlarging mechanism and metric thread transmission route. Set $u_b = \dfrac{18}{45} \times \dfrac{15}{48} = 1/8$, then 8 kinds of fine feeds ranging from 0.028mm/r to 0.054 mm/r can be obtained.

By analyzing the transmission route of longitudinal and transverse feeds, it can be found that the transverse feeds are always the half of longitudinal feeds in the same transmission route, and the number of transverse feeds equals the number of longitudinal feeds.

5. Primary structure of CA6140 lathe

(1) Headstock

Main functions of the headstock are: to carry the spindle and realize its start, stop, reversing, brake and speed changing; to transmit the drive motion from the power source to the spindle and the feed motion from the spindle to the feed system.

1) Sub-assembly of the spindle. As a sub-assembly of the spindle, it should be of higher revolving accuracy, enough rigidity and good anti-vibration. The spindle assembly structure of the CA6140 lathe is shown in Fig. 2-14. The accuracy of the front bearing is one grade higher than that of the back bearing; both the front and the back bearing's clearance can be adjusted. The spindle of the CA6140 lathe is a step shaft with a through hole in which the long rod or other pneumatic, electromotive, hydraulic clamping device can pass through. The taper hole in the front of the spindle is a No.6 Morse taper hole used for mounting the centre or mandrel. The flange-type structure at the front end of the spindle is used to install chucks or driving plates. There is a left-hand helical gear on the right side of the spindle. The axial thrust produced in gearing directs at the front bearing to counteract the axial cutting force, thus reducing the axial force born by front bearing.

2) Unloaded pulley. It is well known that if the pulley is mounted directly on a overhang shaft, the shaft would produce deformation under the action of the pulling force of the V-belt. The unloaded pulley shown in Fig. 2-15 is used to pass the radial load to the box to prevent bending of shaft I from the pull of the V-belt, thus improving the smoothness of the transmission.

The unloading route is: radial force F_r (from V-belt) —screw—spline sleeve—deep groove ball bearing—flange —box.

3) Bidirectional friction clutch, brake and its manipulating mechanism. A bidirectional friction clutch, brake and its manipulating mechanism is illustrated in Fig. 2-16.

The multi-piece bidirectional friction clutch is used to control the start, stop and reversing of

② 大进给量。运动经由正常螺距的英制螺纹传动路线传动时，使用增倍组中 $\frac{28}{35} \times \frac{35}{28}$ 传动路线，可得到 0.86～1.59 mm/r 的 8 种大进给量。

③ 较大进给量。运动经由扩大导程机构及英制螺纹传动路线传动，且主轴处于较低的 12 级转速时，可将进给量扩大 4 倍或 16 倍，得到 1.71～6.33 mm/r 的 16 种较大的纵向进给量。

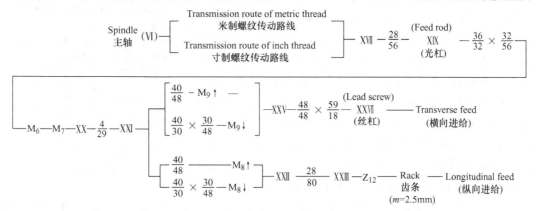

Fig. 2-13　Transmission route of longitudinal and transverse feed　纵向和横向机动进给传动路线

④ 细进给量。运动经由扩大导程机构及米制螺纹传动路线传动，且主轴以高速（450～1400r/min）运转时，增倍变速组使用 $\frac{18}{45} \times \frac{15}{48}$ 传动路线，可得到 0.028～0.054mm/r 的 8 种细进给量。

从传动路线表达式可以分析出，当主轴箱及进给箱中的传动路线相同时，所得到的横向进给量是纵向进给量的一半，而横向进给量的级数与纵向的相同。

5. 机床的主要结构

（1）主轴箱　主轴箱的主要功用是支承主轴并使其实现旋转、起动、停止、变速和换向；把来自动力源的驱动运动传给主轴，同时也把进给运动传给进给装置。

1）主轴组件。主轴组件应具有较高的旋转精度、足够的刚度和优良的抗振性。主轴组件的结构如图 2-14 所示。主轴前端轴承的精度应比后端的轴承高一级；前、后轴承的间隙均可调整。CA6140 型卧式车床的主轴是空心阶梯轴，其内孔是为了穿过长棒料以及气动、液压或电气等夹紧装置。主轴前端的锥孔为莫氏 6 号锥度，用于安装顶尖或心轴。主轴前端部采用短锥法兰式结构，用于安装卡盘或拨盘。主轴右侧有一个左旋斜齿轮。齿轮传动中产生的轴向力指向前轴承以抵消轴向切削力，从而可减小由前轴承承受的轴向力。

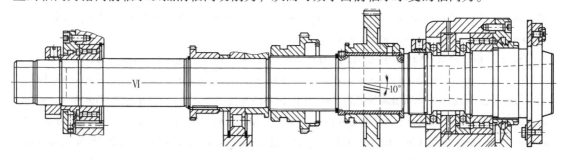

Fig. 2-14　Sub-assembly of the spindle of the CA6140 lathe　CA6140 车床主轴组件

the spindle, as shown in Fig. 2-16a. It consists of left and right parts which have the same structure. The left clutch is used to drive the spindle to rotate counterclockwise, and the right clutch is used to drive the spindle to rotate clockwise. As the counterclockwise rotation of the spindle is used for cutting, whereas the clockwise rotation of the spindle is merely used for the return cutter in the threading, the number of friction plates in the left clutch is more than that in the right clutch. Take the left clutch for example, the inner friction plate 2 and outer friction plate 1 is mounted on the shaft I alternately. Inner friction plates 2 are joined with the shaft I by spline. Outer friction plates 1 are joined with the left dual gear by means of 4 teeth, but assembled to the shaft I by smooth hole. Before pressing the inner and outer friction plates together tightly, the shaft I can't drive the dual gear to rotate. After the inner and outer fric-

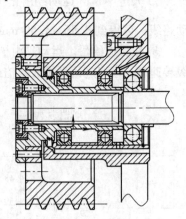

Fig. 2-15 Unloaded pulley
卸荷带轮

tion plates in the left clutch are pressed tightly to the stopping shims 11 and 12, the shaft I can drive the dual gear to rotate by friction force, thus realizing the counterclockwise rotation of the spindle. The right clutch has the same structure and operation principle as the left one. When the sliding bush is in the middle position, the friction plates in both the left and right clutch are loose, the spindle stops rotation.

The bidirectional friction clutch is an overloading insurance mechanism in addition to transmit motion and power. When the machine tool is over loaded, the friction plates would slip, making the spindle stop and preventing the machine tool from damage. The pressure among the friction plates is determined according to the rating torque transmitted by the friction clutch. After the friction plate is worn, the pressure would be decreased. The pressure can be adjusted by turning the nut 3 screwed on the pressing bushs.

The friction clutch is controlled by handle 18, as shown in Fig. 2-16b. Pulling handle 18 up, the handle would swing counterclockwise around shaft 19. The sector gear 17 turns clockwise and drive the rack 22 moving to the right. The movement of the rack 22 is transmitted by a series of mechanisms to pull the rod 8. The movement of the rod 8 to the left brings the pressing bush 5 to the left to compress the inner and outer friction plates, thus realizing the counterclockwise rotation of the spindle. Similarly, pulling the handle 18 down to lower end can compress the clutch on the right and make the spindle rotate clockwise. When the handle 18 is located in the middle position, the clutches disengage and the spindle stops.

The function of the brake is to make the spindle stop rotation rapidly. The same handle 18 is used to control the start, stop, brake and reversing in order to operate the machine conveniently and realize the interlock between the braking and the connecting transmission chains. The brake of the spindle is realized by means of the brake belt 15 and brake plate 16. The tension can be adjusted by the screw 13. After adjusting the screw 13, it should be inspected if the brake belt 15 is loose when the clutch is compressed. Be insure that the transmission chain is in "turn-off" in braking situation,

2）卸荷带轮。众所周知，如果把带轮直接安装在悬臂轴上，该轴就会在带拉力的作用下产生弯曲变形。如图 2-15 所示的卸荷带轮用来把径向力传递到箱体上，以防止 I 轴因带的拉力而产生变形，从而提高传动的平稳性。卸载路线是：来自 V 型带的径向力—螺钉—花键套—深沟球轴承—法兰—主轴箱体。

3）双向片式摩擦离合器、制动器及其操纵机构。CA6140 型车床的双向片式摩擦离合器、制动器及其操纵机构如图 2-16 所示。

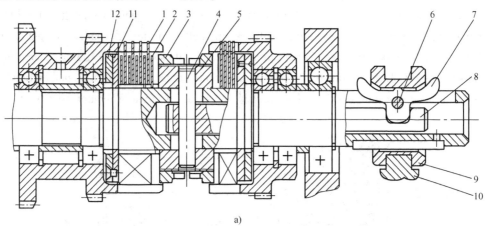

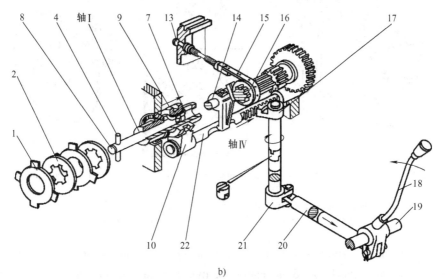

Fig. 2-16　Bidirectional friction clutch, brake and its manipulating mechanism
双向片式摩擦离合器、制动器及其操纵机构
a）Bidirectional friction clutch 双向片式摩擦离合器
b）brake and its manipulating mechanism 制动器及操纵机构
1—Outer friction plate 外摩擦片　2—Inner friction plate 内摩擦片　3—Nut 螺母　4—Pin 圆柱销
5—Pressing bush 压套　6—Hinge pin 销轴　7—Centering block 摆块　8—Pull rod 拉杆
9—Sliding bush 滑套　10—Fork 拨叉　11、12—Stopping shim 止推片　13—Adjusting screw 调节螺钉
14—Lever 杠杆　15—Brake belt 制动带　16—Brake plate 制动盘　17—Sector gear 扇形齿轮
18—Handle 手柄　19—Shaft 操纵轴　20—Rod 杆　21—Crank arm 曲柄　22—Shaft with rack 齿条轴

双向多片式摩擦离合器用于控制主轴的开停和换向，如图 2-16a 所示。它由结构相同的左、右两部分组成。左离合器传动主轴正转，右离合器传动主轴反转。由于主轴正转用于切

and there is no brake when turning on the transmission chain.

4) Speed-change manipulating mechanism. Fig. 2-17 shows a speed-change manipulating mechanism in the headstock of CA6140 lathe. This is a concentrated manipulating mechanism to manipulate a sliding dual-gear on shaft Ⅱand a sliding tri-gear on shaft Ⅲ. One handle is used in this manipulating mechanism to change 6 kinds of drive ratios from shaft Ⅰ to shaft Ⅲ. Turning the handle can make the shaft 4 as well as the cam 3 and crank 2 mounted on it rotate together. There are 6 speed changing positions on the cam 3. Positions 1, 2, 3 are used to move the dual-gear on shaft Ⅱ to the left end, and positions 4, 5, 6 are used to move the dual-gear on shaft Ⅱ to the right end. The rotation of the crank 2 can drive the fork 1 to move via the pin on the crank, thus driving the sliding tri-gear to move along the spline shaft Ⅲ. Each time the handle turns a circle, there are 6 kinds of combinations corresponding to the gear engaging position. The shaft Ⅲ can obtain 6 kinds of rotation speeds.

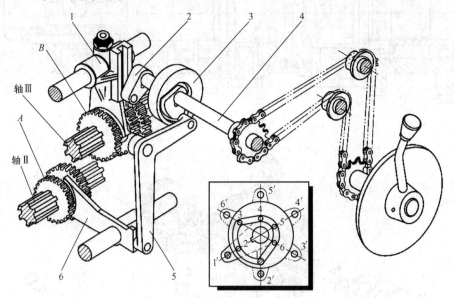

Fig. 2-17 Speed-change manipulating mechanism with sliding gears on shaft Ⅱ and shaft Ⅲ
Ⅱ、Ⅲ轴滑移齿轮变速操纵结构
A—Dual-gear 双联齿轮 B—Tri-gear 三联齿轮 1、6—Fork 拨叉
2—Crank 曲柄 3—Disc cam 盘形凸轮 4—Shaft 轴 5—Lever 杠杆

(2) Apron assembly The functions of the apron assembly are to transform the rotary motion of the feed mechanisms into the longitudinal motion of the carriage and the transverse motion of the cross-slide. The apron assembly is primarily composed of manipulating mechanisms for longitudinal and transverse power feed and fast movement, half nut mechanism, interlock mechanisms, overrunning clutch, safety clutch and so on.

1) Manipulating mechanism for longitudinal and transverse power feed. Fig. 2-18 shows a manipulating mechanism for power feed. The mechanism utilizes a handle 1 to control concentrically the connection, disconnection and reversing. The direction of the pulling handle is the same as the tool-rest moving direction. For example, pulling the handle to left or right can drive the fork 16 to move forward or backwards. The movement of the fork 16 drives the clutch M_8 to engage to one of

削加工，而主轴反转仅用于车螺纹时退刀，因此左离合器摩擦片的数量比右离合器的摩擦片多。以左离合器为例，多个内摩擦片 2 和外摩擦片 1 相间安装。内摩擦片 2 以花键与轴 I 相联接，外摩擦片 1 以及四个凸齿与空套双联齿轮相联接并空套在轴 I 上。内外摩擦片未被压紧时，彼此互不联系，轴 I 不能带动双联齿轮转动。当把左离合器的内外摩擦片压紧在止推片 11 和 12 上时，通过摩擦片间的摩擦力使空套双联齿轮与轴 I 联接，使主轴正向旋转。右离合器的结构和工作原理与左离合器相同。滑套 9 处于中间位置时，左右两离合器的摩擦片都松开，主轴的传动断开，停止转动。

摩擦离合器除了靠摩擦力传递运动和扭矩外，还能起过载保护作用。当机床过载时，摩擦片打滑，主轴停止转动，可避免损坏机床。摩擦片间的压紧力是根据离合器应传递的额定扭矩来确定的。当摩擦片磨损以后，压紧力减小，这时可通过旋转装在压套 5 上的螺母 3 来调整。

离合器由手柄 18 操纵（图 2-16b）。扳动手柄 18 向上扳绕操纵轴 19 逆时针摆动。扇形齿轮 17 顺时针转动使齿条轴 22 向右移动，通过一系列机构的传递使拉杆 8 向左移动（图 2-16a），通过圆柱销 4 带动压套 5 向左压紧内外摩擦片，实现主轴正转。同理，将手柄 18 扳至下端位置时，右离合器压紧，主轴反转。当手柄 18 处于中间位置时，离合器脱开，主轴停止转动。

制动装置的功用是使主轴迅速停止转动。为了操作方便和实现制动与接通传动链这两组动作的互锁，故采用同一手柄 18 操纵。主轴的制动是通过制动带 15 和制动盘 16 来实现的。制动带的拉紧程度由调节螺钉 13 调整。调整后应检查在压紧离合器时制动带是否松开，确保制动时接不通传动链，接通传动链时不能实现制动。

4）变速操纵机构。图 2-17 所示为 CA6140 型车床主轴箱中的一种变速操纵机构。它用一个手柄同时操纵轴 Ⅱ、Ⅲ 上的双联滑移齿轮和三联滑移齿轮，变换轴 I—Ⅲ 间的六种传动比。转动手柄可使传动轴 4 以及装在其上的凸轮 3 和曲柄 2 一起转动。凸轮 3 上有 6 个变速位置。位置 1、2、3 可将轴 Ⅱ 上的双联齿轮移向左端位置，位置 4、5、6 可将双联齿轮移向右端位置。曲柄 2 转动可以通过其上的销子带动拨叉 1 移动，从而拨动三联齿轮在轴 Ⅲ 上移动。手柄每旋转一周，对应齿轮啮合位置有 6 种组合，使轴 Ⅲ 得到 6 种转速。

（2）溜板箱　溜板箱的主要功用是将进给机构的旋转运动转变为大拖板的纵向直线运动和滑鞍的横向直线运动。溜板箱主要由纵、横向机动进给和快速移动的操纵机构、开合螺母及操纵机构、互锁机构、超越离合器和安全离合器等组成。

1）纵、横向机动进给操纵机构。图 2-18 所示为 CA6140 型车床的机动进给操纵机构。它利用一个手柄集中操纵纵向、横向机动进给运动的接通、断开和换向。手柄扳动方向与刀架运动方向一致。例如，向左或向右扳动手柄 1，可使拨叉 16 向前或向后移动，从而拨动离合器 M_8 与轴 ⅩⅫ 上两个空套齿轮之一啮合，接通纵向机动进给运动，使刀架相应地向左或向右移动。

同理，向后或向前扳动手柄 1 可接通横向机动进给运动，使刀架相应地向前或向后移动。

手柄 1 扳至中间直立位置时，离合器 M_8 和 M_9 均处于中间位置，机动进给传动链断开。当手柄扳至左、右、前、后任一位置时，如按下装在手柄 1 顶端的按钮 K，则快速电动机起动，刀架便在相应方向上快速移动。刀架快速移动传动路线表达式为

the two gears mounted on the shaft XXII by smooth hole, thus turning on the longitudinal power feed motion and making the tool-rest move to left or right correspondingly.

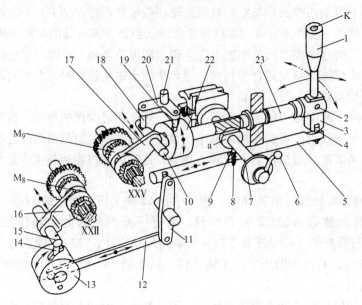

Fig. 2-18 Manipulating mechanism power feed 机动进给操纵机构(CA6140)
1、6—Handle 手柄 2、21—Pin 销轴 3—Handle seat 手柄座 4、9—Ball pin 球头销
5、7、23—Shaft 轴 8—Spring pin 弹簧销 10、15—Fork shaft 拨叉轴 11、20—Lever 杠杆
12—Link rod 连杆 13、22—Cam 凸轮 14、18、19—Round pin 圆销 16、17—Shift fork 拨叉

Similarly, pulling the handle forward or backward can switch on the transverse feed motion and drive the tool rest to move forward or backward accordingly.

When the handle locates in the middle position, two clutches M_8 and M_9 are also in the middle position, the transmission chain for power feed is disconnected. Whether the handle is on the left or right, the front or back, pushing the button K on the top of the handle 1, then the motor for fast movement starts, driving the tool rest to move rapidly in the corresponding direction.

The transmission route expression for the tool rest fast movement is as follows:

$$\text{Motor}(电动机) - \frac{13}{29} - \text{XXII} - \frac{4}{29} - \text{XXIII} - \begin{Bmatrix} M_8 \ldots (\text{longitudinal 纵向}) \\ M_9 \ldots (\text{transverse 横向}) \end{Bmatrix}$$

The rapid moving speed of the tool rest to right in longitudinal is:

$$v_{L \to R} = 2800 \times \frac{13}{29} \times \frac{4}{29} \times \frac{40}{30} \times \frac{30}{48} \times \frac{28}{80} \times \pi \times 2.5 \times 12 \text{m/min} = 4.76 \text{m/min}$$

2) Half nut mechanism. The structure of the half nut mechanism is shown in Fig. 2-19. The half nut is also named as split nut, which consists of the upper half nut 26 and lower half nut 25. The half nut which is installed in a dovetail guideway can be moved up and down. There are two pins 27 mounted respectively on the back of both upper and lower half nut, which are inserted separately into two curve grooves in the grooved plate 28. When threading, turning the handle 6 to make the grooved plate rotate counterclockwise (Fig. 2-19b), the curve grooves force two pins in it to

$$\text{Motor（电动机）} \frac{13}{29} - \text{XXII} - \frac{4}{29} - \text{XXIII} - \begin{Bmatrix} M_8 \cdots \text{（longitudinal 纵向）} \\ M_9 \cdots \text{（transverse 横向）} \end{Bmatrix}$$

刀架向右快速移动的速度为

$$v_{L \to R} = 2800 \times \frac{13}{29} \times \frac{4}{29} \times \frac{40}{30} \times \frac{30}{48} \times \frac{28}{80} \times \pi \times 2.5 \times 12 \text{m/min} = 4.76 \text{m/min}$$

2）开合螺母机构。开合螺母机构的结构如图 2-19 所示。开合螺母由上下两个半螺母 25 和 26 组成，装在溜板箱体后壁的燕尾形导轨中，可上下移动。上下半螺母的背面各装有一个圆销 27，它们分别插入槽盘 28 的两条曲线槽中。需要车螺纹时，扳动手柄 6，经轴 7 使槽盘逆时针转动时（图 2-19b），曲线槽迫使两圆销互相靠近，带动上下半螺母合拢，与丝杠啮合。槽盘顺时针转动时，曲线槽通过圆销使两半螺母相互分离，与丝杠脱开啮合。

3）互锁机构。为了避免损坏机床，在接通机动进给或快速移动时，开合螺母不应合上。反之，开合螺母合上时，不允许接通机动进给或快速移动。

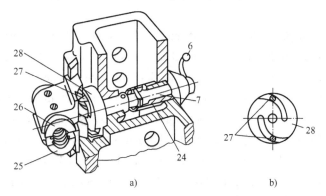

Fig. 2-19　Half nut mechanism　开合螺母机构
6—Handle 手柄　7—Shaft 轴　24—Supporting sleeve 支承套　25—Lower half nut 下半螺母
26—Upper half nut 上半螺母　27—Pin 圆销　28—Grooved plate 槽盘

图 2-20 所示的互锁机构是由开合螺母操纵轴 7 上的凸肩 a，轴 5 上的球头销 9 和弹簧销 8 以及支承套 24（图 2-18、图 2-19）等组成。图 2-18 表示丝杠传动和纵横向机动进给均未接通的情况，此位置称中间位置。此时可扳动手柄 1 至前、后、左、右任意位置，接通相应方向的纵向或横向机动进给，或者扳动手柄 6 使开合螺母合上。

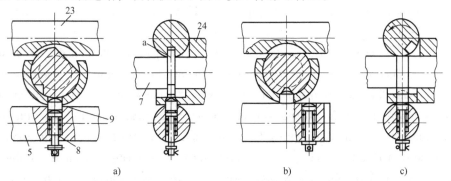

Fig. 2-20　Operation principle of interlocking mechanism 互锁机构工作原理
5、7、23—Shaft 轴　8—Spring pin 弹簧销　9—Ball head pin 球头销
24—Supporting sleeve 支承套

move toward each other, thus driving the two half nuts to come close and engage with the leadscrew. Turning the handle 6 to make the grooved plate rotate clockwise, the curve grooves force the two half nuts to separate via two pins in it and disengage with the leadscrew.

3) Interlock mechanisms

Functions: In order to avoid damaging the machine tool, the half nut should not be closed when switching on powered feed or fast movement. Whereas once the half nut is closed, it is forbidden to connect powered feed or fast movement.

Interlock mechanisms shown in Fig. 2-20 is composed of the shaft shoulder a on the shaft 7, ball stud 9, spring pin 8 and supporting sleeve 24, etc. (see Fig. 2-18 and Fig. 2-19). Fig. 2-18 illustrates the situation without connecting the leadscrew drive and the longitudinal and transverse power feed. This position is called middle position. Now, you can pull the handle 1 freely to connect the longitudinal or transverse power feed, or you can also pull the handle 6 to close the half nuts.

If you pull the handle 6 to close the half nuts, then the shaft 7 would turn clockwise a certain angle. The shoulder a enters the slot on shaft 23 and locks the shaft 23 so that it can't rotate. At the same time, the shoulder pushes the ball stud 9 into the hole of shaft 5. As the half of the ball stud 9 still remains in the supporting sleeve 24, the shaft 5 is locked and can't move along its axis, as shown in Fig. 2-20a. Here both the longitudinal and transverse power feed can't be connected. When the longitudinal power feed is connected, the movement of shaft 5 along its axis brings the hole on it to move together. As the ball stud 9 contacts to the surface of shaft 5, it can't move down, the shaft 7 is locked and can't rotate, as shown in Fig. 2-20b. Similarly, when the transverse power feed is connected, because of the rotation of shaft 23, the shoulder a of shaft 7 can't enter the slot and is supported by the surface of shaft 23. Therefore, the shaft 7 can't turn, as shown in Fig. 2-20c. It can be seen that once the longitudinal or transverse power feed is connected, the half nut mechanism wouldn't close.

4) Safety clutch. In the power feed movement, when the feed resistance is overlarge or when the movement of the tool rest is hindered, a overload protection equipment named as safety clutch is equipped in the feed transmission chain to stop the feed automatically in order to prevent the machine tool from damaging owing to overloading or accident. The structure of the safety clutch is shown in Fig. 2-21.

The safety clutch consists of the left part 5 and the right part 6 with helical teeth on one end. The left part 5 is fitted with key to the star wheel 4 of the overrunning clutch M_6, and the star wheel 4 with internal smooth surface is mounted (without key) on the shaft XX. The right part 6 with a spline hole is fitted to the shaft XX with spline. Under normal conditions, the two parts mesh with each other under the pressure of the spring 7, and the motion from the leadscrew is transmitted to the shaft XX and worm 10 via the gear Z_{56}, overrunning clutch M_6 and safety clutch M_7. If the load acting on the tool rest increases for some reasons, when the axial force is larger than the pressure of the spring 7, the right part 6 would disengage with the left part 5, leading to the skid of two parts. And then the power feed transmission chain is disconnected, and the tool rest stops feeding. When over-

如果向下扳动手柄 6 使开合螺母合上，则轴 7 顺时针转过一个角度，其上的凸肩 a 嵌入轴 23 的槽中，将轴 23 卡住使其不能转动。同时，凸肩又将球头销 9 压入轴 5 的孔中。由于球头销 9 的另一半还留在支承套 24 中，所以就将轴 5 锁住，使其不能轴向移动，如图 2-20a 所示。这时纵、横向机动进给都不能接通。如果接通纵向机动进给，因轴 5 沿轴线的移动使其上的孔也相应移动。球头销 9 被轴 5 的表面顶住不能往下移动，因而轴 7 被锁住而无法转动，如图 2-20b 所示。如果接通横向机动进给时，由于轴 23 转动了位置，轴 7 上的凸肩 a 被轴 23 顶住，使轴 7 无法转动，如图 2-20c 所示。由此可见，接通纵向或横向机动进给后，开合螺母均不能合上。

4）安全离合器。机动进给时，当进给力过大或刀架移动受阻时，为了避免损坏机床，在进给传动链中设置了安全离合器（即过载保护装置）来自动地停止进给。安全离合器的结构如图 2-21 所示。

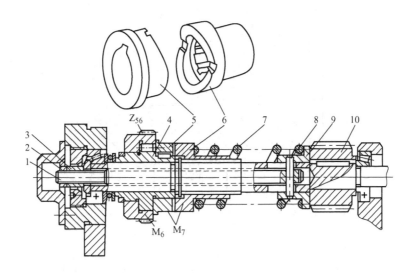

Fig. 2-21　Safety clutch 安全离合器
1—Pull rod 拉杆　2—Locking nut 锁紧螺母　3—Regulating nut 调整螺母
4—Star wheel 星轮　5—Left part 左半部　6—Right part 右半部　7—Spring 弹簧
8—Pin 圆销　9—Seat of spring 弹簧座　10—Worm 蜗杆

安全离合器由端面带螺旋形齿爪的左右两半部 5 和 6 组成，其左半部 5 用键装在超越离合器 M_6 的星轮 4 上，且与轴 XX 空套，右半部 6 与轴 XX 用花键联接。在正常工作情况下，在弹簧 7 的压力作用下，离合器左右两半部分相互啮合，由光杠传来的运动，经齿轮 Z_{56}、超越离合器 M_6 和安全离合器 M_7，传至轴 XX 和蜗杆 10。如果某种原因使刀架上的载荷增大，当进给力超过弹簧 7 的压力时，离合器右半部 6 将与左半部 5 脱开，导致安全离合器打滑。于是机动进给传动链断开，刀架停止进给。过载现象消除后，弹簧 7 使安全离合器重新自动接合，恢复正常工作。机床许用的最大进给力取决于弹簧 7 产生的弹力，而该弹力可在一定范围内进行调整。

loading disappears, the spring 7 makes the clutch mesh again automatically and return to the normal working situation. The maximum feeding resistance endured by machine tool depends on the force produced by the spring 7, which can be adjusted in a certain range.

2.2.3 Lathe cutters

There are many kinds of lathe cutters. Commonly used lathe cutters and their usages are shown in Fig. 2-22. External cylindrical turning tools are used for turning external cylindrical and conical surfaces. There are straight turning tools and bent turning tools. The bent turning tool has higher versatility and can be used for cylindrical turning, facing and chamfering. The lathe cutters with leading cutting edge angle $K_r = 90°$ can be used for turning multi-diameter shafts, shoulders of the shaft and slender shafts with low rigidity.

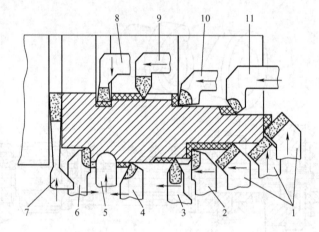

Fig. 2-22 Types and usages of lathe cutters 车刀的类型与用途
1—45°bent turning tool 45°弯头车刀 2—90°external turning tool 90°外圆车刀
3、9—Threading tool 螺纹车刀 4—75°external turning tool 75°外圆车刀
5—Form turning tool 成形车刀 6—90°left-hand turning tool 90°左切外圆车刀
7、8—Grooving tool 切槽刀 10、11— Boring cutter 镗刀

Generally, lathe cutters have 4 types of structural forms: solid tool, brazed insert, clamped insert and indexing insert, as shown in Fig. 2-23. The solid tool is made of high-speed steel (Fig. 2-23a). The lathe cutter with brazed insert means that the cemented-carbide insert is brazed on the shank by means of brazing method (Fig. 2-23b). Fig. 2-23c shows the lathe cutter with clamped insert which is clamped mechanically onto the shank and can be resharpened by grinding if necessary. For many applications it is more economical to clamp the insert to the shank. The lathe cutter with clamped indexing insert (Fig. 2-23d) is similar to the cutter shown in Fig. 2-23c, but there is a little difference in insert. The indexing insert has several cutting edges. When one edge becomes dull, the insert can be indexed and clamped, and put into turning operation again. Commonly used cemented-carbide indexing inserts are shown in Fig. 2-24.

2.2.3 车刀

车刀有多种类型。常用车刀种类及用途如图 2-22 所示。外圆车刀用于加工外圆柱面和外圆锥面，它分为直头和弯头两种。弯头车刀通用性较好，可以车削外圆、端面和倒棱。当外圆车刀的主偏角 $\kappa_r = 90°$ 时，可用于车削阶梯轴、凸肩及刚度较低的细长轴。外圆车刀按进给方向又分为左偏刀和右偏刀。

车刀在结构上可分为整体式、焊接式和机械夹固式整体车刀，如图 2-23 所示。整体式车刀主要是整体高速钢车刀（图 2-23a）。焊接车刀是将硬质合金刀片用焊接的方法固定在普通碳钢刀体上（图 2-23b）。机夹重磨车刀（图 2-23c）是采用普通硬质合金刀片，用机械夹固的方法将其夹持在刀柄上使用的车刀，切削刃用钝后可以重磨，经适当调整后仍可继续使用。机夹重磨车刀在许多应用场合显得更加经济有效。机夹可转位车刀（图 2-23d）是采用机械夹固的方法将可转位刀片固定在刀体上。刀片制成多个刀刃，当一个刀刃用钝后，只需将刀片转位、重新夹固，即可使新的刀刃投入工作。常用的硬质合金可转位刀片如图 2-24 所示。

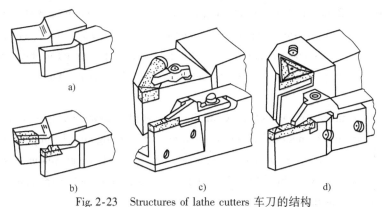

Fig. 2-23　Structures of lathe cutters　车刀的结构
a) Solid tool 整体式　b) Brazed insert 焊接式　c) Clamped insert 机夹式　d) Indexing insert 转位式

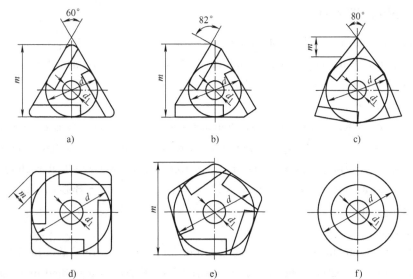

Fig. 2-24　Cemented carbide indexing inserts 硬质合金可转位刀片
a)、b)、c) Triangular inserts 三角形刀片　d) Square insert 正方形刀片
e) Pentagon insert 五角形刀片　f) Circular insert 圆形刀片

2.3 Grinding wheels and grinding machines

2.3.1 Introduction

Grinding is a process carried out with a grinding wheel made up of abrasive grains for removing very fine quantities of material from the workpiece surface. The required size of the abrasive grains are thoroughly mixed with the bonding material and then pressed into a disc shape with given diameter and thickness. This can be comparable to a milling process with an infinite number of cutting edges.

Grinding is a process used for the following purposes.

1) Machining materials which are too hard for other machining processes to cut;
2) Close dimensional accuracy of the order of 0.3 to 0.5 μm;
3) High surface finish such as Ra = 0.15 to 1.25 μm.

This accounts for 25% of all the machining processes used for roughing and finishing.

The abrasive grains are basically spherical in shape with large sharp points which act as cutting edges. All grains are of random orientations, and the rake angles presented to the work material can vary from positive to a large negative value. Sometimes grits slide rather than cut because of its orientation.

The depth of cut taken by each of the grain is very small. However a large number of grits act simultaneously and hence the material removed is large. The cutting speeds employed are also high. As a result the chips produced are very small and red hot. They often get welded easily to the abrasive grain or to the workpiece. Thus the grinding process is a very inefficient one compared to the conventional metal cutting processes.

2.3.2 Grinding wheels

A grinding wheel is also a kind of cutting tool, which is made by mixing the appropriate grain size of the abrasive with the required bond and pressed into shape. The characteristics of the grinding wheel depend upon a number of variables.

1. Abrasive types

These are hard materials with adequate toughness which act as cutting edges for a sufficiently long time. They also have the ability to fracture into smaller pieces when the force increases, which is termed as friability. This property gives the abrasives the necessary self-sharpening capability. The commonly used abrasives are aluminum oxide (Al_2O_3), silicon carbide, cubic boron nitride (CBN), diamond, etc.

2. Grain size

The grain size means the size of an abrasive grain. Based on the stipulation in GB/T 2481.1—1998 and GB/T 2481.1—2009, the coarse grain size (F4 – F220) is identified by a number which is based on the sieve size used. The sieve number is specified in terms of the number of openings per square inch. Thus the larger the grain number, the finer the grain size. The fine grain size (F230 –

2.3 砂轮与磨床

2.3.1 概述

磨削是指利用由磨料制成的磨具从工件表面上去除很少量材料所进行的工艺过程。砂轮是用规定粒度大小的磨粒与结合剂充分拌匀后压制成具有一定直径和厚度的盘状工具。磨削可以看作具有无限多刀刃的铣削过程。磨削用于以下目的：

1) 可以加工别的加工方法难以加工的硬材料。
2) 加工尺寸误差可达 $0.3 \sim 0.5 \mu m$。
3) 可以达到很小的表面粗糙度，如 $Ra = 0.15 \sim 1.25 \mu m$。

据统计，磨削约占所有加工方法的 25%，既可用于粗加工，也可用于精加工。

磨粒的形状大致呈球形，具有相当于切削刃的大而锋利的棱角。所有磨粒呈无序排列，加工时所出现的前角可以从很大的正前角到很大的负前角。有时磨粒不是在切削而是在工件表面上划擦。

每个磨粒的切削深度很小。但是由于众多的磨粒在同时切削，因此材料去除率还是比较可观的。由于采用的切削速度很高，因此形成的切屑很小且呈现红热状态。这些红热的切屑很容易粘接到磨粒上或工件表面上。因此，与传统金属切削加工方法相比，磨削是一种加工效率很低的加工方法。

2.3.2 砂轮

砂轮是磨削加工使用的切削工具，它是用结合剂将磨粒粘结在一起并压制成形的。砂轮的特性取决于许多因素。

1. 磨料类型

磨料是具有足够韧性而硬度很高的材料，以便能够充当切削刃并具有足够长的使用寿命。当磨削力增大到一定程度时，磨料也具有破碎成小片的能力，称为易碎性。这种性能使得磨粒具有必要的自锐性。常用的磨料有氧化铝（Al_2O_3）、碳化硅（SiC）、立方氮化硼（CBN）和金刚石（TR）等。

2. 粒度

粒度表示磨料颗粒尺寸的大小。GB/T 2481.1—1998 和 GB/T 2481.2—2009 规定，粗磨粒粒度 F4～F220 用筛分法区别，F 后面的数字大致为每英寸筛网长度上筛孔的数目；微粉粒度 F230～F2000 用沉降法区别。

表面加工后的粗糙度与所用的磨料粒度有关。细的磨粒采用很小的切削深度，因而可以获得较高的表面粗糙度。细的磨粒可用来制造成形砂轮。粗的磨粒有利于提高材料去除率。但粗磨粒易碎，因此不适合用于断续磨削。当工件材料较软、塑性大时，为避免砂轮堵塞，常采用由颗粒较粗的磨料制作的砂轮。

3. 结合剂

结合剂的作用是将磨料粘结在一起以共同抵挡磨削力。常用的结合剂如下：

(1) 陶瓷　这是最常用的结合剂。它具有强度高、耐热性好、气孔率大、耐腐蚀，适

F2000), called micro powder, is identified by sedimentation method.

The surface finish generated after grinding depends upon the grain size of grinding wheel. Fine grains take a very small depth of cut and hence provide a better surface finish. Fine grains are also used for making form grinding wheels. Coarse grains are good for higher material removal rates. These have better friability and as a result are not good for intermittent grinding where they are likely to chip easily. When grinding soft materials with good plasticity, coarse grains are generally used to make the grinding wheel in order to avoid jamming grinding wheel.

3. Bond

The function of the bond is to keep the abrasive grains together under the action of the grinding forces. The commonly used bond materials are as follows:

(1) Vitrified　This is the most commonly used bond. It is strong, rigid and porous, and not affected by fluids. It is suitable for various grinding operations.

(2) Synthetic resin　It has good strength and is more elastic than vitrified bonds. It is generally used for rough grinding, parting off and high speed grinding (50 to 65m/s).

(3) Rubber　Of all the bonds used, this is the most flexible. The bond is made up of natural or synthetic rubber. It is generally used for making the cutting off wheels, the regulating wheels in centreless grinding and the polishing wheels.

(4) Metal　This is used in the manufacture of diamond and CBN wheels. This bond has very high strength, good wear resistance, but poor self-sharpening capability.

4. Grade

The grade of the grinding wheel is also called the hardness of the wheel. This designates the force holding the grains, or it means the difficulty of abrasive falling off the grinding wheel in grinding. The grade of a wheel depends on the kind of the bond, structure of the wheel and amount of abrasive grains. Greater bond content and a strong bond result in a harder grinding wheel. The hardness of the grinding wheel has nothing to do with the hardness of the abrasive grain. Even the same abrasives can make the grinding wheels with different hardness. Harder wheels hold the abrasive grains till the grinding force increases to a great extent. In grinding, if the hardness of the grinding wheel is over large, the dull abrasives are difficult to fall off the wheel, which would lead to the increase of the grinding temperature and furthermore cause the burn of the work surface; if the hardness of the grinding wheel is over low, the abrasives would fall off the grinding wheel rapidly, which will be difficult to give full scope to the abrasives.

Soft wheels are generally used for grinding hard materials, and hard wheels are used for grinding soft materials. Softer wheels should be selected when grinding the workpiece with thin wall or poor thermal conductivity. Harder wheels should be selected in finish grinding or form grinding.

The grade and code of the grinding wheel are listed in Tab.2-2. In the grinding process, the grades from L to N are used for grinding the steels without hardening, and the grades H to K are used for grinding the hardened alloy steels. The grinding wheel with grades K and L should be selected when grinding the surfaces with high surface quality; the grinding wheel with grades H and J should be selected when grinding the cemented-carbide cutters. The grades from H to N are most commonly used.

用于各类磨削加工。

（2）树脂　它具有强度高，弹性比陶瓷结合剂好，通常用于粗磨、切断以及磨削速度为 50～65m/s 的高速磨削。

（3）橡胶　这是所用的结合剂中弹性最好的。它可以由天然橡胶制成，也可以用合成橡胶。通常用于切断砂轮、无心磨削的导轮以及抛光轮。

（4）金属　金属结合剂用于制作金刚石砂轮和 CBN 砂轮。金属结合剂具有高强度，良好的耐磨性，但自锐性差。

4. 等级

砂轮的等级也称砂轮硬度。它表示在磨削时保持磨粒不从砂轮表面脱落的力的大小，或者说它是指磨削时磨粒从砂轮表面脱落的难易程度。砂轮的等级取决于结合剂的种类、砂轮的组织以及磨粒所占比例的多少。结合剂的含量多、强度高，砂轮的硬度就高。砂轮的硬度与磨料的硬度无关。同一种磨料，可以做出不同硬度的砂轮。砂轮硬度高可以保证磨粒在很大磨削力作用下也不会从砂轮上脱落。磨削时，若砂轮太硬，则磨钝了的磨料不能及时脱落，会使磨削温度升高而造成工件烧伤；若砂轮太软，则磨料脱落过快而不能充分发挥磨料的磨削效能。

软砂轮常用来磨削硬度高的工件材料，而硬砂轮则用来磨削硬度低的工件材料。磨削薄壁件及导热性差的工件时，选用较软的砂轮；精磨与成形磨时，应选用硬些的砂轮，以利于保持砂轮的廓形。

砂轮等级及代号见表 2-2。在磨削加工中，砂轮等级 L～N 用于磨削未淬硬钢，H～K 用于磨削淬火合金钢。磨削表面质量要求高的表面时选用 K、L 等级的砂轮，刃磨硬质合金刀具选用 H、J 等级的砂轮。但是最常使用的砂轮等级是 H～N。

5. 组织

砂轮组织表示砂轮磨料的颗粒配比与分布。砂轮组织级别分为紧密、中等、疏松三大类，如图 2-25 所示。组织等级用数字来表示，见表 2-3。磨粒在砂轮总体积中所占比例越大，孔隙越小，砂轮的组织越紧密，反之，则组织疏松。

Table 2-2　Grade and code of a grinding wheel　砂轮的等级及代号

Grade 等级	Super-soft 超软		Soft 软			Mid-soft 中软		Medium 中		Mid-hard 中硬		Hard 硬		Super-hard 超硬
Code 代号	D	E	F	G	H	J	K	L	M	N	P	Q	R	S	T	Y

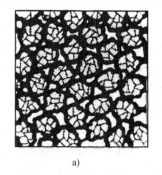  

a)　　　　　　　　　　　b)　　　　　　　　　　　c)

Fig. 2-25　Structure of a grinding wheel　砂轮的组织
a) dense 紧密　b) medium 中等　c) open 疏松

5. Structure

The structure of the grinding wheel represents the grain spacing. It can be classified into dense, medium and open and is shown in Fig. 2-25 conceptually. It is generally denoted by numbers as shown in Tab. 2-3. The spaces among the grains allows for collecting chips and for carrying cutting fluid into grinding region. The larger the proportion of abrasive grains in the total wheel volume, and the smaller the proportion of pores, the denser the structure of a grinding wheel. Contrarily, the opener the structure is.

Open structures are used for high stock removal and consequently produce a rough finish. Dense structures are used for precision grinding and profile grinding. The grinding wheel with open structure should be selected when grinding the workpiece with thin wall or in face grinding.

The grinding wheel marking system should be able to specify the abrasive, grain size, grade, structure and bond. Grinding wheels are marked with symbols that designate their properties. Generally, there is a sign on the end surface of the grinding wheel to indicate the properties of the grinding wheel. For example, 1 - 300 × 50 × 75 - AF60L5V - 35m/s. "1" stands for the plane grinding wheel; "300 × 50 × 75" stand for the outside diameter, thickness and inside diameter (mm) respectively; "A" denotes that the abrasive is brown corundum; "F60" denotes the No. of the grain size; "L" denotes that the grade is mid-soft; "5" denotes the No. of the structure; "V" denotes the bonding material, and "35m/s" means the maximum grinding speed.

2.3.3 Grinding machines

Grinding operations are generally classified based on the type of the surface produced. The grinding operations can be classified as cylindrical grinding, surface grinding and centerless grinding. All the machine tools which use the abrasive tools (such as grinding wheel, abrasive belt, oil stone and other abrasive materials) to machine workpieces are called grinding machines. There are many types of grinding machines. The primary types are cylindrical grinding machines, internal cylindrical grinding machines and surface grinding machines.

1. Cylindrical grinding machines

There are two types of cylindrical grinding operations. One is the external cylindrical grinding, the other is the internal cylindrical grinding.

Center-type cylindrical grinding machines are used generally for grinding straight and conical cylindrical surfaces. The typical movements in a cylindrical grinding machine are shown in Fig. 2-26. The grinding wheel A is installed on a device similar to the tool post and is driven by an independent power at a high speed suitable for the grinding operation. The work E which is normally held between two centers D is rotated at very low speed in the direction opposite to that of the grinding wheel. Infeed motion 4 is provided by moving the wheel head at right angle to the longitudinal axis of the table.

Fig. 2-27 shows a universal cylindrical grinding machine which can be used for grinding the internal and external cylindrical surface, conical cylindrical surface, end face of the shaft shoulder and so on. The worktable 3 is driven by hydraulic power or manually and moves along the longitudinal guideways on the bed 1. There are an upper worktable and a lower worktable. The upper workta-

疏松的砂轮具有高的材料切除率，但磨削过的表面粗糙。紧密的砂轮适用于成形磨削和精密磨削。磨削区面积大的（如端磨）或薄壁件磨削时，应选择组织疏松的砂轮。

Table 2-3　Structure of a grinding wheel 砂轮的组织等级

No. of structure 组织号	0	1	2	3	4	5	6	7	8	9	10	11	12
Abrasive rate 磨料率/(%)	62	60	58	56	54	52	50	48	46	44	42	40	38
Density 疏密程度	Dense 紧密				Medium 中等				Open 疏松				

砂轮的标注系统应能够指明所使用的磨料、粒度、硬度、组织和结合剂。砂轮通常用表示它们性能的符号来标注。通常在砂轮的端面上都印有标志，用以表示砂轮的特性。例如，1—300×50×75—AF60L5V—35m/s。其中，1 表示平面砂轮，其余则分别表示：外径为 300mm，厚度为 50mm，内径为 75mm，磨料为棕刚玉（A），粒度号为 F60，硬度为中软（L），组织号为 5，结合剂为陶瓷（V），最高线速度为 35m/s。

2.3.3　磨床

磨削加工通常根据所加工表面的类型进行分类。磨削加工可以分为外圆磨削、平面磨削、无心磨削等。凡是以磨料磨具（如砂轮、砂带、油石、研磨料等）为工具对工件进行切削加工的机床统称为磨床。磨床的种类很多，其中主要类型有外圆磨床、内圆磨床、平面磨床等。

1. 外圆磨床

有两种类型的圆柱面磨削操作，一种是外圆磨削，另一种是内圆磨削。

顶尖型外圆磨床通常用于磨削圆柱面和圆锥面。外圆磨床所具有的运动如图 2-26 所示。砂轮 A 装在一个类似于刀架的装置上，由独立动力源单独驱动以适合于磨削加工的速度旋转。工件 E 通常安装在两顶尖之间以非常低的速度转动，其旋转方向与砂轮相反。横向进给运动是通过砂轮架在垂直于工作台纵向轴线方向上的移动实现的。

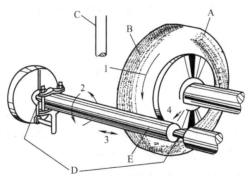

Fig. 2-26　Cylindrical grinding operation 外圆磨削
1—Primary motion 主动
2—Rotary feed motion 旋转进给运动
3—Longitudinal feed motion 纵向进给运动
4—Infeed motion 径向进给运动

图 2-27 所示为一万能外圆磨床的外形图，它适用于单件小批生产中磨削内外圆柱面、圆锥面、轴肩端面等。工作台 3 能以液压或手轮驱动，在床身 1 的纵向导轨上作进给运动。工作台由上、下两层组成，上工作台可相对于下工作台在水平面内回转一个不大的角度

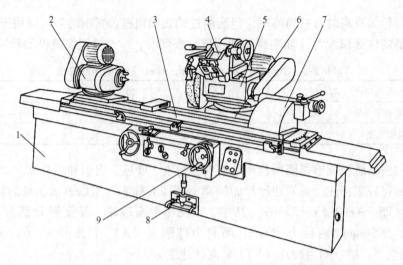

Fig. 2-27　Universal cylindrical grinding machine 万能外圆磨床
1—Bed 床身　2—Headstock 头架　3—Worktable 工作台　4—Internal grinding head 内圆磨头
5—Wheel head 砂轮架　6—Slide 滑鞍　7—Tailstock 尾座　8—Pedal switch 脚踏开关　9—Handwheel 手轮

ble can swivel an angle (±10°) relative to the lower in horizontal plane in order to grind a long conical surface, as shown in Fig. 2-28b. The headstock 2 attached to the worktable is used for mounting the workpiece and driving it to rotate. The headstock 2 can be swiveled to a certain angle in a horizontal plane in order to grind a short taper hole, as shown in Fig. 2-28d. The tailstock 7 is located to the proper position on the worktable and used to support the workpiece. The wheel frame 5 and internal grinding head 4 are installed on the slide 6 which can be moved transversely by turning the handwheel 9. The wheel head 5 can also be swiveled on the slide 6 to a certain angle (±30°) to grind the short taper with a steep angle, as shown in Fig. 2-28c.

Several basic grinding methods are longitudinal grinding (Fig. 2-28a, b, d) and plunge grinding (Fig. 2-28c).

2. Surface grinding machine

Surface grinding machines are generally used for generating flat surfaces. These machines are similar to milling machines in construction as well as motion. In light of grinding method and layout, there are basically four types of surface grinding machines as shown in Fig. 2-29. Fig. 2-29a shows the type with a horizontal spindle and reciprocating table; Fig. 2-29b shows the type with a horizontal spindle and rotating table; Fig. 2-29c shows the type with a vertical spindle and reciprocating table; Fig. 2-29d shows the type with a vertical spindle and rotating table.

The table in the case of reciprocating machines is generally moved by hydraulic power. A cross feed motion is given to the wheel head at the end of each table motion. Vertical spindle machines are generally of a bigger capacity. The diameter of the wheel is wider than the workpiece, and as a result, no traverse feed is required. They are suitable for grinding very flat surfaces in production.

Compared with the rotating table, the reciprocating table has a broader machining range, but there is a time lost in the reversing table. The rotating table has higher grinding efficiency, but it is not suitable for grinding the longer plane or the plane with step.

（±10°）以磨削长锥面，如图 2-28b 所示。头架 2 固定在工作台上，用来安装工件并带动工件旋转。为了磨短的锥孔，头架在水平面内可转动一个角度，如图 2-28d 所示。尾座 7 可在工作台的适当位置上固定，以顶尖支承工件。滑鞍 6 上装有砂轮架 5 和内圆磨头 4，转动横向进给手轮 9，通过横向进给机构能使滑鞍和砂轮架作横向进给运动。砂轮架也能在滑鞍上调整一定角度（±30°），以磨削锥度较大的短锥面，如图 2-28c 所示。

万能外圆磨床的基本磨削方法有纵向磨削法（图 2-28 a、b、d）和切入磨削法（图 2-28c）。

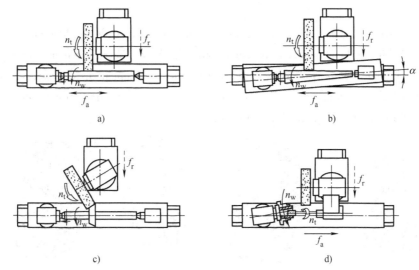

Fig. 2-28 Schematic diagram of universal cylindrical grinding 万能外圆磨床加工示意图

2. 平面磨床

平面磨床主要用于磨削各种平面。这类机床无论在结构还是在运动方面都和铣床相似。根据磨削方法和机床布局，平面磨床主要有 4 种类型，其磨削方式如图 2-29 所示。其中，图 2-29a 为卧轴矩台型；图 2-29b 为卧轴圆台型；图 2-29c 为立轴矩台型；图 2-29d 为立轴圆台型。

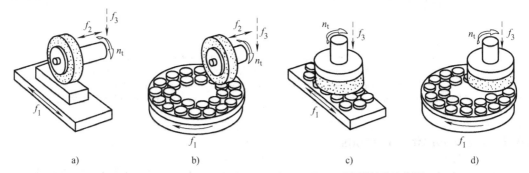

Fig. 2-29 Schematic diagram of surface grinding 平面磨削示意图

往复运动的工作台通常由液压驱动。砂轮架在工作台每次移动到头时作横向进给运动。立式磨床的加工能力较大，其砂轮的直径大于工件的宽度，因此无需横向运动，在生产中适合加工很大的平面。

Compared with the peripheral grinding, the face grinding has higher grinding efficiency. But because of large contact surface between the grinding wheel and the workpiece, it is difficult to remove the grinding heat by cooling and chips from the grinding area. Therefore, the face grinding generally has lower machining accuracy and poor surface finish. The peripheral grinding is likely to acquire good machining quality.

2.4 Gear cutting machines and cutting tools

As the gear transmission has many advantages, it has found wide applications in mechanical drive systems. With the development of transmission technology, the requirements for gears are getting higher and higher. There are many kinds of gears in mechanical drive systems, as shown in Fig. 2-30. In order to machine various kinds of gears and meet machining requirements, many gear cutting methods and gear cutting machines are developed.

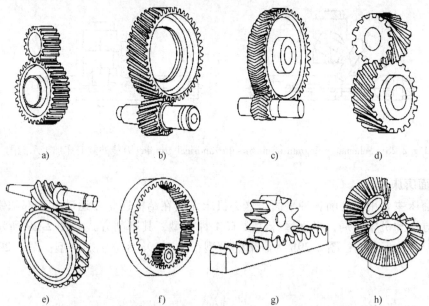

Fig. 2-30 Commonly used gear transmission types 常用的齿轮传动类型
a) Spur gear transmission 直齿轮传动   b) Helical gear transmission 斜齿轮传动
c) Herringbone gear transmission 人字齿轮传动   d) Spiral gear transmission 螺旋齿轮传动
e) Worm gear transmission 蜗轮传动   f) Internal gear transmission 内齿轮传动
g) Rack and gear transmission 齿轮齿条传动   h) Straight bevel gear transmission 直齿锥齿轮传动

2.4.1 Gear cutting methods

There are many kinds of gear cutting methods. Based on the tooth machining principle, gear cutting methods are classified as gear forming and gear generation.

1. Gear forming

In the process of gear forming, a form cutter with the shape of the gear tooth is used to cut gears. The gear form milling, gear form grinding, and gear broaching belong to the gear cutting method.

矩形工作台与圆形工作台相比较，前者的加工范围较宽，但有工作台换向的时间损失；后者为连续磨削，生产率较高，但不能加工较长的或带台阶的平面。

砂轮端面磨削与周边磨削相比，端面磨削的生产率较高，但因砂轮和工件接触面积大，冷却困难，且切屑不易排除，所以加工精度较低，表面粗糙度较大；而周边磨削可获得较高的加工质量。

2.4 齿轮加工机床及其刀具

齿轮传动有很多优点，因而在机械传动装置中得到广泛应用。随着传动技术的发展，对齿轮的要求越来越高。在机械传动装置中有各种各样的齿轮，如图 2-30 所示。为了加工各种各样的齿轮并满足各自的加工要求，许多加工方法和加工机床应运而生。

2.4.1 齿轮加工方法

齿轮加工方法很多。但按齿形加工原理，齿轮加工方法可分为成形法和展成法。

1. 成形法

成形法所用刀具的切削刃形状与被加工齿轮的齿槽形状相同，这种方法有成形铣齿、成形磨齿、拉齿等。

成形铣齿可以在普通铣床上进行。如图 2-31 所示，成形铣刀有两种。盘形齿轮铣刀通常用于加工模数 $m \leqslant 8\text{mm}$ 的齿轮，而指形齿轮铣刀通常用于加工模数 $m > 8\text{mm}$ 的齿轮。

成形铣刀沿工件径向进给直到全齿高，而工件沿其轴线进给通过铣刀以加工一个齿槽。在一个齿槽加工完后，工件退离刀具利用分度头进行分度。分度后继续重复前述的加工过程。

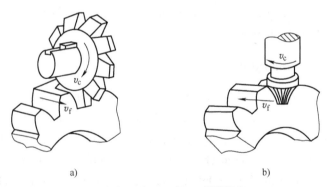

Fig. 2-31　Gear milling 成形铣齿
a) Disc-type gear milling cutter 盘形齿轮铣刀
b) Finger-type gear milling cutter 指形齿轮铣刀

成形铣齿的精度和生产率都较低，仅在单件小批生产中采用。成形磨齿是用与齿形相同的成形砂轮磨削齿槽的加工方法。它是一种用于加工硬齿面和高精度齿轮的精加工方法。拉齿仅用于大量生产中加工内齿轮。

2. 展成法

为了获得更精确的齿轮，齿轮通常利用一种刀具展成加工，该刀具类似于按照齿轮啮合

The gear form milling can be done by means of general purpose milling machines. There are two types of gear milling cutters, as shown in Fig. 2-31. A disc-type gear milling cutter is generally used to cut the gears with module $m \leqslant 8$. A finger-type gear milling cutter is generally used to cut the gears with module $m > 8$.

The gear milling cutter is fed radially into the workpiece till the full depth is reached. Then the workpiece is fed past the cutter to complete the machining of one tooth space. After a tooth space is cut, the workpiece is returned apart from the cutter, and then is indexed by means of dividing head. The machining process is repeated after indexing the gear blank by one tooth.

The gear milling has low machining efficiency and accuracy, and it is only suitable for job production. The gear form grinding means using the form grinding wheel with the same profile as a gear tooth to grind the gear space.

The gear form grinding is a gear finish machining method used for grinding the gears with hard face and higher accuracy. The gear broaching is only used to cut internal gears in large volume production.

2. Gear generating

To obtain more accurate gears, the gear is generally generated using a cutter, which is similar to the gear with which it meshes by following the general gear theory. If the module and pressure angle of the gear cutter are the same as that of the gears to be cut, one cutter theoretically can generate all gears regardless of the number of teeth. All the teeth have the correct profile and mesh correctly. The gears produced by generation are more accurate and the manufacturing process is also fast. These are used for cutting gears in large volume production.

There are several gear generating methods, such as gear shaping, gear hobbing, gear shaving, gear honing and gear generation grinding, and so on. Here the gear hobbing is mainly introduced as it is the most commonly used method.

2.4.2 Gear cutting machines

The machine tools which are used to machine the gear tooth are named as gear cutting machines. There are many kinds of gear cutting machines, such as gear hobbing machine, gear shaping machine, gear shaving machine, gear honing machine, gear grinding machine, gear planner, gear milling machine, gear broaching machine, and so on. The gear hobbing machine is one of the most widely used gear cutting machines. Only the knowledge about the gear hobbing machine is introduced in this book.

1. Gear hobbing principle

Based on the gear generation principle, gear hobbing simulates the meshing process of a pair of helical gears, as shown in Fig. 2-32a. If one of the two meshing gears has only several teeth or even one tooth with a very small pitch angle, the helical gear becomes a worm, as shown in Fig. 2-32b. After slotting and relieving the flank on the worm, the worm becomes a hob, as shown in Fig. 2-32c.

2. Transmission principle of gear hobbing

(1) Transmission principle of straight gear hobbing

理论进行啮合的齿轮。因此，只要刀具与被加工齿轮的模数和压力角相同，一把刀具理论上可以加工同一模数、不同齿数的齿轮，而且齿形准确，能够正确啮合。展成法加工的齿轮生产率和加工精度都比较高，常用于大量生产。

展成法加工齿轮的方法有多种，如插齿、滚齿、剃齿、珩齿、展成法磨齿等。因滚齿最常用，故本节仅对滚齿加工方法进行介绍。

2.4.2 齿轮加工机床

齿轮加工机床是加工齿轮轮齿的机床。齿轮加工机床的种类很多，如滚齿机、插齿机、剃齿机、珩齿机、磨齿机、刨齿机、铣齿机、拉齿机等。滚齿机是齿轮加工机床中应用最广的一种，本书仅介绍有关滚齿机的知识。

1. 滚齿原理

根据齿轮展成原理，滚齿模拟了一对螺旋齿轮的啮合过程，如图 2-32a 所示。将这对啮合传动副中的一个齿轮的齿数减少到几个或一个，并使螺旋升角很小，它就成了蜗杆，如图 2-32b所示。再对蜗杆开槽并铲背，就成了齿轮滚刀，如图 2-32c 所示。

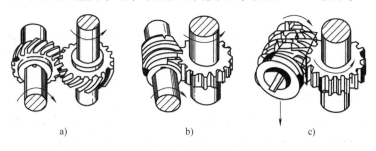

Fig. 2-32　Gear hobbing principle 滚齿原理

2. 滚齿传动原理

（1）滚切直齿轮的传动原理　图 2-33 所示为滚切直齿轮的传动原理图。由图 2-33 可知，它有三条传动链。

1）主运动传动链。电动机—1—2—u_v—3—4—滚刀，属于外联传动链。滚齿主运动是滚刀的旋转运动 B_{11}。u_v 是用来调整渐开线成形运动的速度参数。

2）展成运动传动链。滚刀—4—5—u_x—6—7—工件。靠滚刀的旋转运动 B_{11} 和工件的旋转运动 B_{12} 组成的复合运动渐开线齿廓，属于内联传动链。展成运动用来保证滚刀与被加工齿轮之间

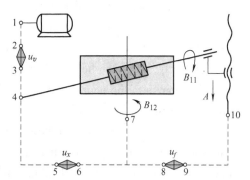

Fig. 2-33　Transmission principle of straight gear hobbing 滚切直齿轮传动原理

的啮合关系。u_x 是用来调整渐开线成形运动的轨迹参数，它影响渐开线的形状。

3）轴向进给运动传动链。工件—7—8—u_f—9—10—刀架，属于外联传动链。轴向进给运动是为了确保能切出全齿宽。工件转一转，刀架沿齿宽方向移动一定距离。调整 u_f 得到所需要的轴向进给量。进给量影响齿面粗糙度。

（2）加工斜齿圆柱齿轮的传动原理　图 2-34a 所示为滚切斜齿轮的传动原理图。由图

The transmission principle of straight gear hobbing is illustrated in Fig. 2-33. It can be seen from Fig. 2-33 that there are three transmission chains. They are:

1) Transmission chain of primary motion: Motor—1—2—u_v—3—4—hobbing cutter. It belongs to "external connection transmission chain". The primary motion of gear hobbing is the rotation of the hobbing cutter B_{11}. u_v is used for adjusting the speed parameter of the involute forming motion.

2) Transmission chain of generation motion: Hobbing cutter—4—5—u_x—6—7—workpiece. The formation of the involute depends on a compound motion composed of B_{11} and B_{12}. It belongs to "internal-connection transmission chain". Generation motion is used to maintain the meshing relation between the hob and the gear to be cut. u_x is used for adjusting the track parameter of the involute forming motion. It has an effect on the shape of the involute.

3) Transmission chain of axial feed motion: Workpiece—7—8—u_f—9—10—Tool rest. It belongs to "external connection transmission chain". The axial feed motion is used for hobbing the whole tooth width. When the workpiece turns a circle, the hob moves along the tooth width a certain distance. The required axial feed can be obtained by adjusting u_f. It has an effect on the surface roughness.

(2) Transmission principle of helical gear hobbing The transmission principle of helical gear hobbing is illustrated in Fig. 2-34a. It can be seen from Fig. 2-34a that there are four transmission chains. But three transmission chains are the same as that of straight gear hobbing. The different one is the transmission chain of the differential motion. The transmission chain is: Tool rest—12—13—u_y—14—15—Σ—6—7—u_x—8—9—Work table. It belongs to "internal connection transmission chain". This is an additional motion which is used to generate helix. When the workpiece rotates a circle accurately, it happened that the cutter moves a lead of helix.

An additional motion can be illustrated by Fig. 2-34b. Suppose the work to be cut is a right hand gear. When the hob moves from the point a to the point b, in order to cut a helical, the point b' should be turned to the point b. That means the work should turn an additional bb' besides original B_{12}. Similarly, when the hob reaches the point c, the work turn an additional cc'. In this way, when the hob moves to the point p, the point p' on the work should be turned to the point p, i. e. the work should turn an additional circle.

3. Installation of the gear hobbing cutter

In gear hobbing, it is required that the helix direction of the hobbing cutter be consistent with tooth alignment of the gear. Before hobbing, the installation angle of the hobbing cutter must be set. Suppose that the hobbing cutter is located in the front of the gear, ω stands for the helical angle of the cutter, δ stands for the installation angle of the cutter.

When hobbing a straight gear, as shown in Fig. 2-35, there is $\delta = \omega$.

When hobbing a helical gear, as shown in Fig. 2-36, there is $\delta = \beta \pm \omega$, where β is the helical angle of the gear to be cut. When the helix direction of the helical gear is the same as that of the hobbing cutter, take " $-$ "; When the helix direction of the helical gear is oppose to that of the hobbing cutter, take " $+$ ".

可知，它有四条传动链，其中三条与滚切直齿轮的相同。所不同的是差动运动传动链。其传动链是：刀架—12—13—u_y—14—15—Σ—6—7—u_x—8—9—工作台，属内联系传动链。这是一条附加运动传动链，即滚刀移动一个工件螺旋线导程时，工件应准确地附加转过一转。

对此可用图 2-34b 来说明。设工件为右旋，当滚刀从 a 点移到 b 点时，为了能切出螺旋线齿线，应使工件的 b' 点转到 b 点，即在原来 B_{12} 的基础上再多转个 bb'。同理，当滚刀进给到 c 点时，工件应附加转动 cc'。以此类推，当滚刀进给到 p 点时，应使工件上的 p' 点转到 p 点，即工件应附加转一转。

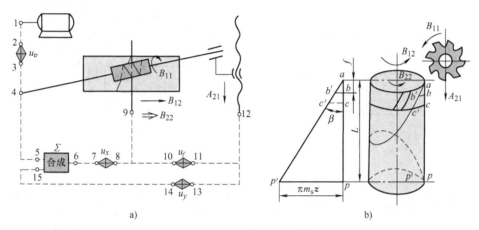

Fig. 2-34　Transmission principle of helical gear hobbing 滚切斜齿轮传动原理

3. 滚刀的安装

安装滚刀时必须使滚刀的螺旋线方向与工件齿长方向一致，因此，加工前必须正确调整滚刀的安装角。图 2-35 所示为滚切直齿圆柱齿轮时滚刀的安装角。其中，滚刀位于工件前面，滚刀的螺旋升角为 ω。滚刀的安装角 $\delta = \omega$。

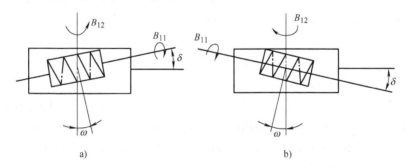

Fig. 2-35　Installation angle of hobbing cutter in hobbing straight gear 滚切直齿圆柱齿轮时滚刀的安装角
　　　　a) Hobbing straight gear by using right-hand hobbing cutter 右旋滚刀滚切直齿轮
　　　　b) Hobbing straight gear by using left-hand hobbing cutter 左旋滚刀滚切直齿轮

用滚刀加工斜齿圆柱齿轮时，由于滚刀和工件的螺旋方向都有左、右方向之分，因此就有四种不同的组合，如图 2-36 所示。那么 $\delta = \beta \pm \omega$，其中 β 为被加工齿轮的螺旋角。当滚刀与被加工斜齿轮的螺旋线方向相同时取"－"号，方向相反时取"＋"号。

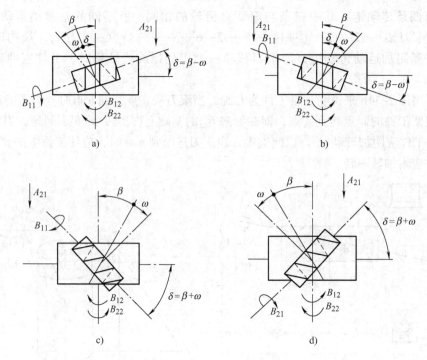

Fig. 2-36 Installation angle of the hobbing cutter in the hobbing helical gear 滚切斜齿圆柱齿轮时滚刀的安装角
a) Hobbing a left-hand helical gear by using a left-hand hobbing cutter 左旋滚刀滚切左旋齿轮
b) Hobbing a right-hand helical gear by using a right-hand hobbing cutter 右旋滚刀滚切右旋齿轮
c) Hobbing a right-hand helical gear by using a left-hand hobbing cutter 左旋滚刀滚切右旋齿轮
d) Hobbing a left-hand helical gear by using a right-hand hobbing cutter 右旋滚刀滚切左旋齿轮

4. Model Y3150E hobbing machine

(1) Usage and layout of the Y3150E hobbing machine   The layout of the Y3150E hobbing machine is shown in Fig. 2-37. It is primarily used to cut the spur gear and helical gear. If the worm gear hob is used, the worm gear can also be cut through the manual feed in the radial direction. The column 2 is installed on the bed 1, and the carriage 3 together with the tool rest can be moved up and down along the guideways on the column 2 to realize the axial feed motion. The toolbar 4 used for mounting the hob is assembled into the spindle located in the toolrest housing 5. The toolrest housing 5 can turn around its axis a certain angle whose value has to do with the helix angle of the hob. The workpiece is located on the mandrel 7 of the rotary table 9 and rotates with the table. The back pillar 8 is tied to the table 9, which can move along the guideways on the bed 1 in order to adjust the radial position of the workpiece relative to the hob. The support 6 is used to support the mandrel 7 by means of center in order to increase the rigidity of the mandrel.

(2) Main technical parameters of the Y3150E hobbing machine

Max. diameter of the gear	500mm
Max. width of the gear	250mm
Max. module of the gear	8mm
Min. teeth number	$5 \times k$ (threads of hob)

4. Y3150E 型滚齿机

（1）Y3150E 型滚齿机的用途与布局　Y3150E 型滚齿机外形如图 2-37 所示。它主要用来加工直齿轮和斜齿轮。如果使用蜗轮滚刀，也可以通过手动径向进给加工蜗轮。立柱 2 固定在床身 1 上，刀架溜板 3 可沿立柱 2 上的导轨作垂直方向的直线移动（轴向进给运动）。安装滚刀的刀杆 4 固定在刀架体 5 中的刀具主轴上。刀架体能绕自身轴线倾斜一个角度，其大小与滚刀的螺旋升角大小及旋向有关。工件安装在回转台 9 的心轴 7 上，并随回转台一起旋转。后立柱 8 和回转台 9 连成一体，可沿床身的导轨作水平移动，用于调整工件与滚刀间的径向位置，以适应不同直径的工件或加工蜗轮时作径向进给运动。支架 6 可用轴套或顶尖支承工件心轴 7，以增加心轴的刚性。

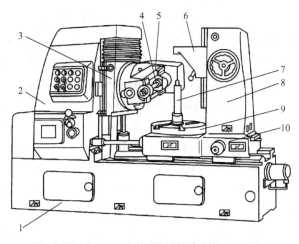

Fig. 2-37　Layout of the Y3150E hobbing machine
Y3150E 型滚齿机的外观图
1—Bed 床身　2—Column 立柱　3—Carriage 溜板
4—Toolbar 刀杆　5—Toolrest housing 刀架体
6—Support 支架　7—Mandrel 心轴　8—Back pillar 后立柱
9—Rotary table 回转台　10—Saddle 床鞍

（2）Y3150E 型滚齿机的主要技术参数

工件最大直径	500mm
工件最大加工宽度	250mm
工件最大模数	8mm
工件最小齿数	$5 \times k$（滚刀线数）
滚刀主轴转速	40，50，63，80，100，125，160，200，250r/min
刀架轴向进给量	0.4，0.56，0.63，0.87，1，1.16，1.41，1.6，1.8，2.5，2.9，4mm/r
机床轮廓尺寸	2439mm×1272mm×1770mm

5. Y3150E 型滚齿机的传动系统分析

（1）Y3150E 型滚齿机的传动系统图　Y3150E 型滚齿机的传动系统如图 2-38 所示。

（2）滚切直齿轮的传动链及换置计算　在滚切直齿圆柱齿轮时需要两个成形运动和三条传动链。一个是用于生成渐开线的展成运动（$B_{11} + B_{12}$），它是利用确定运动轨迹（渐开线）的展成运动传动链和确定运动速度的主运动传动链来实现的。另一个是用于形成直线（导线）的刀架轴向直线运动，它只包括一条确定运动速度的轴向进给运动传动链。

1）主运动传动链。

a. 查找首、末两端件：滚刀和主轴。

b. 计算位移：n_m(r/min)——n_t(r/min)。

c. 运动平衡式

$$n_m \times \frac{115}{165} \times \frac{21}{42} \times u_{2-3} \times \frac{A}{B} \times \frac{28}{28} \times \frac{28}{28} \times \frac{28}{28} \times \frac{20}{80} = n_t \quad (2\text{-}17)$$

Rotation speeds of the hob spindle　　40, 50, 63, 80, 100, 125, 160, 200, 250 r/min

Axial feed of the tool slide　　0.4, 0.56, 0.63, 0.87, 1, 1.16, 1.41, 1.6, 1.8, 2.5, 2.9, 4mm/r

Total dimension　　2439mm × 1272mm × 1770mm

5. Analysis of the transmission system of the Y3150 E hobbing machine

(1) Transmission system diagram of the Y3150 E hobbing machine　Transmission system diagram of the Y3150 E hobbing machine is shown in Fig. 2-38.

(2) Transmission chain in the spur gear hobbing and displacement calculation　Two forming motions and three transmission chains are required in the spur gear hobbing. One is the generation motion ($B_{11} + B_{12}$) used to generate the involute. It is realized by means of a transmission chain of generation motion used to determine its track (involute) and a transmission chain of primary motion used to determine its motion speed. The other is the axial linear motion of the tool slide used to produce a straight line (directrix). It only includes a transmission chain of axial feed motion used to determine the kinematic velocity.

1) Transmission chain of primary motion.

a. To find the first and last part: Motor— Hob

b. Displacement calculation: n_m r/min ——n_t r/min

c. Motion balance equation:

$$n_m \times \frac{115}{165} \times \frac{21}{42} \times u_{2\text{-}3} \times \frac{A}{B} \times \frac{28}{28} \times \frac{28}{28} \times \frac{28}{28} \times \frac{20}{80} = n_t \quad (2\text{-}17)$$

d. Displacement formula:

$$u_v = u_{2\text{-}3} \times \frac{A}{B} = \frac{n_t}{124.583} \quad (2\text{-}18)$$

where n_m stands for the rotation speed of the electric motor; n_t stands for the rotation speed of the hob; u_{2-3} stands for the drive ratio of the shaft Ⅱ to the shaft Ⅲ in the speed gear box.

It can be seen from the above formula that after n_t is determined, u_v can be calculated. Thereby, the pairs of meshing gears in the gear box and the tooth number of interpolating gears can be determined. The main speed change gears A/B provided by the Y3150E hobbing machine are 22/44, 33/33, 44/22.

2) Transmission chain of generation motion. Transmission chain of generation motion can be found easily from Fig. 2-38. Its transmission route is:

Hob (Ⅷ)—80/20—Ⅶ—28/28—Ⅵ—28/28—Ⅴ—28/28—Ⅳ—42/56—Ⅸ—Synthesizer— $\frac{e}{f} \times \frac{a}{b} \times \frac{c}{d}$ — XⅢ—1/72—XXV (Workpiece)

a. To find the first and last part

Hob— Workpiece (Worktable)

b. Calculating displacement

$1/K$ $r_{(\text{hob})}$ ——$1/Z_w$ $r_{(\text{work})}$

c. Motion balance equation

d. 计算换置公式

$$u_v = u_{2-3} \times \frac{A}{B} = \frac{n_t}{124.583} \tag{2-18}$$

式中 n_m——电动机的转速；
n_t——滚刀的转速；
u_{2-3}——变速箱中Ⅱ轴和Ⅲ轴之间的传动比。

由上述公式可知，当 n_t 确定后就可以算出 u_v，并由此确定变速箱内啮合的齿轮对和交换齿轮的齿数。Y3150E 型滚齿机提供的主变速交换齿轮 A/B 分别为 22/44、33/33、44/22。

2）展成运动传动链。从图 2-38 中不难找出展成运动传动链。其传动路线表达式为

滚刀主轴（Ⅷ）—80/20—Ⅶ—28/28—Ⅵ—28/28—Ⅴ—28/28—Ⅳ—42/56—Ⅸ—合成机构—$\frac{e}{f} \times \frac{a}{b} \times \frac{c}{d}$—ⅩⅢ—1/72—ⅩⅩⅤ（工件）

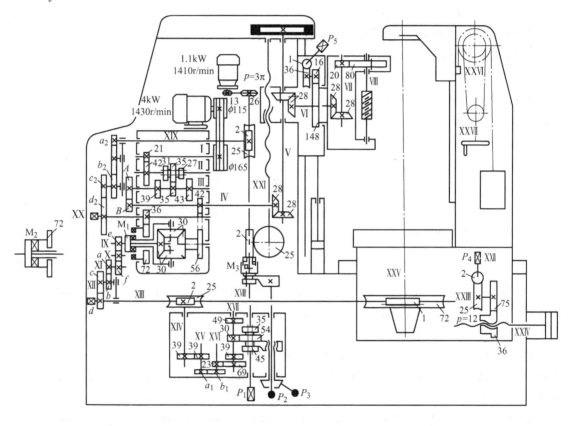

Fig. 2-38 Transmission system diagram of Y3150E Y3150E 型滚齿机传动系统图

a. 查找首、末两端件：滚刀和工件（工作台）。
b. 计算位移：$1/K$ r$_{(滚刀)}$——$1/Z_w$ r$_{(工件)}$。
c. 运动平衡式

$$\frac{1}{K} \times \frac{80}{20} \times \frac{28}{28} \times \frac{28}{28} \times \frac{28}{28} \times \frac{42}{56} \times u_\Sigma \times \frac{e}{f} \times \frac{a}{b} \times \frac{c}{d} \times \frac{1}{72} = \frac{1}{Z_w} \tag{2-19}$$

$$\frac{1}{K} \times \frac{80}{20} \times \frac{28}{28} \times \frac{28}{28} \times \frac{28}{28} \times \frac{42}{56} \times u_\Sigma \times \frac{e}{f} \times \frac{a}{b} \times \frac{c}{d} \times \frac{1}{72} = \frac{1}{Z_w} \qquad (2\text{-}19)$$

where, u_Σ is the transmission ratio of the synthesizer. $u_\Sigma = 1$ in hobbing straight spur gears.

d. Displacement formula

$$u_x = \frac{a}{b} \times \frac{c}{d} = \frac{f}{e} \cdot \frac{24K}{Z_n} \qquad (2\text{-}20)$$

where, e and f are the structure interpolating gears which are selected according to the tooth number of the gear to be cut. It is used to adjust u_x to prevent u_x overlarge or over-small so as to choose interpolating gears conveniently. $\frac{a}{b} \times \frac{c}{d}$ is the transmission ratio of generating interpolation gears.

When $5 \leqslant Z_n/K \leqslant 20$, $e = 48$, $f = 24$;

When $21 \leqslant Z_n/K \leqslant 142$, $e = 36$, $f = 36$;

When $Z_n/K \geqslant 143$, $e = 24$, $f = 48$.

3) Transmission chain of axial feed motion

a. To find the first part and the last part

Workpiece (worktable)—Toolrest

b. Calculating displacement

1 r_(worktable)——$f$mm_(tool rest)

c. Motion balance equation

$$1 \times \frac{72}{1} \times \frac{2}{25} \times \frac{39}{39} \times \frac{a_1}{b_1} \times \frac{23}{69} \times u_{17\text{-}18} \times \frac{2}{25} \times 3\pi = f \qquad (2\text{-}21)$$

where u_{17-18} stands for the drive ratio of the shaft XVII to the shaft XVIII in the feed gear box.

d. Displacement formula

$$u_f = \frac{a_1}{b_1} \times u_{17\text{-}18} = \frac{f}{0.4608\pi} \qquad (2\text{-}22)$$

(3) A transmission chain in helical gear hobbing and displacement calculation The difference between the spur gear hobbing and the helical gear hobbing lies only in the difference of the shape of the guide line. The former is a straight line, and the latter is a helical line. This means they are different in the track parameters. It is clear that there is still a transmission connection in between the linear movement of the tool rest and the rotation of the workpiece in order to generate the helical line.

1) Transmission chains of the primary motion and the axial feed motion. The two transmission chains are the same exactly as that of the spur gear hobbing.

2) The transmission chain of the generation motion is also the same as that of the spur gear hobbing except for $u_\Sigma = -1$.

When the helical spur gear is cut, the clutch M_1 with short teeth is replaced by M_2 with long teeth. The teeth length of M_2 is long enough to engage with the end teeth of the shell (bar H) and the gear Z_{72} mounted on the shell simultaneously. In this way, they linked each other to form a whole body. The bar is linked to outside. Motions from the transmission chain of the generation mo-

其中，u_Σ 是合成机构的传动比。在滚切直齿圆柱齿轮时，$u_\Sigma = 1$。

d. 计算换置公式

$$u_x = \frac{a}{b} \times \frac{c}{d} = \frac{f}{e} \cdot \frac{24K}{Z_\text{工}} \tag{2-20}$$

其中，e 和 f 是结构交换齿轮，根据被加工齿轮的齿数选取，用于调整 u_x 的数值，使其不会过大或过小，以便于交换齿轮的选择。$\frac{a}{b} \times \frac{c}{d}$ 称为分齿交换齿轮。

当 $5 \leqslant Z_\text{工}/K \leqslant 20$ 时，取 $e = 48$，$f = 24$；当 $21 \leqslant Z_\text{工}/K \leqslant 142$ 时，取 $e = 36$，$f = 36$；当 $Z_\text{工}/K \geqslant 143$ 时，取 $e = 24$，$f = 48$。

3）轴向进给运动传动链。

a. 查找首、末两端件：工件（工作台）—刀架。

b. 计算位移：$1\text{r}_{(\text{工作台})}$——$f\text{mm}_{(\text{刀架})}$。

c. 运动平衡式

$$1 \times \frac{72}{1} \times \frac{2}{25} \times \frac{39}{39} \times \frac{a_1}{b_1} \times \frac{23}{69} \times u_{17-18} \times \frac{2}{25} \times 3\pi = f \tag{2-21}$$

其中，u_{17-18} 代表进给箱中 XVII 轴到 XVIII 轴之间的传动比。

d. 计算换置公式

$$u_f = \frac{a_1}{b_1} \times u_{17-18} = \frac{f}{0.4608\pi} \tag{2-22}$$

（3）滚切斜齿圆柱齿轮的传动链及换置计算　滚切直齿轮和滚切斜齿轮的差别仅在于导线的形状不同。前者是直线，而后者是螺旋线。这就意味着它们的轨迹参数不同。显然，为了生成螺旋线，在刀具的直线运动和工件的转动之间存在传动联系。

1）主运动传动链和轴向进给运动传动链与滚切直齿轮时完全相同。

2）展成运动传动链和滚切直齿轮时也相同，只是其合成机构的传动比 $u_\Sigma = -1$。

在滚切斜齿圆柱齿轮时，短齿离合器 M_1 更换成长齿离合器 M_2，长齿离合器 M_2 与合成机构的壳体（即转臂 H）以及装在壳体上的齿轮 Z_{72} 同时啮合。于是，它们连接成一个整体。转臂 H 与外部相连。来自展成运动传动链的运动与来自差动运动传动链的运动分别通过 Z_{56} 和 Z_{72} 进入合成机构。合成后的运动经 IX 轴输出。

3）差动运动传动链。差动运动传动链的功用是当滚刀沿工件轴向进给时使工件产生相应的附加转动。

a. 查找首、末两端件：刀架—工件（工作台）。

b. 计算位移：$L\text{mm}_{(\text{刀架})}$——$1\text{r}_{(\text{工作台})}$。

这就意味着当刀架移动一个螺旋线导程 L 时，工件就会比原来的正常转动多转或少转一圈。

c. 运动平衡式

$$\frac{L}{3\pi} \times \frac{25}{2} \times \frac{2}{25} \times \frac{a_2}{b_2} \times \frac{c_2}{d_2} \times \frac{36}{72} \times u_\Sigma \times \frac{e}{f} \times u_x \times \frac{1}{72} = 1\text{r} \tag{2-23}$$

其中，$L = \frac{\pi m_n z}{\sin\beta}$。$L$ 代表被加工斜齿轮的螺旋线导程；m_n 代表齿轮的法向模数；β 是齿轮的

tion and the transmission chain of the differential motion are input to the synthesizer through Z_{56} and Z_{72} separately. Synthesis motions are output through the shaft IX.

3) A transmission chain of the differential motion. The function of a transmission chain of the differential motion is to have the workpiece produce an additional rotation when the hob is fed along the axis of the workpiece.

a. To find the first and the last part: Tool rest —Workpiece (Worktable)

b. Calculating displacement: L mm$_{(tool\ rest)}$ ——1 r$_{(workpiece)}$

That means when the tool rest moves a lead of helical line, the workpiece would turn 1 circle adding to or subtracting from the normal rotation.

c. Motion balance equation

$$\frac{L}{3\pi} \times \frac{25}{2} \times \frac{2}{25} \times \frac{a_2}{b_2} \times \frac{c_2}{d_2} \times \frac{36}{72} \times u_\Sigma \times \frac{e}{f} \times u_x \times \frac{1}{72} = 1\text{r} \quad (2\text{-}23)$$

where $L = \frac{\pi m_n z}{\sin\beta}$. L is the helix lead of the helical gear to be cut; m_n is the normal module of the gear; β is the helical angle of the gear; $u_x = \frac{a}{b} \times \frac{c}{d} = \frac{f}{e} \cdot \frac{24K}{Z_n}$; $u_\Sigma = u_{\Sigma 2} = 2$ for transmission chain of differential motion.

d. Displacement formula

Substituting L, u_Σ and u_x into the Eq. (2-23), then

$$u_y = \frac{a_2}{b_2} \times \frac{c_2}{d_2} = 9 \cdot \frac{\sin\beta}{m_n K} \quad (2\text{-}24)$$

where $\frac{a_2}{c_2} \times \frac{c_2}{d_2}$ is the transmission ratio of differential interpolating gears.

It can be seen from Eq. (2-24) that u_y does not contain the tooth number Z. As the choice of differentia interpolating gears a_2, b_2, c_2, d_2 has nothing to do with Z, when cutting a pairs of helical gears, though they have a different tooth number, they can be machined with the exact equal helical angle, and are not affected by the calculation error of u_y.

2.4.3 Gear cutting tools

Gear cutting tools are used to cut gear tooth. There are many kinds of gear cutting tools with complex structure in order to machine different gears. Gear cutting tools are categorized into gear forming cutters and gear generating cutters.

(1) Gear milling cutters   Commercial gear milling cutters are available as a set for a given module. These can be used on either horizontal axis or vertical axis milling machines. There are two types of gear milling cutters, as shown in Fig. 2-39.

(2) Gear generating cutters   There are many kinds of gear generating cutters, but the most often used gear generating cutters in production practice are hobs, gear shaping cutters, and gear shaving cutters, as shown in Fig. 2-40. The cutter itself is equivalent to a gear, which meshes with the gear to be cut in gear cutting process. This meshing motion is named as generation motion.

螺旋角；$u_x = \dfrac{a}{b} \times \dfrac{c}{d} = \dfrac{f}{e} \cdot \dfrac{24K}{Z_\text{工}}$；对于差动运动传动链，$u_\Sigma = u_{\Sigma 2} = 2$。

d. 计算换置公式。

将 L、u_Σ 和 u_x 代入式（2-23），那么有

$$u_y = \dfrac{a_2}{b_2} \times \dfrac{c_2}{d_2} = 9 \cdot \dfrac{\sin\beta}{m_n K} \tag{2-24}$$

其中，$\dfrac{a_2}{c_2} \times \dfrac{c_2}{d_2}$ 是差动交换齿轮的传动比。

由式（2-24）可知，u_y 中并不包含齿数 Z。由于差动交换齿轮 a_2、b_2、c_2、d_2 的选择与齿数 Z 无关，当加工一对螺旋齿轮时，尽管它们的齿数不同，但都可以加工成具有相同的螺旋角，而且不受 u_y 计算误差的影响。

2.4.3 齿轮刀具

齿轮刀具是用来加工齿形的。为满足各种齿轮的加工需要，齿轮刀具种类很多，结构复杂。齿轮刀具可分为成形法切齿刀具和展成法切齿刀具两大类。

（1）成形铣齿刀具　市场上销售的成形铣齿刀具是针对某一模数成套供应的。它们要么是在普通卧式铣床上使用，要么是在普通立式铣床上使用。两种类型的成形铣齿刀具如图2-39所示。

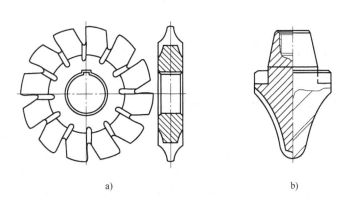

Fig. 2-39　Gear milling cutters 成形齿轮铣刀
a）Disc-type gear milling cutter 盘形齿轮铣刀
b）Finger-type gear milling cutter 指形齿轮铣刀

（2）展成法切齿刀具　展成法切齿刀具有许多种，在生产中最常用的有齿轮滚刀、插齿刀、剃齿刀等，如图2-40所示。刀具本身相当于一个齿轮，切齿时刀具与被加工齿轮之间有相对的啮合运动，即展成运动。

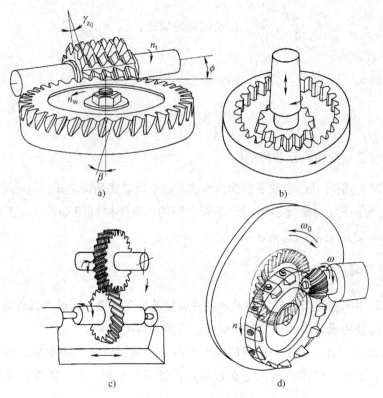

Fig. 2-40　Gear generating cutters 展成法切齿刀具
a) Hob 齿轮滚刀　b) Gear shaping cutter 插齿刀
c) Gear shaving cutter 剃齿刀　d) Spiral gear milling cutter 弧齿锥齿轮铣刀

2.5　Other machine tools and cutting tools

2.5.1　Hole-making machine tools and cutting tools

Observing various products around us, we find that the vast majority have several holes in them. Hole-making in the workpiece is one of the most common tasks in the manufacturing industry. It is estimated that of all the machining operations performed, there are about 1/4 hole making operations. Drilling machines and boring machines are the most commonly used hole making machines.

1. Drilling machines

All machines which are mainly used to drill holes by means of drills are named as drilling machines. In addition to drilling, drilling machines can also be used for core drilling, reaming, tapping, countersinking, counterboring, spot facing and so on, as shown in Fig. 2-41.

Drilling machines come in a variety of shapes and sizes. The types of drilling machines range from bench-type drill presses, used to drill small-diameter holes, to large radial drilling machines, which can accommodate large workpieces. The most commonly used drilling machines in hole making operations are as follows.

(1) Upright drill press　A typical drill press is shown in Fig. 2-42. It is composed of a base

2.5 其他类型的机床与刀具

2.5.1 孔加工机床与刀具

观察周围的产品,不难发现绝大多数产品中都会有几个孔。在工件上进行孔加工是机械制造业中最平常的工作。据估计,孔加工约占所有加工任务的 1/4。钻床和镗床都是最常用的孔加工机床。

1. 钻床

所有主要用钻头在工件上加工孔的机床统称为钻床。除钻孔外,在钻床上还可完成扩孔、铰孔、攻螺纹、锪孔、锪平面等工作,如图 2-41 所示。

有各种形状和规格的钻床。钻床的类型有用于钻削小孔的台钻和能在大型工件上钻孔的摇臂钻床。在孔加工方面最常用的钻床如下。

(1) 立式钻床 图 2-42 所示为一典型的立式钻床。它由底座 1、工作台 2、主轴箱 3 和立柱 4 等部件组成。主轴箱内有主运动及进给运动的传动机构。对于直柄的小直径钻头借助于钻头卡爪安装在主轴上,而带有莫氏锥的钻头可直接装在主轴的锥孔内,由主轴带动作旋转主运动,主轴套筒可以手动或机动作轴向进给运动。工作台可沿立柱上的导轨作调位运动。工件用工作台上的台虎钳夹紧,或用压板直接固定在工作台上加工。立式钻床只适用于在单件、小批生产中加工中、小型工件。

(2) 摇臂钻床 摇臂钻床的外形及其主要部件和具有的运动如图 2-43 所示。加工时,

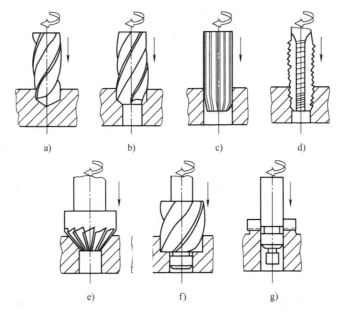

Fig. 2-41 Functions of the drilling machine 钻床的功用
a) Drilling 钻削 b) Core drilling 扩孔 c) Reaming 铰孔 d) Tapping 攻螺纹
e) Countersinking 锪锥孔 f) Counterboring 锪平底孔 g) Spot facing 锪平面

(1), a worktable (2), a spindle head (3), a column (4) and other components. There are transmission mechanisms for primary motion and feed motion in the spindle head. The drill bit is mounted into the spindle either by means of drill chuck for the small size drills that are straight shank type, or by the Morse taper hole in the end of the spindle. The spindle that drives the cutting tools is located inside a quill, which can be fed up and down by means of manual operation or by means of power feed. The workpiece is normally placed on an adjustable table, either by clamping it directly into the slots and holes on the table or by using a vise, which in turn can be clamped onto the table. The upright drill press is only suitable for the small and medium sized workpiece in job and low volume production.

(2) Radial drilling machine The schematic diagram of a radial drilling machine illustrating the major parts and movements is shown in Fig. 2-43. The workpiece is placed on the worktable or the base. There are two columns, the inner one is fixed to the base, and the outer one together with the arm and the spindle head can be rotated around the inner column. The spindle head can move along the radial arm to any position while the position of the radial arm can be adjusted on the outer column in the vertical direction, thus allowing the drill bit to reach any position in the radial range of the arm. The spindle control mechanisms such as on/off, speed changing, reversing, brake etc. are all arranged in the spindle head. Radial drilling machines are widely used for medium and large sized workpieces in job, low and medium volume production.

2. Boring machines

Boring machines are generally used for enlarging the existed holes on a large and heavy workpieces. In addition to boring a hole, boring machines can also be used for drilling, core-drilling, reaming, tapping, surface milling and so on. There are many types of boring machines, of which horizontal boring machines are most widely used.

(1) Horizontal boring machine Horizontal boring machines are very versatile and useful in machining large parts. The construction of these machines are illustrated in Fig. 2-44. The headstock 11 can be moved along the head column 10 which is mounted on the bed 1. The workpiece is mounted on the table 6 that can move longitudinally in both the axial direction and radial direction. In addition to boring bar, other cutting tools, such as drills, reamers and milling cutters etc. can also be mounted on the spindle that rotates in the headstock. The spindle is also provided with longitudinal power feed so that drilling and boring can be done over a considerable distance without moving the table.

When using a boring bar and a single point cutting tool to machine a hole, the major problem is the lack of rigidity of the boring bar. The end support 4 mounted on the end column provides rigid support for the long boring bar and permits very accurate work to be done.

(2) Jig borer A jig borer is a kind of universal boring machine with high precision. Jig boring machines are costly because they are built to ensure the utmost accuracy through extra rigidity, low thermal expansion, ultra-true geometry, and refined measuring means. The spindles of most machines run in preloaded anti-friction bearings.

A vertical jig boring machine is depicted in Fig. 2-45. The workpiece is mounted on the table 2 whose position can be ensured by controlling the longitudinal and transverse movement of the saddle 5. The headstock 3 can be moved up and down along the guideways on the column 4 so as to adapt to the workpiece with different heights.

工件安装在工作台或底座上。内立柱固定在底座上,外立柱连同摇臂和主轴箱可绕内立柱旋转摆动,摇臂可沿外立柱升降以便于加工不同高度的工件,主轴箱能在摇臂的导轨上作径向移动,使主轴与工件孔中心对正,然后用夹紧装置将内外立柱、摇臂与外立柱、主轴箱与摇臂间的位置分别固定。主轴的旋转运动及主轴套筒的轴向进给运动的开停、变速、换向、制动机构,都布置在主轴箱内。摇臂钻床广泛应用在单件和中小批生产中加工大、中型零件。

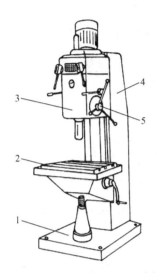
Fig. 2-42 Upright drill press 立式钻床
1—Base 底座 2—Worktable 工作台
3—Spindle head 主轴箱 4—Column 立柱
5—Hand lever 手柄

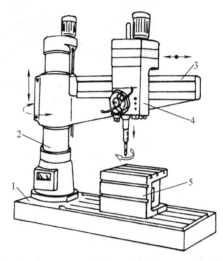
Fig. 2-43 Radial drilling machine 摇臂钻床
1—Base 底座 2—Column 立柱 3—Radial arm 摇臂
4—Spindle head 主轴箱 5—Worktable 工作台

2. 镗床

镗床通常用于加工大型工件上的预制孔。除了镗孔以外,还可以车端面、车外圆、车螺纹、车沟槽、钻孔、铣平面等。有各种类型的镗床,其中以卧式镗床应用最为广泛。

(1) 卧式镗床 卧式镗床的工艺范围很广,尤其适用于加工大型工件。卧式镗床的外形如图 2-44 所示。主轴箱 11 可沿装在床身 1 上的前立柱 10 的导轨上、下移动。工作台 6 用于安装工件,它可以作纵向和径向进给。除了镗杆外,其他刀具如钻头、铰刀、铣刀等也可以安装在旋转的主轴上。主轴可以纵向机动进给,这样在工作台不移动的情况下也可以在很大的距离范围内钻孔和镗孔。

当使用镗杆和单刃刀具镗孔时,存在的主要问题是镗杆的刚度不足。为此装在后立柱上后支架为长的镗杆提供刚性支承,以保证加工出精确的工件。

(2) 坐标镗床 坐标镗床是一种通用性很强的高精密机床。坐标镗床之所以价格昂贵,是因为它具有极高的刚度、低的热膨胀系数、极准确的几何形状和精密的测量装置,从而可以确保极高的加工精度。大多数坐标镗床的主轴都用经过预紧的滚动轴承支承。

立式坐标镗床如图 2-45 所示。工作台 2 用于安装工件,其位置可以通过控制床鞍 5 的纵向和横向移动来保证。主轴箱 3 可以在立柱 4 的竖直导轨上调整上下位置,以适应不同高度工件的加工要求。

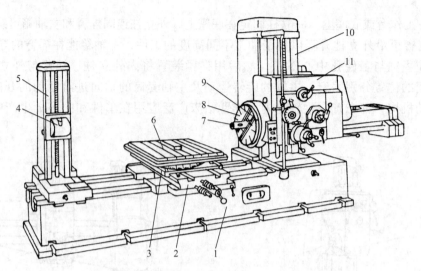

Fig. 2-44 Horizontal boring machine 卧式镗床
1—Bed 床身 2—Lower slide 下滑座 3—Upper slide 上滑座 4—End support 后支架
5—End column 后立柱 6—Work table 工作台 7—Spindle 镗轴 8—Rotatable flat plate 平旋盘
9—Radial tool post 径向刀座 10—Head column 前立柱 11—Headstock 主轴箱

Although jig boring machines are constructed for precision work on jigs and fixtures, they are now replaced by modern CNC machining centers.

3. Hole-making cutting tools

There are various kinds of hole-making cutting tools. In light of their usage, they fall into two categories. One is used for making a hole in solid material, such as twist drill, center drill, spade drill and deep hole drill; the other is used for enlarging the existing hole or improve the hole quality, such as core drill, counterboring and countersinking drill, reamer and boring tool and so on.

(1) Twist drills Twist drills are the most widely used hole-making cutting tools, which are especially suitable for roughing holes with the diameter $d \leqslant 30$mm. The structure of a standard twist drill is illustrated in Fig. 2-46a. It basically consists of two parts: one is the operation part composed of guiding part and cutting part, the other is the shank with either straight or taper shape which is used for holding the drill and transmitting the cutting torque. There are two major cutting edges, two minor cutting edges and one chisel edge in the cutting part, as shown in Fig. 2-46b. Most of the twist drills have two helix flutes which serve as channels through which the chips are withdrawn from the hole and allow the cutting fluid to reach the cutting edges. The drills with straight shank generally have a small diameter ($d_0 \leqslant 13$mm), and are held in the spindle with the help of a drill chuck. The drills with taper shank usually have a large diameter ($d_0 > 13$mm), and are directly held in the spindle with the help of a self-holding taper. The tang at the end of the taper shank fits into a slot in the spindle. Its functions are to drive the drill, prevent it from slipping and help in removing it from the spindle.

To reduce the friction between the drill and the hole, each land is reduced in diameter except at the major edge, leaving a narrow margin with a complete diameter to aid in supporting and guiding the drill. The chisel edge is formed by the intersection of the two flanks. In order to provide strength to the drill, the cutting edge is thickened gradually from the bottom. Two cutting lips of the twist drill are connected by the core of the drill. The core of the drill is made with a positive cone in order to provide strength and rigidity to the drill, as shown in Fig. 2-46c.

The major technical parameters of the twist drill are the helix angle β, point angle 2ϕ (cutting edge an-

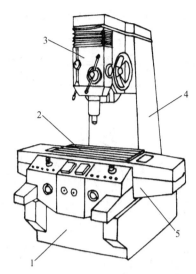

Fig. 2-45　Vertical jig boring machine with
single column 立式单柱坐标镗床
1—Bed 床身　2—Worktable 工作台　3—Headstock 主轴箱
4—Column 立柱　5—Saddle 床鞍

尽管坐标镗床是为加工钻镗模和夹具上的精密零件设计制造的,但如今已被现代数控加工中心所替代。

3. 孔加工刀具

孔加工刀具种类很多。按其用途可分为两大类:一类是在实心材料上加工孔的刀具,如麻花钻、中心钻、扁钻和深孔钻等;另一类是对工件上已有孔进行再加工的刀具,如扩孔钻、锪钻、铰刀和镗刀等。

(1) 麻花钻　麻花钻是应用最广泛的孔加工刀具,一般用于在实心材料上孔径 $d \leqslant 30\text{mm}$ 的粗加工。标准麻花钻的结构如图 2-46a 所示。它基本上包括两部分:一是由引导部分和切削部分组成的工作部分;另一部分是用来定心和传递扭矩的刀柄(有直柄与锥柄两种)。标准麻花钻的切削部分有两条主切削刃、两条副切削刃和一个横刃,如图 2-46b 所示。大多数麻花钻都有两条对称的螺旋槽,螺旋槽可用作排屑和输送切削液的通道。通常直柄麻花钻的直径 d_0 较小 ($d_0 \leqslant 13\text{mm}$),而且借助于钻头卡爪安装在主轴上。锥柄麻花钻的直径 d_0 较大 ($d_0 > 13\text{mm}$),而且借助于自锁锥度直接装在主轴的锥孔内。麻花钻锥柄顶端的扁尾可装入主轴锥孔内的槽里,不仅可用来传递扭矩、防止打滑,还可以辅助从主轴锥孔中取钻头。

为了减小钻头与孔壁的摩擦,刃带有很小的倒锥,刃带也起支承和导向作用。两后面交线形成横刃。为了保证钻头的强度,切削刃从底部开始逐渐变厚。麻花钻的两个刃瓣由钻芯连接在一起,为了增加钻头的强度和刚度,钻芯制成正锥,如图 2-46c 所示。

麻花钻的主要几何参数有:螺旋角 β、顶角 2ϕ (主偏角 $\kappa_\text{r} \approx \phi$)、前角 γ_o、后角 α_o、横刃长度 b_ψ、横刃斜角 ψ 等。

(2) 其他钻的类型　扩孔钻是对工件已有孔进行再加工,以扩大孔径和提高加工质量的加工工具。扩孔钻的刀齿一般有 3~4 个。与麻花钻相比,扩孔钻有许多优点,如导向性

gle $\kappa_r \approx \phi$), rake angle γ_o, clearance angle α_o, chisel edge length b_ψ, chisel edge angle Ψ and so on.

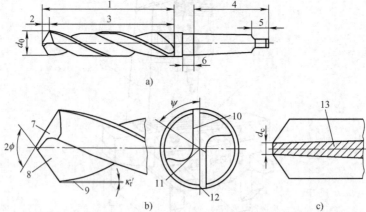

Fig. 2-46 Structure of a standard twist drill 标准麻花钻的结构
1—Operation part 工作部分 2—Cutting part 切削部分 3—Guiding part 引导部分
4—Shank 刀柄 5—Tang 扁尾 6—Neck 颈部 7—Flank 后面 8—Rake face 前面
9—Minor edge 副切削刃 10—Major edge 主切削刃 11—Chisel edge 横刃
12—Land or margin 刃带 13—Core of drill 钻芯

(2) Other types of drills Core drills are used for enlarging the existing holes such as those in castings. They are either of the three-flute or four-flute type. Compared with twist drills, they have many advantages, such as good guidance and cutting smoothness, tool body with high strength and rigidity, good cutting condition without a chisel edge and so on. These help to improve the machining quality and efficiency.

The primary types of core drills are shown in Fig. 2-47, where the drilling diameters of solid core drills are $\phi10 - \phi32$mm, and the diameters of shell core drills are $\phi25 - \phi80$mm.

As shown in Fig. 2-48a, counterboring can be carried out by means of the tool similar to the shell end mill with the cutting edges along the side as well as the end while a pilot portion is present for the tool to enter the hole machined already to keep the concentricity with the hole. In the counterboring operation, the hole is enlarged with a flat bottom to provide a smooth bearing surface that is normal to the axis of the hole for the bolt head, or a nut.

Countersinking tools , as shown in Fig. 2-48b, Fig. 2-48c, are similar to counterboring tools except that their blades are made with a specific conical angle so as to accommodate the counter sunk machine screw head. The most common angles are 60°, 90°and 120°. The depth of countersinking should be deep enough to accommodate the screw head to fully flush with the surface.

Spot facing tools are similar to counterboring tools, but they remove only a very small amount of material around the existing hole to provide a flat surface perpendicular to the axis of the hole. This is normally done to provide a bearing surface for a washer, or a nut, or a bolt head. This has to be done only in case where the existing surface is not smooth. The tool used can be the same as that for counterboring or the tool as shown in Fig. 2-48d.

(3) Reamers A reamer is a multi-tooth cutter which rotates and moves linearly into an already existing hole. The previous operation could be drilling or preferably boring. Reaming provides a smooth surface as well as close tolerance on the diameter of the hole. Generally the reamer follows the already existing hole and therefore will not be able to correct the hole misalignment.

Generally, there are two kinds of reamers: hand reamers and machine reamers, as shown in Fig. 2-49. Machine reamers are divided into shank reamers (Fig. 2-49a) and shell reamers (Fig. 2-

好，切削平稳，刀体强度和刚性较好，没有横刃从而使切削条件得到改善。这些都有助于提高切削效率和加工质量。

扩孔钻的主要类型如图 2-47 所示，其中整体式扩孔钻的扩孔范围为 $\phi 10 \sim \phi 32 \text{mm}$；套式扩孔钻的扩孔范围为 $\phi 25 \sim \phi 80 \text{mm}$。

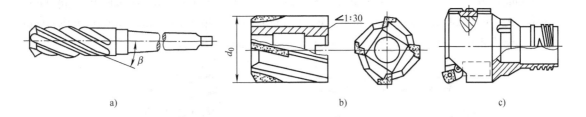

Fig. 2-47　Types of core drills　扩孔钻的类型

如图 2-48a 所示，锪圆柱形沉头孔是利用带有侧刃和端面刃，结构上类似于套式面铣刀的刀具对已有孔进行孔口加工。锪钻端部的引导部分进入已加工的孔内以保证与孔的同轴度。已有孔被锪成平底沉头孔以便为螺栓头或螺母提供一个垂直于孔轴线的光滑支承面。

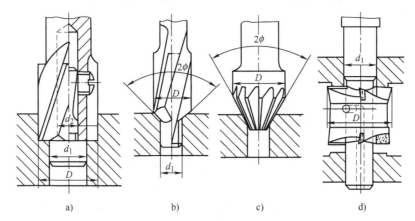

Fig. 2-48　Cutting tools for counterboring, countersinking and spot facing
锪圆柱孔、锥孔和孔口平面的刀具
a) Counterboring tool 锪圆柱孔刀具　b)、c) Countersinking tool 锪锥孔刀具
d) Spot facing tool 锪孔口平面刀具

锪锥孔刀具与锪圆柱形沉头孔的刀具结构相似，只是刀刃做成特定的圆锥角以便能够放进沉头螺钉，如图 2-48b、c 所示。最常用的角度有 60°、90°、120° 三种。锥孔的深度应足够，以便使装入的螺钉头与表面平齐。

锪平面刀具与锪圆柱形沉头孔的刀具结构相似，但其只在孔口表面切除很少量的材料以获得一个与孔垂直的平面。这种操作通常是为垫片、螺母或螺栓头提供一个支承平面而进行的，而且是在已有平面不够光滑的情况下才做的。所用刀具可以是锪圆柱形沉头孔的刀具，或图 2-48d 所示刀具。

（3）铰刀　铰刀是一种边旋转边直线移动进入已有孔中进行精加工的多刃刀具。先前

49b). The straight shank reamers are used for reaming the hole with $\phi 1 - \phi 20$mm, and tapered shank reamers are used for reaming the hole with $\phi 10 - \phi 32$mm. A shell reamer, which is hollow and mounted on an arbor, is generally used for holes larger than $\phi 20$mm. A hand reamer has a straight shank with a square tang for a wrench, and used for reaming holes with $\phi 1 - \phi 50$mm. The reamer flutes are either straight or helical (Fig. 2-49c). Taper holes can be machined by taper reamers. As the larger reaming allowance exists in the reaming taper hole, there are roughing taper reamers and finishing taper reamers, usually with 2 or 3 reamers in set (Fig. 2-49d).

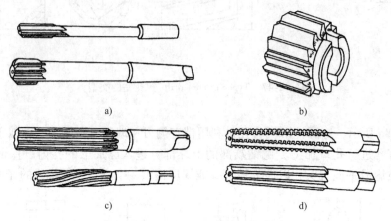

Fig. 2-49 Types of reamers 铰刀类型
a) Reamers with straight and taper shank 直柄与锥柄铰刀   b) Shell reamer 套式铰刀
c) Reamers with straight and helix flutes 带有直槽和螺旋槽的铰刀
d) Roughing and finishing taper reamer 粗、精锥铰刀

A reamer consists of operation part, neck and shank. The operation part is divided into cutting part and calibrating part, as shown in Fig. 2-50. The cutting edge angle κ_r has very deep influence on the machining accuracy, the surface roughness of the hole and the axial cutting force. Over-large κ_r would decrease the centering accuracy and increase the axial cutting force; Over-small κ_r would increase the cutting width and go against the chip removal. Generally, $\kappa_r = 0.5° - 1.5°$ for hand reamers, and $5° - 15°$ for machine reamers.

(4) Boring tools   Boring tools are generally classified into the single point tool that is similar to those used in the turning operation and the tool bit with two cutting edges.

Single point boring tools are more widely used for boring holes because they are simple in structure, easy to fabricating and versatile. Most of single point boring tools have the dimension adjusting device, as shown in Fig. 2-51. The boring tools with fine tuning device are used on the precising boring machines in order to improve the dimension adjusting accuracy, as shown in Fig. 2-52.

The boring tool block with two blades has cutting edges on both sides of the boring tool, as shown in Fig. 2-53. This cutter can eliminate the influence of the radial force on the boring bar. The diameter and its accuracy of the hole to be bored are ensured by the diametric dimension of the boring tool. As the two blades can be adjusted in radial direction, the cutter can be used for boring the hole in a certain range of diameters. Two-blade boring tools adopt floating connection structure. That is, the boring tool block is put into the radial slot of the boring bar, and balanced automatically rel-

的操作可以是钻孔或更好一点的镗孔。铰孔可以进一步把孔加工的更加光洁，直径误差更小。通常铰刀是用已有的空导向，因此它不能修整孔的直线度。

铰刀一般分为手用铰刀及机用铰刀两种，如图 2-49 所示。机用铰刀可分为带柄的（图 2-49a）和套式的（图 2-49b）。直柄铰刀加工直径为 $\phi1 \sim \phi20$mm，锥柄铰刀加工直径为 $\phi10 \sim \phi32$mm。手用铰刀柄部为直柄，带有供扳手扭动的方形扁尾，用于加工直径为 $\phi1 \sim \phi50$mm 的孔。铰刀形式有直槽式和螺旋槽式两种（图 2-49c）。可以用锥度铰刀加工锥孔。铰制锥孔时由于铰制余量大，锥铰刀常分为粗铰刀和精铰刀，一般做成 2 把或 3 把为一套（图 2-49d）。

铰刀由工作部分、颈部及柄部组成。工作部分又分为切削部分与校准（修光）部分，如图 2-50 所示。铰刀的主偏角 κ_r 对孔的加工精度、表面粗糙度和铰削时进给力的大小影响很大。κ_r 值过大，切削部分短，铰刀的定心精度低，还会增大进给力；κ_r 值过小，切削宽度增加，不利于排屑；手用铰刀的 κ_r 值一般取为 $0.5° \sim 1.5°$，机用铰刀的 κ_r 值取为 $5° \sim 15°$。

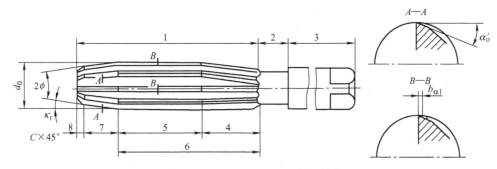

Fig. 2-50 Reamer structure 铰刀结构
1—Operating par 工作部分　2—Neck 颈部　3—Shank 刀柄　4—Back taper 倒锥　5—Cylindrical part 圆柱部分
6—Calibrating part 校准部分　7—Cutting part 切削部分　8—Chamfer length 倒棱长度

（4）镗刀　镗刀一般分为单刃镗刀与多刃镗刀两大类。

单刃镗刀结构简单，制造容易，通用性好，故使用较多。单刃镗刀一般均有尺寸调节装置，如图 2-51 所示。在精镗机床上常采用微调镗刀（图 2-52）以提高调整精度。

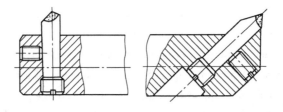

Fig. 2-51 Single point boring tools 单刃镗刀

双刃镗刀两边都有切削刃，如图 2-53 所示。工作时可以消除径向力对镗杆的影响，工件的孔径与精度由镗刀径向尺寸保证。镗刀上的两个刀片径向可以调整，因此，可以加工一定尺寸范围的孔。双刃镗刀多采用浮动连接结构，镗刀片插在镗杆的槽中，依靠作用在两个切削刃上的径向力自动平衡其位置，可消除因镗刀安装误差或镗杆偏摆引起的加工误差。

ying on the radial cutting force acting on the cutting edges. This structure can eliminate the machining error caused by the tool installation error or runout of the boring bar.

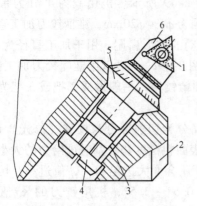

Fig. 2-52　Fine tuning boring tool 微调镗刀
1—Indexable insert 转位刀片　2—Boring bar 镗杆
3—Pilote key 定向键　4—Clamping screw 夹紧螺钉
5—Fine tuning nut 微调螺母　6—Tool holder 刀夹

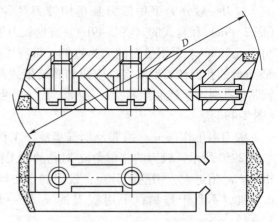

Fig. 2-53　Boring tool with two blades 双刃镗刀

2.5.2　Milling machines and milling cutters

1. Introduction

After lathes, milling machines are the most widely used for manufacturing applications. Whether the surface is flat or curved, inside or outside, almost all shapes and sizes can be machined by milling. The commonly used milling operations are shown in Fig. 2-54. In milling, the workpiece is fed into a rotating milling cutter, which is a multi-point tool, unlike a lathe, which uses a single point cutting tool.

2. Milling machines

To satisfy various requirements milling machines come in a number of sizes and varieties. In view of the large material removal rates milling machines come with a very rigid spindle and a large power. The varieties of milling machines available are: knee and column type milling machines, production (bed) type milling machines, plano millers, and other special type milling machines.

(1) Horizontal knee and column type milling machines　The layout of a horizontal knee and column type milling machine is shown in Fig. 2-55. It is composed of a base 8, a column 1, an arbor 3, an overarm 2, an outboard support 6, a knee 7, a worktable 4, a saddle 5 and other components. The column 1 which is fixed to the base 8 is used for mounting and supporting each component of the machine.

The feed mechanism is arranged in the knee 7. The saddle 5 and worktable 4 are mounted on the knee. The worktable has the longitudinal T-slots for the purpose of work holding. The table moves along the X-axis on the saddle while the saddle moves along the Y-axis on the guide ways provided on the knee. The feed is provided either manually with a hand wheel or connected for automatic by a lead screw, which in turn is coupled to the main spindle drive. The knee can move up and

2.5.2 铣床与铣刀

1. 概述

铣床是继车床后应用最广的机床。无论是平面还是曲面，是内表面还是外表面，几乎所有形状和规格的表面都可以在铣床上铣削。常见的铣削操作如图 2-54 所示。在铣削中工件向着旋转的铣刀进给。铣刀不是车刀那样的单刃刀具，而是多刃刀具。

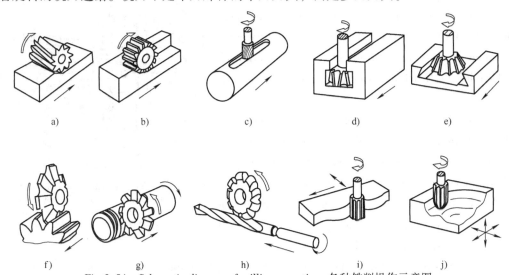

Fig. 2-54　Schematic diagram of milling operations　各种铣削操作示意图
a) Slab or plain milling 铣平面　b) Side milling 铣台阶　c) Keyway milling 铣键槽　d) T-slot milling 铣 T 形槽
e) Dovetail milling 铣燕尾槽　f) Gear form milling 成形铣齿　g) Thread milling 铣螺纹
h) Spiral grooving 铣螺旋槽　i) Profile milling 仿形铣　j) Surface contouring 面轮廓铣削

2. 铣床

铣床的种类和规格繁多，以满足各种加工需要。鉴于铣床具有很高的材料切除率，因此铣床的主轴刚性好、功率大。目前可用的铣床类型有升降台铣床、床身式铣床、龙门铣床以及其他特殊用途的铣床。

（1）卧式升降台铣床　图 2-55 所示为卧式升降台铣床的外形。它由底座 8、立柱 1、铣刀轴（刀杆）3、悬梁 2、悬梁支架 6、升降台 7、工作台 4 及床鞍 5 等主要部分组成。床身固定在底座 8 上，用于安装和支承机床的各个部件。

在升降台 7 内有进给机构，而在升降台上装有床鞍 5 和工作台 4。工作台上有供安装工件或夹具的纵向 T 形槽。工作台可在床鞍 5 上沿 X 轴移动，而床鞍则在升降台的导轨上沿 Y 轴移动。机床进给可以利用

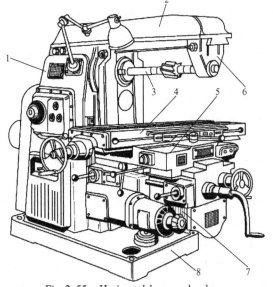

Fig. 2-55　Horizontal knee and column type milling machine 卧式升降台铣床
1—Column 立柱　2—Overarm 悬梁
3—Arbor 铣刀轴　4—Worktable 工作台
5—Saddle 床鞍　6—Outboard support 悬梁支架
7—Knee 升降台　8—Base 底座

down (Z-axis) on a dovetail provided on the column 1.

(2) Vertical knee and column type milling machines

Another type of knee and column milling machine is the vertical axis type, as shown in Fig. 2-56. Its construction is very similar to the horizontal axis type, except for the spindle type and location. The spindle 2 is located in the vertical direction and is suitable for using the shank to mount the milling cutters such as end mills. In view of the location of the tool, the setting up of the workpiece and observing the machining operation are more convenient.

A vertical axis milling machine is relatively more flexible and suitable for machining complex cavities such as die cavities in tool rooms. The vertical head 1 is provided with a swiveling facility in a horizontal plane whereby the cutter axis can be swiveled. This is useful for tool rooms where more complex milling operations are carried out. The universal machine has a table which can be swiveled in a horizontal plane at about 45° to either the left or the right. This makes the universal machine suitable for milling spur and helical gears as well as worm gears and cams.

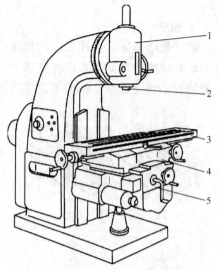

Fig. 2-56 Vertical knee and column type milling machine 立式升降台铣床
1—Vertical head 立铣头 2—Spindle 主轴
3—Worktable 工作台 4—Saddle 床鞍
5—Knee 升降台

3. Milling cutters

There are a large variety of milling cutters available to meet specific requirements. The versatility of the milling machine is contributed to a great extent by the variety of milling cutters that are available.

(1) Slab milling cutters These cutters as shown in Fig. 2-57, generally made of HSS, have a number of teeth along its circumference, each tooth acting as a single-point cutting tool.

Cutters for slab milling may have straight or helical teeth resulting in, respectively, orthogonal or oblique cutting action. The helical teeth on the cutter are preferred over the straight teeth because the load acting on the tooth is lower, resulting in a smoother operation and reducing tool forces and chatter.

There are coarse-tooth milling cutters and fine-tooth milling cutters. The coarse-tooth milling cutters come with a small number of teeth to allow for more chip space. Therefore, the coarse-tooth milling cutters are used for rough milling operations. The fine-tooth milling cutters generally have straight teeth and large number of teeth, which is suitable for finishing. The milling cutters with larger diameters are generally made into tipped milling cutters.

(2) Face milling cutters Face milling cutters, as shown in Fig. 2-58, are used for machining larger flat surfaces. They have cutting edges on the periphery (major cutting edges) and face (minor cutting edges). The axis of the face milling cutter is perpendicular to the workpiece surface. Most of face milling cutters are made into tipped shell face milling cutters. The tooth material is either HSS or cemented carbide. They are widely used on both horizontal and vertical milling machines with a high material removal rate.

手轮手动操作，也可以通过连接主电动机的丝杠自动控制。升降台可沿着立柱上的燕尾导轨上下移动。

（2）立式升降台铣床　另一种升降台铣床是立式结构，如图 2-56 所示。其结构与卧式铣床极其相似，只是主轴类型和安装方位不同。主轴 2 位于垂直方向，并且适合于使用能够安装像立铣刀或面铣刀之类刀具的刀柄。鉴于刀具所处位置，这类机床更便于工件的安装调整和加工情况的观察。

立式铣床比卧式铣床更加灵活，可用于加工类似于工具车间模具型腔的复杂型腔表面。立铣头 1 具有可在垂直平面内旋转的装置，因而铣刀轴也可以旋转。这尤其适用于在工具车间进行更复杂的铣削加工。万能铣床具有可在水平面内左右旋转 45° 的工作台，从而使得它能够加工直齿轮、斜齿轮以及蜗轮和凸轮等。

3. 铣刀

铣刀种类很多，以满足不同的加工需求。事实上铣床的通用性在很大程度上正是由各式各样的铣刀所赋予的。

（1）圆柱铣刀　圆柱铣刀如图 2-57 所示，一般都是用高速钢制成的，许多切削刃分布在其圆柱表面上，每一个刀齿都相当于一把单刃刀具。圆柱铣刀有直切削刃或螺旋形切削刃之分，分别产生正切削或斜切削的效果。由于螺旋形切削刃的切削负荷小，从而使切削过程较平稳、减小切削力和振动，因而性能比直切削刃优越。

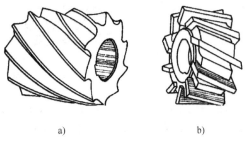

Fig. 2-57　Slab milling cutters 圆柱铣刀
a) Solid type 整体式　b) Tipped type 镶齿式

圆柱铣刀有粗齿、细齿之分。粗齿铣刀的刀齿数少，容屑槽大，因此用于粗加工。细齿铣刀通常做成直刃，刀齿数多，因而适用于精加工。铣刀外径较大时，常制成镶齿的。

（2）面铣刀　面铣刀如图 2-58 所示，主要用于加工较大平面。面铣刀的圆周和端面都分布有切削刃，其中圆周上的是主切削刃，端面上的是副切削刃。铣刀的轴线垂直于被加工表面。大多数面铣刀都做成镶齿套式的。刀头材料不是高速钢就是硬质合金。面铣刀具有高的材料切除率，主要用在立式铣床或卧式铣床上。

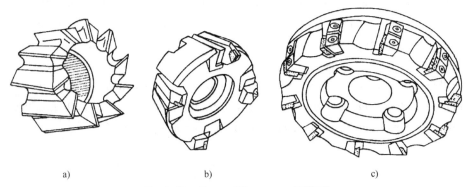

Fig. 2-58　Face milling cutters 面铣刀
a) Solid tooth 整体式　b) Brazed inserts 焊接式　c) Clamped inserts 机夹式

（3）立铣刀　如图 2-59 所示，立铣刀通常在立式铣床上用于加工那些别的铣刀不能加

(3) End milling cutters   End milling cutters as shown in Fig. 2-59 are generally used in vertical axis milling machines. They are used for milling slots, key ways and pockets where other type of milling cutters cannot be used. The end mills have the cutting edge running through the length of the cutting portion as well as distributing on the face in radial to a certain length. The helix angle of the cutting edge promotes smooth and efficient cutting even at high cutting speeds and feed rates.

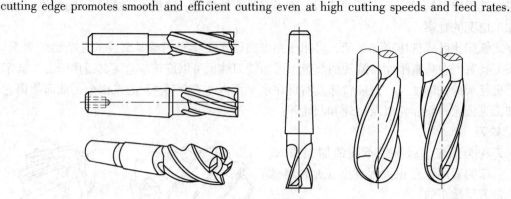

Fig. 2-59   End milling cutters 立铣刀

The end teeth of the end mills may be terminated at a distance from the cutter centre or may proceed till the centre. Those with the cutting edge up to the centre are called slot drills or end cutting end mills since they have the ability to cut into the solid material. The other type of end mills which have a larger number of teeth cannot cut into the solid material and hence require a pilot hole drilled before a pocket is machined. Further, the end face can be a ball end shape to be used for milling three dimensional contours such as in die cavities.

(4) Side and face milling cutters. Side and face milling cutters as shown in Fig. 2-60 have cutting edges not only on the face like the slab milling cutters, but also on both the sides. As a result, these cutters have a better cutting condition and higher cutting efficiency. They are more versatile since they can be used for side milling as well as for slot milling. Side and face milling cutters are classified into straight tooth (Fig. 2-60a) and staggered tooth (Fig. 2-60b) side milling cutters. Staggered tooth side milling cutters are generally used for milling deep slots since their staggering teeth provide greater chip space. When the solid side and face milling cutters are resharpened, their width will be varied greatly. Tipped side and face milling cutters (Fig. 2-60c) do not have this problem.

(5) Other milling cutters   In addition to the milling cutters mentioned above, there are other types of milling cutters such as slitting saws, angular milling cutters, form milling cutters, T-slot milling cutters, dovetail milling cutters, and so on, as shown in Fig. 2-61.

工的槽、键槽和凹槽。立铣刀在整个切削部分的全长上都有切削刃，而在端面上也分布着一定长度的径向切削刃。螺旋切削刃有利于在高速和大进给情况下实现平稳高效铣削。

立铣刀的端面切削刃可以不通过中心，也可以通过中心。端面切削刃延伸至中心的立铣刀称为键槽铣刀，或者称为能进行端面切削的立铣刀，因为这种铣刀具有切入实体材料的能力。其他具有多个刀齿的立铣刀不能直接切入实体材料，因此在加工凹槽前需要提前加工一个引导孔。此外，立铣刀的端部呈球形，以便铣削像模具型腔那样的三维曲面。

（4）三面刃铣刀　图 2-60 所示的三面刃铣刀不仅在圆周上有切削刃，而且在两侧面也有切削刃。因此，这种铣刀不仅有良好的切削条件，而且切削效率也高。三面刃铣刀的通用性好，不仅可以用于铣侧面，还可以用来铣沟槽。整体三面刃铣刀有直齿（图 2-60a）和错齿（图 2-60b）两种。由于错齿三面刃铣刀具有更大的容屑空间，因此通常用来铣深槽。整体三面刃铣刀重磨后宽度尺寸变化较大，镶齿三面刃铣刀（图 2-60c）不存在这一问题。

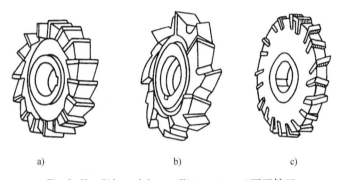

Fig. 2-60　Side and face milling cutters 三面刃铣刀
a）Straight tooth 直齿　b）Staggered tooth 错齿　c）Tipped type 镶齿

（5）其他铣刀　除了前面介绍的几种铣刀外，还有其他类型的铣刀，如成形铣刀、T 形槽铣刀、燕尾槽铣刀、锯片铣刀、角度铣刀等，如图 2-61 所示。

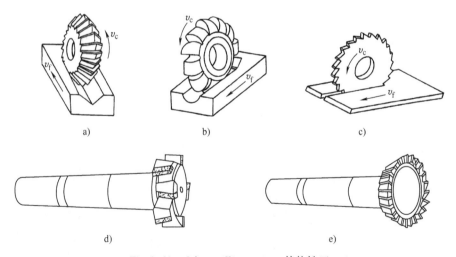

Fig. 2-61　Other milling cutters 其他铣刀
a）Angular milling cutter 角度铣刀　b）Form milling cutter 成形铣刀　c）Slitting saw 锯片铣刀
d）T-slot milling cutter T 形槽铣刀　e）Dovetail milling cutter 燕尾槽铣刀

Chapter 3
Machining Process Planning

第3章
机械加工工艺规程的制订

3.1 Basic concepts

3.1.1 Production course and process course of the machine

1. Production course

Manufacturing means transformation of materials and information into goods for the satisfaction of human needs. A series of interrelated labor processes through which raw materials are changed into finished products is referred to as production course. In the production of mechanical products, production course consists of product design, production preparation, transportation and storage of raw materials, blank manufacturing, machining, heat treatment, assembly and adjustment, check and trial run, painting and package etc.

2. Process and process course

A process is simply a method or technique by which finished or semi-finished products can be manufactured from raw materials. Process course is one part of production course in which the shape, dimensions and performance of objects to be machined are directly changed into desired ones. It involves blank manufacturing, machining, heat treatment, assembly etc.

In general, machine manufacturing technological process includes the machining process and the assembly process. Machining process contains the whole process through which the dimensions, shape and surface quality of a blank are directly changed into a qualified part step by step by means of the metal cutting methods. The process involved in assembling different workparts into a qualified product is called the assembly process.

3.1.2 Elements of process course

Machining technological process of a part consists of one or several operations in sequence. A blank can be fabricated into a finished or semi-finished product by these sequential operations.

1. Working operation

The term working operation or operation can be defined as a part of process course which is completed continuously by one operator (or a group of operators) who machines one workpiece (or several workpieces simultaneously) on a definite working place (or a definite machine tools). Working operation is the basic element of the technological process as well as of the production planning and of the cost accounting. It can be seen from the definition of the operation that there are four key factors (same operator(s), same workpiece(s), same working place, continuous) which are used to partition the working operation. If one of the four factors varies in machining process, it means the working operation has been changed. An equipment is a working place. The vital sign to distinguish operations is to see whether the operation place (i.e. equipment) has been changed.

Fig. 3-1 is a sketch map of a shaft with shoulders. The volume of the part to be machined is small. The machining process of the shaft is shown in Tab. 3-1.

3.1 基本概念

3.1.1 机器的生产过程和工艺过程

1. 生产过程

制造指的是把材料和信息转换成满足人类需求的商品的过程。生产过程实质上是指由原材料转变为产品的各个相互关联的劳动过程总和。在机械产品的生产中，生产过程包括产品设计、生产准备、原材料的运输和储存、毛坯制造、机械加工、热处理、装配与调整、检验与试车、喷漆以及包装等一系列过程。

2. 工艺和工艺过程

工艺指的是把原材料制成成品或半成品的方法或技术。工艺过程是生产过程的一部分。凡是用来直接改变生产对象的形状、尺寸、相对位置和性质等，使其成为成品或半成品的过程称为工艺过程。它涉及毛坯制造、机械加工、热处理和装配等。

机械制造工艺过程一般是指机械加工工艺过程和机械装配工艺过程。采用机械加工的方法，按照一定的顺序逐步地改变毛坯的形状、尺寸、各表面间相互位置及表面质量，使其成为合格零件的过程，称为机械加工工艺过程。将各种零件装配成为合格产品的工艺过程称为机械装配工艺过程。

3.1.2 工艺过程的组成

某零件的机械加工工艺过程通常由一个或若干个按顺序排列的工序组成。通过这一连串的工序可以把毛坯做成成品或半成品。

1. 工序

由一个（或一组）工人，在同一台机床（或同一个工作地）对一个（或同时几个）工件所连续完成的那一部分工艺过程称为工序。工序是组成工艺过程的基本单元，也是生产计划和成本核算的基本单元。从工序的定义可以发现，划分工序的关键要素有四个，即同一工人，在同一工作地（或机床），对同一工件，连续完成。在加工过程中，只要有其中一个要素发生变化，即换了一个工序。一台设备就是一个工作地。区分工序的重要标志是看设备是否发生改变。

图 3-1 所示为一双向阶梯轴的示意图。该零件为小批量生产。其工艺过程见表 3-1。

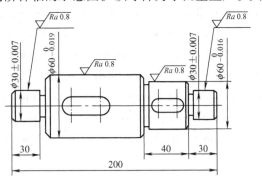

Fig. 3-1　A shaft with shoulders 阶梯轴

2. Working step

The working step can be defined as the part of the operation completed continuously under the condition that the machining surface, cutting tool and cutting variables, including the cutting speed, feed are kept unchanged. When any of the factors is changed, it would be considered to be another working step. In the operation 1 listed in Tab. 3-1, there are four working steps: facing one end, centering one end, facing the other end and centering the other end.

Generally speaking, several identical working steps that happened successively can be seen as one working step, which is called continuous working step. On the contrary, it would become a compound working step if several surfaces of a part are machined simultaneously by several cutting tools or a compound cutting tool. For example, four identical holes shown in Fig. 3-2 are drilled one by one, the process is a continuous working step. While four drills are used to machine these holes simultaneously, the step is named compound working step. Another example that can be classified as compound working step is illustrated in Fig. 3-3.

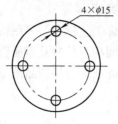

Fig. 3-2 Continuous working step
连续工步

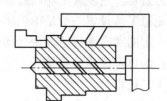

Fig. 3-3 Compound working step
复合工步

3. Working stroke

Working stroke is the manufacturing process that is completed by cutting the surface one time with the same cutting tool and the same cutting parameters. A working step may consist of a number of working strokes if the machining allowance is quite large. Fig. 3-4 is a schematic diagram which shows the relationship between the working step and the working stroke.

4. Setting-up

In order to achieve satisfying machining effects, a workpiece must be located and clamped in a fixture before it is machined. The process to locate and clamp the workpiece on fixture or worktable is called setting-up. A working operation may have either single setting-up or several setting-ups. For example, two setting-ups are needed in the first working operation shown in Tab. 3-1 because two ends are needed to be machined separately. And each setting-up includes two working steps, that is, facing and centering. But it is very important to remember that setting-up times should be arranged as few as possible in one operation, or else errors produced because of several times of setting-up would be increased and the production efficiency would be reduced greatly as well.

Table 3-1　Machining process of the shaft　阶梯轴的工艺过程

Operation No.（工序号）	Operation name（工序名称）	Machine tool 机床
1	Facing（车端面），center drilling（钻中心孔）	Engine lathe（车床）
2	Cylindrical Turning（车外圆），grooving（车槽），chamfering（车倒角）	Engine lathe（车床）
3	Keyway milling（铣键槽），deburring（去毛刺）	Vertical milling machine（铣床）
4	Cylindrical grinding（磨外圆）	Grinding machine（磨床）

2. 工步

在加工表面、切削工具不变以及切削用量中的切削速度和进给量均不变的情况下所连续完成的那部分工序称为一个工步。当其中任何一个因素改变后就变为另一工步。如表 3-1 所示的工序 1 就包括四个工步：车一端面，钻中心孔；车另一端面，钻中心孔。

一般来讲，连续进行的几个相同的工步可视为一个工步，称为连续工步。相反，如果同时用几把刀具或复合刀具加工工件上不同的几个表面，那就视为一个复合工步。如图 3-2 所示的四个直径相同的孔一个接一个钻削，该过程就是一个连续工步。但是当用四把钻头同时钻削这四个孔时，就成为一个复合工步。图 3-3 所示的多刀加工就是复合工步的另一个实例。

3. 工作行程

用同一把刀具和同一切削参数（转速和进给量）在加工表面上切削一次所完成的工艺过程称为一次工作行程（走刀）。如果要切去的金属层很厚，需要几次走刀，一个工步可包括数次走刀。图 3-4 所示为工步和工作行程之间的关系示意图。

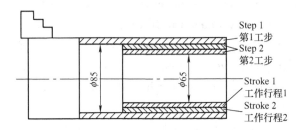

Fig. 3-4　Working step and working stroke　工步与工作行程

4. 安装

为了获得满意的加工效果，工件在加工前必须在机床或夹具上进行定位和夹紧。完成工件一次定位并夹紧的过程称为安装。在同一加工工序中，工件或许只安装一次，但也可能需要安装多次。例如，表 3-1 中的第一道工序需要经过两次安装，因为轴的两端需要分别加工。而每一次安装包括两个工步，即车端面、钻中心孔。值得强调的是，工件在加工中应尽量减少安装次数，因为多次安装不仅会增加安装误差，还会大大降低生产效率。

5. 工位

为了减少安装次数，常采用回转工作台、回转夹具或移位夹具等，使工件在一次安装中先后处于不同的位置进行加工。工件在一次装夹后，工件与夹具或机床的可动部分一起相对

5. Working position

In order to reduce the mounting times, a rotary table, or rotary fixture, or movable fixture is often used to make the workpiece occupy different positions successively in one setting-up. After the workpiece is clamped, each position occupied by the workpiece and fixture or the movable part of the machine tool together relative to the cutter or the fixed part of the machine tool is called one working position. Fig. 3-5 illustrates an example of a working operation with multi working positions. A rotary table with four positions is used to fulfill loading/unloading, drilling, core drilling and reaming sequentially in an aggregate drilling machine.

3.1.3 Production program and production type

Production program and production type have an influence on the production organization and management, the layout of the workshop, the processing method, the manufacturing equipments to be used and the technical requirements for operators. So it is necessary to learn something about the two terminologies before set out to make process.

1. Production program

A production program N of a workpart refers to the production volume of the workpart per year, including spare parts and rejected parts, which can be calculated by Eq. (3-1).

$$N = Qn(1+\alpha)(1+\beta) \qquad (3\text{-}1)$$

where Q is the annual yield of a product; n is the number of the parts used in one product; α is the average percentage of the spare parts, β is the average percentage of the rejected parts.

2. Production type and its technological feature

Production type means the specialized classification of the enterprise production capacity. Based on the production program, size and complexity of a product, the production of an enterprise can be generally classified as job production, batch production, and mass production, that is, three different kinds of production types.

(1) Job production There are many kinds of products, but each product is manufactured in small quantity, and sometimes even individual product is manufactured. The products to be manufactured are often changed and rarely repeated. The production of heavy duty machines, special-purpose machines and the trial-manufacture of new products are all organized in this way.

(2) Batch production There are still more kinds of products, but there are a certain numbers for each kind of product. The same products are repeatedly manufactured in batches in a year. The number of the products fabricated in one batch is called batch size. According to the batch size and the characteristics of the product to be manufactured, batch production can be further classified into small batch production, medium batch production and large batch production. The production of both general-purpose machine tools and electric motors belongs to the batch production.

(3) Mass production The output of a product is very large. The same operation of a part is done repeatedly in the most of working places. The production of automobiles, tractors, bicycles, and bearings belongs to the mass production.

于刀具或机床的固定部分所占据的每一个位置,称为工位。图 3-5 所示为一个具有多工位的加工实例。在一台组合钻床上利用回转工作台在一次安装中依次完成工件的装卸、钻孔、扩孔、铰孔等内容。

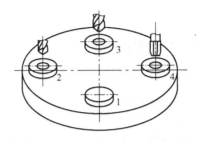

Fig. 3-5　Schematic diagram of multi-position machining 多工位加工示意图
1—Loading/unloading workpiece 装卸工件　2—Drilling 钻孔
3—Core drilling 扩孔　4—Reaming 铰孔

3.1.3　生产纲领和生产类型

生产纲领和生产类型对生产组织和管理、车间布局、加工方法、所使用的装备以及对操作工人的技术要求等都有影响。

因此在编制工艺之前有必要了解下面两个术语。

1. 生产纲领

零件的生产纲领是指该零件包括备品和废品在内的年产量,可按式(3-1)计算。

$$N = Qn(1 + \alpha)(1 + \beta) \tag{3-1}$$

式中　Q——某产品的年产量;
　　　n——每台产品中该零件的数量;
　　　α——平均备品率;
　　　β——平均废品率。

2. 生产类型及其工艺特征

生产类型指的是一个企业生产能力的专业化分类。根据产品的生产纲领、尺寸大小和复杂程度,企业生产一般分为单件生产、成批生产和大量生产三种类型。

(1) 单件生产　产品的种类多,同一产品生产数量很少(仅制造一个或少数几个),加工对象经常变换且很少重复。重型机械、专用设备的制造和新产品试制都属于单件生产。

(2) 成批生产　产品的种类较多,每种产品均有一定的数量,加工对象周期性地重复。同一产品(或零件)每批投入生产的数量称为批量。根据产品的特征和批量的大小,成批生产可分为小批生产、中批生产和大批生产。通用机床和电动机等生产属于成批生产。

(3) 大量生产　产品的产量很大,大多数工作地长期重复地进行某种零件的某一工序的加工。汽车、拖拉机、自行车、轴承等生产都属于大量生产。

生产类型不同,产品制造工艺、所用设备和工艺装备以及生产的组织方式均不同。表 3-2 列出了三种生产类型的工艺特征。

Different production types have different manufacturing processes, machine tools, toolings and production organization modes. The characteristics of different production types are summarized in Tab. 3-2.

3.2 Procedure of process planning

3.2.1 Process rule and its functions

Often there are several manufacturing plans which can meet the design requirements of a product or a component. But there must be a relative optimal manufacturing plan under a specific production condition. The relevant contents to the manufacturing process of a product are specified in the forms of document expressed by charts, tables and words. This document is called process rule (also called technological document).

Process rule is an essential technological document for manufacturing factories or workshops. Its functions can be summarized in the following four aspects.

(1) Main technological documents used to guide production course As a rule, process rule is formulated on the basis of process theories, necessary process tests and summarization of practical experiences. Only when the whole production process is performed according to these rules, can high production quality and normal production order be guaranteed.

(2) Basis of production preparation The supply of raw materials or blanks, design and manufacturing of process equipment, the purchase of new machine tools, the planning of production schedule and the calculation of production costs are all completed on the basis of a process route.

(3) Important technical reference to build new factory or rebuild former factory When building or rebuilding factories or workshops, the types and numbers of machine tools required in production, workshop area, and proper operators can be determined according to the process rule.

(4) Main means used to store and exchange technology Often, the technical level of an enterprise can be reflected partly by its process level. As the carrier of process level, process rule is the footstone of the development of an enterprise. It can be used to exchange and spread advanced process technologies among the factories in the same industry.

Process rule is composed of a series of technological files. These files are often presented in the form of sheets, such as process procedure sheet, process sheet, operation sheet, inspection sheet and so on. Generally, a process procedure sheet is a kind of simplified technological file and is often used in job or small volume production. The process sheet is often used in medium volume production. The operation sheet is used in large volume production and mass production.

Table 3-2 Main technological characteristics of three production types 三种生产类型的主要工艺特征

Types 类型 Features 特征	Job production 单件生产	Batch production 成批生产	Mass production 大量生产
Interchangeability and assembling method 互换性与装配方法	No interchangeability, individual fitting method. 无互换,修配法装配	Interchangeable for most parts. 大部分零件具有互换性	Full interchangeability, interchangeable assembly. 完全互换,互换法装配
Blank type & machining allowance 毛坯类型及加工余量	Manual modeling for casting, free forgings, low precision, large machining allowance. 铸件用手工造型;自由锻件,毛坯精度低,加工余量大	Die casting, or die forging, moderate blank precision and machining allowance. 部分毛坯用金属型铸件或模锻件,毛坯精度与加工余量中等	Blank manufacturing with high accuracy and high efficiency, low machining allowance. 采用高精度、高生产率的毛坯制造方法,加工余量小
Equipment 机床设备	General purpose machine tool part of CNC machine. 通用机床,部分采用数控机床	CNC machine, FMC, part of special purpose machine tools. 数控机床、柔性制造单元、部分专用机床	Dedicated transfer line, FNL, or CNC machines. 专用生产线、柔性生产线或数控机床
Tooling (fixture, cutter and measuring tool). 工艺装备（夹具、刀具和量具）	Standard attachments, cutters and measuring tools. 标准附件、标准刀具和万能量具	Fixtures are widely used, special-purpose cutters and measuring tools. 广泛采用夹具、专用刀具和量具	Fixtures, cutters and measuring tools with high efficiency are widely used. 广泛采用高效夹具、刀具和量具
Requirements for worker. 对工人的要求	Skilled workers are needed. 需要技术熟练的工人	Moderate skilled workers and programmers are needed. 需要一定熟练程度的工人和编程技术人员	Low-skill operators, but skilled maintenance workers are needed. 低技能的操作工人,而设备维护人员的技术要熟练
Process rule 工艺规程	Simple process procedure sheet. 简单的工艺过程卡	process sheet, and operation sheets for key operations. 有工艺卡,对关键工序还有工序卡	Detailed process rule is provided. 有详细的工艺规程

3.2 工艺规程制订的步骤

3.2.1 工艺规程及其作用

达到一种产品或零部件设计要求的制造方案往往不止一个，但在具体生产条件下总会存在一个相对最优的制造方案。把这种制造产品或零部件方案下的工艺过程和操作方法等内容以图、表、文字的形式规定下来的工艺文件称为工艺规程。

工艺规程是制造厂或车间的重要工艺文件。其作用可以概括为以下四个方面。

（1）用于指导生产　通常，工艺规程是在工艺理论、必要的工艺试验以及实践经验总结的基础上制订出来的。只有整个工艺过程都按此规程执行，才能确保产品质量和正常的生产秩序。

（2）各项生产准备工作的依据　原材料和毛坯的供应，工艺装备的设计、制造及选购，新设备的购置，制订生产计划，计算生产成本等，都需要根据机械加工工艺规程来进行。

（3）新建或扩建工厂（或车间）的重要资料　在新建或扩建工厂（或车间）时，根据工艺规程可以确定生产所需机床的种类与数量、车间面积、所需工人的工种、技术等级和数量、各辅助部门的安排等。

（4）技术储存交流的主要手段　工艺规程体现了一个企业的工艺技术水平。作为工艺水平的载体，工艺规程是企业发展的基石。它可以用于同行业之间的技术交流和推广。

工艺规程是由一系列工艺文件所构成的。工艺文件一般以卡片的形式来体现，如工艺过程卡、工艺卡、工序卡、检验卡等。通常，工艺过程卡是一种简化的工艺文件，常用在单件小批生产中。在中批生产中多采用工艺卡。在大批大量生产中采用工序卡。

3.2.2 Original information required in process planning

In most cases, the manufacturing process for a given part or product is not unique. The task of process planning is to find the most reasonable process route among those candidate ones. For this reason, process planning technicians must collect original materials related to this part and/or product as detail as possible. These materials are listed as follow:

① Part drawings and assembly drawing of the product.
② Production program and production type.
③ Quality standards used to inspect products.
④ Blank production situation.
⑤ Production condition of the factory.
⑥ Development of current production technology at home and abroad.

3.2.3 Procedure of process planning

1) Technological analysis of the workpart to be machined. The primary tasks of this step are to analyze the structural manufacturability, technical requirements, production type, and so on.

2) Selection of blank. The types of blanks and their quality have great influence on process planning. Therefore, the fabrication method and the accuracy of blank should be specified based on the function, the production program, the mechanical properties and the structural features of the part before process planning. The common used blank types in manufacturing industry are castings, forgings, sectional materials, welding parts, pressing parts, etc. Also, current production condition of the blank workshop should be taken into account when selecting a blank.

In large volume production, high-precision and high-efficiency blank processing methods, such as precision casting, die forging, cold stamping and powder metallurgy, are used extensively since the geometry shapes of these blanks are close to the finished parts and large amount of cutting works could be saved. Hence, manufacturing cost can be reduced for a great deal.

3) Drawing up process route. This is a critical step in process planning. It includes selecting location datum and processing methods, dividing machining stages, arranging machining sequence of each surface, determining the degree of process concentration and process dispersal, arranging heat treatment, inspection and other auxiliary operations.

4) Determining equipment and tooling needed in each working operation. Tooling is the general name of cutting tools, fixtures, measuring instruments and auxiliary tools. The production type, the structural features and the requisite machining quality of the part should be considered in selecting machine tools and tooling. Sometimes, the design task of a special purpose equipment and tooling is put forward if necessary.

5) Specifying technical requirements and inspection means used in the key operations.

6) Determining machining allowances, calculating operation dimensions and their tolerances for each operation.

7) Specifying cutting variables for each operation. In order to simplify process document and production management, the specific cutting variables are not specified clearly in process sheets but decided by operators themselves according to specific production conditions in job and small volume

3.2.2 制订工艺规程所需的原始资料

在大多数情况下,所给定零件或产品的制造工艺不是唯一的。制订工艺规程的任务就是从候选的工艺路线中找出最合理的一个。为此,工艺人员必须尽可能详细地收集与所加工零件有关的原始资料。所要收集的原始资料包括:

① 产品的整套装配图和零件图。
② 产品的生产纲领和生产类型。
③ 产品的质量验收标准。
④ 毛坯生产情况。
⑤ 本厂的生产条件和技术水平。
⑥ 国内外生产技术发展情况。

3.2.3 工艺规程制订的步骤

1) 零件的工艺性分析。这一步的主要任务是分析零件的结构工艺性、技术要求、生产类型等内容。

2) 确定毛坯。毛坯的种类和质量对工艺规程的编制影响很大。因此,在编制工艺规程之前,应根据零件在产品中的作用、零件的生产工艺规程、力学性能以及零件本身的结构特点,确定毛坯的制造方法、精度等内容。常用的毛坯种类有铸件、锻件、型材、焊接件、冲压件等。在选择毛坯时也应考虑毛坯车间的现有生产条件。

在大量生产条件下,一些如精密铸造、模锻、冲压和粉末冶金等高精度、高效率的毛坯制造方法被广泛采用,因为这类毛坯的几何形状接近于成品零件的形状,从而可以节省大量的切削加工。因此,可以大大减少零件的制造成本。

3) 拟订机械加工工艺路线。这是制订工艺规程的关键一步。主要内容有:选择定位基准,确定加工方法,划分加工阶段,安排加工顺序,决定工序的集中与分散,安排热处理、检验和其他辅助工序等。

4) 确定各工序所采用的设备和工艺装备。工艺装备包括刀具、夹具、量具、辅具等。在选择机床和工艺装备时应考虑零件的生产类型、结构特点和零件所要求的加工质量。有时对于专用设备和工艺装备,必要时还要提出设计任务书。

5) 确定主要工序的技术要求和质量验收标准。

6) 确定各工序的余量,计算工序尺寸和公差。

7) 确定各工序的切削用量。在单件、小批生产中,为简化工艺文件和生产管理,一般不明确规定具体的切削用量,而由操作者结合具体生产情况自行决定。在中批生产,特别是在大批大量生产时,为了保证生产节拍和加工质量,在工艺规程中对切削用量有明确的规定,并且不得随意改动。

8) 确定时间定额。

9) 技术经济分析。

10) 填写工艺文件。

production. In medium volume, especially in large volume and mass production, in order to ensure production balance and machining quality, cutting variables are specified clearly in operation sheets, and not allowed to modify them arbitrarily.

8) Identifying time standards.

9) Performing technical and economic analysis.

10) Filling in process sheets and other documents that would be used by operators.

3.3 Selection of location datum

3.3.1 Datum and its classification

Point, line, or surface used to determine the position of other points, lines or surfaces on the same workpiece is called datum. According to the function of datum and used situation, datum can be classified into design datum and process datum.

1. Design datum

Design datum refers to the points, lines, or surfaces that are used as references by designers to mark dimensions and relative positions of other graphic elements in part drawings. In other words, design datum is the starting position used to label the dimensions of the workpiece. Fig. 3-6 shows a drawing of a spur gear. The centerline of the gear is the design datum of the cylindrical surface with diameters D and D_0, and the inner cylindrical surface with diameter d, while the end face A is the design datum of the surface B and C since the positions of B and C are determined by the surface A.

2. Process datum

Datum used in the machining or assembly process are called process datum. According to the function of the process datum, process datum can be further divided into operation datum, location datum, measuring datum and assembly datum.

(1) Operation datum The datum used in operation drawing or process file to determine the positions of surfaces to be machined in one operation is called operation datum. The dimensions marked on the operation drawing are referred to as operation dimensions. As shown in Fig. 3-7, a hole is to be machined on a sleeve. Both L_1 and L_2 are called operation dimension for hole drilling, but the surface A is the operation datum in Fig. 3-7a, and the surface B is the operation datum in Fig. 3-7b.

(2) Location datum The datum used to determine the correct position of the workpiece on the machine tool or fixture is called location datum. For example, when the gear teeth shown in Fig. 3-6 are cut on hobbing machine, the surface A and the hole with diameter d are often the location datum (or location surface). In process files, location surfaces are often indicated with the sign "∨". Fig. 3-8 illustrates how the location sign "∨" is used, where the apex of the sign "∨" is attached directly to the location surface.

Location datum can be further classified into rough datum and finish datum. The location surface which has not been machined is called the rough datum surface or rough datum, whereas the location surface which has been machined is called the finish datum surface or finish datum. The location datum used in the first machining operation must be a rough datum. Finish datums are also di-

3.3 定位基准的选择

3.3.1 基准及其分类

用来确定在同一工件上其他点、线、面的位置所依据的点、线、面统称为基准。根据基准的功用和使用场合,基准可以分为设计基准和工艺基准。

1. 设计基准

设计基准是设计人员在零件图上标注尺寸或相互位置关系时所依据的那些点、线、面。换言之,设计基准是零件图上标注尺寸的起始位置。图 3-6 所示为一直齿轮的零件简图。齿轮的中心线即为直径为 D 和 D_0 的外圆柱面以及直径为 d 的内圆柱面的设计基准。由于表面 B、C 的位置是由端面 A 确定的,所以端面 A 是表面 B、C 的设计基准。

2. 工艺基准

零件在加工和装配过程中所采用的基准称为工艺基准。根据工艺基准的功用,工艺基准又可进一步分为工序基准、定位基准、测量基准和装配基准。

(1) 工序基准 在工序图或工艺文件中用来确定本道工序加工表面的尺寸和位置关系的基准,称为工序基准。在工序图上所标注的尺寸称为工序尺寸。如图 3-7 所示,欲在一套筒上钻一个孔,图中的 L_1 和 L_2 都是钻孔的工序尺寸,但在图 3-7a 中,表面 A 是工序基准,而在图 3-7b 中,表面 B 是工序基准。

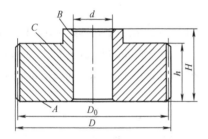

Fig. 3-6　Design datum 设计基准

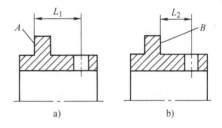

Fig. 3-7　Operation datum 工序基准

(2) 定位基准 在加工时用于确定工件在机床或夹具上正确位置的基准称为定位基准。在滚齿机上加工图 3-6 所示的齿轮时,端面 A 和直径为 d 的孔常作为定位基准(定位基面)。在工艺文件中,定位基面上常标有符号"∨"。图 3-8 给出了如何使用"∨"符号的示例,其中,符号"∨"的顶点直接接触定位表面。

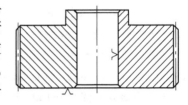

Fig. 3-8　Location datum 定位基准

定位基准又可进一步分为粗基准和精基准。尚未经过机械加工的定位表面称为粗基面,或粗基准,而经过机械加工的定位表面称为精基面,或精基准。在第一道工序使用的定位基准必然是粗基准。而精基准又有主要精基准和辅助精基准之分。主要精基准是零件结构本身就有,且具有一定功能的表面,而辅助精基准是零件结构本身没有,纯粹是为了定位需要而专门加工的。例如,轴上的顶尖孔就是辅助精基准。

vided into major finish datum and auxiliary finish datum. The major finish datum is just the functional surface of the workpiece, whereas the auxiliary finish datum has not any function in assembly drawing, and is machined specially for the location of the workpice. For instance, the center holes at the ends of the shaft are the auxiliary finish datum.

(3) Measuring datum The point, line or surface on which the measurement of the position of machined surfaces is based is called the measuring datum. Usually, the design datum can be selected as the measuring datum. But when it is inconvenient or impossible for the design datum to be used as the measuring datum, we should choose other surfaces existed practically on the workpiece as the measuring datum. For example, the center O of a circle is the design datum of the surface A in Fig. 3-9a and L is a design dimension. But after the surface A is machined, the point C (the lower generatrix of the cylindrical surface B) is selected as the measuring datum to measure the dimension L_1, as shown in Fig. 3-9b.

(4) Assembly datum The datum used to define the relative position of a part or subassembly in a product in assembling process is called the assembly datum. The assembly datum is generally consistent with the major design datum of the workpart. For example, when a gear is to be coupled with a shaft, the hole of the gear is used as the assembly datum. However, if the shaft is to be coupled with a box, the journal of the shaft becomes assembly datum. When the spindle box is to be installed into a bed, the base surface of the box would act as assembly datum.

3.3.2 Selection of location datum

To select location datum is the first step in drawing up the machining process route. The selection of the location datum is of primary importance for drawing up reasonable process route and for designing fixture or jigs with proper structure. Also, it affects the machining accuracy, productivity and production costs of the workpiece to be machined.

In the selection of the location datum, two issues should be considered simultaneously.

1) Which surfaces on the workpiece to be chosen as the finish location surfaces can effectively ensure the machining accuracy of the workpiece? Whether there is any individual operation which requires the second finish datum?

2) In order to machine the finish datum mentioned above, which rough surfaces should be selected as the rough datum?

As the function of the rough datum is different from that of the finish datum, the emphasis considered in specific selection is different.

1. Principles of selecting the rough datum

When selecting the rough datum, the emphasis is to consider how to ensure the mutual position relation between the machined surface and the unmachined surface, and how to ensure each surface to be machined has an adequate machining allowance. Therefore, the principles to select the rough datum are as follows:

(1) Giving priority to the guarantee of the mutual position requirement between two surfaces
If the relative position between the machined surface and unmachined surface must be met, the un-

（3）测量基准　在工件上用来测量已加工表面位置时所依据的点、线、面称为测量基准。通常可以选择设计基准为测量基准。但是当设计基准不便或不可能作为测量基准时，就应选择零件上实际存在的别的表面作为测量基准。如图 3-9a 中圆心 O 是待加工表面 A 的设计基准，L 是设计尺寸。而当 A 面加工后测量时，则以外圆的下母线 C 为测量基准测量尺寸 L_1，如图 3-9b 所示。

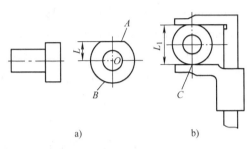

Fig. 3-9　Measuring datum　测量基准

（4）装配基准　在装配时用来确定零件或部件在产品中的相对位置所采用的基准称为装配基准。装配基准一般与零件的主要设计基准相一致。例如，装在轴上的齿轮以内孔为装配基准；装在箱体孔上的轴，其轴颈是装配基准；装在床身上的主轴箱，箱体的底面是装配基准。

3.3.2　定位基准的选择

拟订机械加工工艺路线的第一步是选择定位基准。定位基准的选择对于合理拟订工艺路线以及正确设计夹具是头等重要的。另外，它对零件的加工精度、加工效率以及加工成本都有很大的影响。

在选择定位基准时，需要同时考虑两个问题：

1）究竟选择零件上的哪一个（组）表面作为精基准可以有效保证零件的精度要求？是否有个别工序需要第二个（组）表面作为精基准？

2）要加工上述的精基准，应选择哪一个（组）毛坯面作为粗基准？

由于粗、精基准的用途不同，因此在选择时考虑的侧重点也不同。

1. 粗基准的选择原则

粗基准选择考虑的重点是如何保证非加工面与加工面的位置关系以及保证各加工表面应有足够的加工余量。粗基准的选择原则如下：

（1）优先保证表面间相互位置要求　如果首先要求保证零件上非加工表面与加工表面之间的相互位置要求，则应选择该非加工表面作为粗基准。当零件上有多个不加工表面时，应选择其中与加工表面有较高位置精度要求的不加工表面作为粗基准。

如图 3-10a 所示的铸件，外圆表面 1 为不加工表面。为保证孔壁厚均匀，车削孔 2 时应采用外圆表面 1 作为粗基准；再如图 3-10b 所示的拨杆，虽然不加工面很多，但由于要求 $\phi 22H9$ 孔与 $\phi 40mm$ 外圆同轴，因此在钻 $\phi 22H9$ 孔时应选择 $\phi 40mm$ 外圆作为粗基准，利用自定心夹紧机构使 $\phi 40mm$ 外圆与钻孔中心同轴，如图 3-10c 所示。

（2）优先保证加工余量　如果首先保证工件重要表面的加工余量足够且均匀时，应选

machined one should be chosen as the rough datum. And when there are several unmachined surfaces in the workpiece, the unmachined surface which has higher position requirements relative to the machined one should be chosen as the rough datum.

As shown in Fig. 3-10a, the cylindrical surface 1 is an unmachined surface. In order to achieve even wall thickness, the surface 1 should be chosen as the rough datum when turning the hole 2. A hole with $\phi22H9$ in a driving lever shown in Fig. 3-10b is to be drilled. Although there are several unmachined surfaces, the outer cylindrical surface which is 40mm in diameter should be selected as the rough datum so that the coaxiality of the hole and outer cylindrical surface can be ensured. In drilling hole, self-centering clamp device with three jaws shown in Fig. 3-10c is often used to drill the hole.

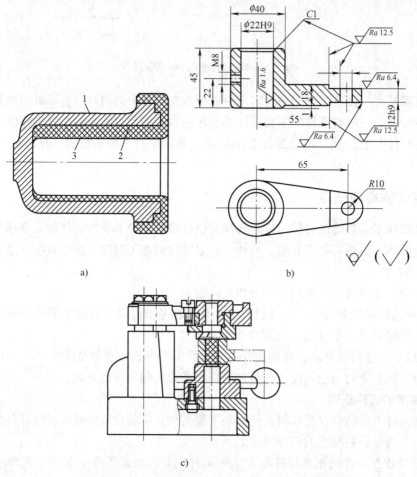

Fig. 3-10   Selection of the rough datum 粗基准的选择
1—毛坯外圆表面   2—待加工孔的加工余量   3—毛坯内圆表面

（2）Giving priority to the guarantee of the machining allowance   If the machining allowance on an important surface must be adequate and uniform, the surface should be selected as the rough location surface.

The surface of the lathe guideway is a main working surface. In order to keep uniform metallur-

择该加工表面为粗基准。

车床床身导轨面是最重要的表面，不仅精度要求高，而且要求导轨面有均匀的金相组织和较高的耐磨性，因此希望加工时导轨面的加工余量要小而均匀。图 3-11 所示车床床身导轨面加工中，有两种定位方案。图 3-11a 以导轨面为粗基准，先加工床腿底面，然后再以床腿底面为精基准加工导轨面；图 3-11b 的加工顺序与图 3-11a 的加工顺序相反。试分析这两种方案中，哪一种正确？不正确的定位方案会导致什么问题？

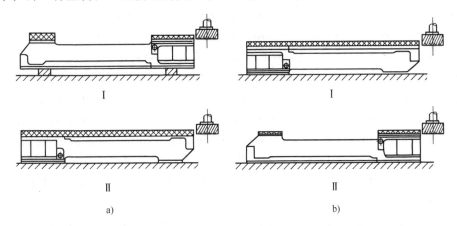

Fig. 3-11　Selection of the rough datum in machining the guideway 床身加工中的粗基准选择

当零件上各个表面都需要加工时，应选择毛坯误差较小、加工余量较少的表面作为粗基准。

（3）可靠定位　为了保证定位准确，夹紧可靠，尽量选用平整、光洁和有足够大面积的表面作为粗基准，不应选择有锻造飞边、铸造浇口、冒口或其他缺陷的表面作为粗基准。

（4）不重复使用粗基准　有了精基准尽量使用精基准，粗基准在同一尺寸方向上一般不应被重复使用。否则，在两次定位中很难保证表面间的相互位置精度。

如图 3-12a 所示零件，其内孔 $\phi16$mm、端面及 $3\times\phi7$mm 孔都需要加工。如果按图 3-12b、c 所示工艺方案，即第一道工序以 $\phi30$mm 外圆为粗基准车端面、镗孔；第二道工序仍以 $\phi30$mm 外圆为粗基准钻 $3\times\phi7$mm 孔，这样就可能使钻出的孔轴线与端面不垂直。如果用图 3-12b、d 所示工艺方案，就可以避免上述问题。

2. 精基准的选择原则

选择精基准时应重点考虑如何减少定位误差，保证加工精度，并使工件装夹准确、可靠、方便。具体选择原则如下：

（1）基准重合原则　尽量选择被加工面的工序基准（或设计基准）作为定位基准称为基准重合原则。采用这一原则可以避免因基准不重合而产生的定位误差。在工序加工精度要求较高的场合应遵循基准重合原则。

图 3-13a 所示为钻孔工序简图。左侧面 C 为工序基准。图 3-13b 表明侧面 C 也是定位基准，工序基准与定位基准重合。因此工序尺寸 B 及其公差可以直接获得。但是，如果选择右侧面 D 作为定位基准，如图 3-13c 所示，那么，工序尺寸 B 就会受到尺寸 A 的影响，而且存在基准不重合误差（$\pm T_A/2$）。

gical structure formed in casting to enhance its wear resistance and good surface quality, it is expected that the guideway surface should have a small and even machining allowance in machining. There are two location plans in the machining of the guideway, as shown in Fig. 3-11. In Fig. 3-11a, the guideway surface is used as the rough location surface to machine the legs of a bed firstly; and then, the bottom of legs is used as the finish location surface to machine the guideway. The machining sequence shown in Fig. 3-11b is contrary to that shown in Fig. 3-11a. Which one is the correct machining plan? What problem would the incorrect location plan cause?

When all the surfaces of a workpiece are to be machined, the surface with smaller blank error and machining allowance should be selected as the rough location surface.

(3) Reliable location It is preferable to choose a flat, smooth surface with larger contact area as a rough location surface to guarantee the accurate position and reliable clamping. Surfaces with casting gates, risers and flashes are not allowed to be chosen as the rough location surface.

(4) Don't use the rough datum repeatedly Finish datums should be used as far as possible if there have been. Generally speaking, it is not allowed to use the rough datum repeatedly in the same dimensional direction. Otherwise, it is hard to ensure the mutual position accuracy of surfaces in two locations.

Fig. 3-12a is a part drawing in which the end face, a hole with $\phi 16$mm in diameter and 3 holes with $\phi 7$mm in diameter are the surfaces to be machined. If the cylindrical surface ($\phi 30$mm) is used as the rough datum in the first operation to machine the end face and the $\phi 16$mm-hole as shown in Fig. 3-12b, and the same rough datum is used again in the subsequent operation to drill 3 $\phi 7$mm-holes, as shown in Fig. 3-12c, the axes of 3 $\phi 7$mm-holes would not be perpendicular to the end face. If the part is machined in light of the machining plan shown in Fig. 3-12b and Fig. 3-12d, the problem mentioned above can be avoided.

2. Principles of selecting the finish datum

When selecting the finish datum, the emphasis is to consider how to reduce the location error, ensure the machining accuracy, and mount the workpart correctly, reliably and conveniently. As a rule, the principles to select the finish datum are as follows:

(1) Datum coincidence principle Datum coincidence means generally to select the design datum or operation datum as the location datum. To use this principle can avoid the location error caused by the non-coincidence of datums. It is better to follow this principle when the operation machining accuracy is very strict.

Fig. 3-13a shows a simplified operation drawing for drilling a hole. The left side C is the operation datum. Fig. 3-13b illustrates that the side C is also the location datum, i.e. the operation datum coincides with the location datum. Therefore, the operation dimension B and its tolerance can be acquired directly. But if the right side D is selected as the location datum, as shown in Fig. 3-13c, the dimension B would be affected by the dimension A, and there is datum non-coincidence error($\pm T_A/2$).

(2) Datum unification principle This principle means selecting a group of surfaces as the location datum to machine most or all surfaces of a workpiece.

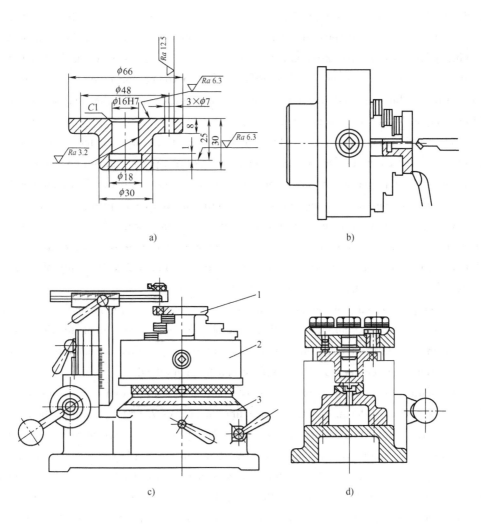

Fig. 3-12 Application of rough datum 粗基准应用
1—Workpart 工件 2—Three-jaw chuck 自定心卡盘 3—Indexing table 分度工作台

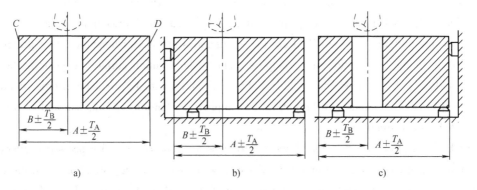

Fig. 3-13 Examples of datum coincidence principle 基准重合原则举例

Usually, there are several design datums in a workpiece to be machined. If the datum coincidence principle is followed in selection of finish datum, there will be several location datums and several kinds of fixtures. If there are a group of surfaces which can be selected as the finish datum in each operation to machine most or all other surfaces of a workpiece conveniently, the group of surfaces should be machined to certain precision as early as possible so that they can be used as the unified finish datum for subsequent operations.

The datum unification principle is widely used in the machining practice. For example, two center holes are used as the unified datum of shaft-type parts; a plane and two holes are used as the unified datum of box-type parts; an end surface and a hole are used as the unified datum of disc-type parts. The datum unification has many advantages. ①It can avoid effectively the errors caused by datum conversion. ②Many surfaces can be machined in a setting, which is in favor of ensuring the mutual position accuracy of different surfaces and reducing the setting error and the time for loading/unloading the workpiece. ③It can simplify the fixture design and realize the unity of fixture structure. ④It is in favor of machining workpieces in transfer line.

It should be pointed out that the datum coincidence principle is often violated when selecting location datum according to the datum unification principle. In these cases, great superiority should be given to machining accuracy. Therefore, the datum unification principle is generally used on condition that the machining accuracy of the workpiece can be guaranteed.

(3) The mutual location datum principle The mutual location datum here means that the surface A is used as the finish location datum to machine the surface B firstly, and then the surface B is used as the location datum to machine the surface A. This principle is used to machine two surfaces which have the higher mutual position accuracy. As shown in Fig. 3-14, there is very strict coaxiality between the journal of the spindle and the conical surface. In order to ensure this requirement, the journal 1 and the journal 2 are used as the finish location datum to machine the conical surface 3, and then, the conical surface 3 is used as the finish location datum to grind the two journals. They are machined again and again in this way until the coaxiality requirement is met.

(4) Self-location datum principle The self-location datum means using the surface to be machined as its own finish location datum. In finish machining, the surface to be machined itself is selected as the finish datum in order to remove a small and uniform machining allowance from the surface. This is the self-location datum principle. Finish reaming, broaching, float boring, finish grinding of the guideway and so on all are the applications of the self-location datum principle.

(5) Convenient for location and clamping The finish location surface, especially the primary location surface to be selected should have a large area and adequate precision in order to ensure the correct and reliable location, and be helpful for simplifying clamping mechanism and convenient operation at the same time.

Each of the principles mentioned above only considers one of several aspects that affect the position of a part in machining. In practice, it is nearly impossible to pay attention to all these aspects involved. The appropriate method is to focus on the key problems to be solved and then apply these principles flexibly to the selection of the rough and finish datums.

(2) 基准统一原则 该原则指的是选择零件上的一组表面作为定位基准去加工该零件上的大多数乃至全部表面。

通常,在要加工的零件上会有几个设计基准。如果按基准重合原则选择精基准,就会有多个定位基准和多种夹具结构。当工件以某一组表面作为精基准定位时,可以比较方便地在各工序中加工大多数(或所有)其他表面,则应尽早地把这一组基准面加工出来,并达到一定精度,以后工序均以它为精基准加工其他表面。

在实际生产中,基准统一原则被广泛使用。例如,轴类零件常使用两顶尖孔定位;箱体类零件常使用一面两孔定位;盘类零件常使用端面和内孔定位等。基准统一的好处很多,①可以有效避免因基准转换过多而产生的误差;②有可能在一次安装中加工许多表面,有利于保证表面间的位置精度,还可以减少安装误差和装卸工件的辅助时间;③可简化夹具设计,实现夹具的统一;④可减少工件搬动和翻转的次数,有利于工件的流水作业或自动化加工。

应当指出,采用基准统一原则常会带来基准不重合问题。此时,应优先考虑保证零件的加工精度。在能够保证加工精度的前提下,一般采用基准统一原则。

(3) 互为基准原则 互为基准在这里指的是以 A 面为精基准加工 B 面,然后再以 B 面为精基准加工 A 面。该原则常用于加工两个相互位置精度要求很高的表面。如图 3-14 所示,主轴颈和圆锥面之间有很严格的同轴度要求。为保证此项要求,在加工圆锥面 3 时以主轴颈 1 和 2 为精基准,然后再以圆锥面 3 为精基准磨削主轴颈 1 和 2。如此反复加工,直到满足同轴度要求为止。

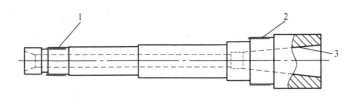

Fig. 3-14 Examples of mutual location datum 互为基准举例
1、2—Journal of spindle 主轴颈 3—Conical surface 圆锥面

(4) 自为基准原则 所谓自为基准指的是以待加工表面自身为精基准加工自身。在精加工工序中,为保证加工面有很小且均匀的余量,常用加工表面本身作为精基准进行加工,这就是"自为基准"原则。精铰孔、拉孔、浮动镗、床身导轨的精磨等都是自为基准原则的应用实例。

(5) 便于装夹原则 所选择的精基准,尤其是主要定位面,应有足够大的面积和精度,以保证定位准确可靠,同时还应使夹紧机构简单,操作方便。

上述基准选择原则中的每一条原则都只说明一个方面的问题。实际上要同时兼顾到方方面面几乎是不可能的。最好是抓住要解决的主要矛盾,然后把上述原则灵活地应用于粗、精基准的选择中。

3.4 Drawing up the process route

Process route refers to the sequencing of parts across the related departments or operations from the blank preparation to the finished product packaging storage in the production process. It includes selecting the location datum and processing methods of each surface, dividing machining stages, determining the degree of process concentration and process dispersal, arranging machining operations sequence by operations merging according to the related rules, specifying clamping methods, as well as arranging heat treatment, inspection and other auxiliary operations. Finally a series of reasonable processing procedure is formed.

Not only does machining process route influence the processing quality of the part and production efficiency, but also it will affect the equipment investment, production area and production cost. So the drawing up process route is a key step to formulate technical procedure, thus we should select the best one through analysis and comparison from the different schemes.

3.4.1 Selection of surface machining methods

When drawing up machining process route, the machining methods of each surface should be firstly determined according to quality requirements for each surface on the part drawing. In the selection of a machining method of a surface, the factors that should be considered include the design and processing features, machining accuracy, surface roughness, workpiece material and heat treatment, production type, economic machining accuracy and surface roughness obtained by different machining methods, available production conditions and technical level in the workshop, etc.

1. Economic machining accuracy and surface roughness obtained by different machining methods

The machining accuracy and surface roughness obtained by different machining methods are quite different. Even using the same machining method, the machining accuracy and surface roughness achieved might be different under different machining conditions. The main reason is that machining quality would be influenced by many factors involved in machining process, such as operator's skill, cutting variables, quality of cutter sharpening and adjustment of machine tools.

Economic machining accuracy is defined as the accuracy achieved under normal production conditions, which mean using the equipment conforming with the quality standard and the operators with standard technique grade, and not prolonging the machining time. It is the basis of determining the surface processing methods.

The economic machining accuracy achieved by different methods can be looked up from the relevant handbooks.

2. Selection of machining methods and machining schemes

1) Firstly, the final machining method should be determined according to the technical requirements for the surface. Then, taking the machining schemes of typical surfaces listed in Tab. 3-3, Tab. 3-4 and Tab. 3-5 for reference, the previous operations can be selected.

3.4 工艺路线的拟订

工艺路线指产品或零、部件在生产过程中,由毛坯准备到成品包装入库,经过企业各部门或工序的先后顺序。工艺路线的拟订包括定位基准和零件各表面加工方法的选择,划分加工阶段,确定工序集中和分散程度,按照一定的规则进行工序的合并,得到一系列的工序序列,选择安装方法,插入必要的热处理、检验工序及其他辅助工序等,最终形成了一系列合理的加工工序的顺序。

机械加工工艺路线不但影响零件的加工质量和生产效率,而且影响企业的设备投资、生产面积和生产成本。拟订工艺路线是制订工艺规程中关键性的一步,通常应多提一些方案,通过分析比较,选出最佳的方案。

3.4.1 表面加工方法的选择

拟订工艺路线时,首先要根据零件图上各表面的加工要求确定各表面的加工方法。选择每一表面加工方法时需要考虑的因素主要包括:零件该表面的设计特征、加工特征、加工精度、表面粗糙度、零件的材料和热处理要求、零件的生产类型、各种加工方法所能达到的加工经济精度和表面粗糙度、车间现有设备和工艺能力等。

1. 各种加工方法的经济加工精度和表面粗糙度

不同加工方法,如车削、铣削、刨削、磨削、钻削、镗削、拉削等,所能达到的加工精度和表面粗糙度也大不相同。即使是同一种加工方法,在不同的加工条件下所达到的加工精度和表面粗糙度也不相同,这是因为影响加工质量的因素有很多,如工人的技术水平、切削用量、刀具的刃磨质量、机床的调整质量等。

经济加工精度指在正常加工条件下(采用符合质量标准的设备、工艺装备和标准技术等级的工人,不延长加工时间)所能保证的加工精度。它是确定各表面加工方法的依据。

各种加工方法所能达到的经济加工精度和表面粗糙度可查阅相关手册。

2. 表面加工方法和加工方案的选择

1)首先根据被加工面的技术要求,确定该表面的最终加工方法。然后参照表 3-3、表 3-4 和表 3-5 列出的外圆表面、内孔表面及平面的加工方案,选定有关的前工序。

Table 3-3 Machining scheme and its economic machining accuracy of external cylindrical surfaces
外圆表面的加工方案及其经济加工精度

Machining scheme 加工方案	Economic accuracy 经济精度	Roughness 表面粗糙度 /μm	Application 应用
Rough turning 粗车	IT11~13	Ra50~100	Widely used for cutting all parts without hardening 广泛用于非淬火零件的加工
→Semi-finishing turning 半精车	IT8~9	Ra3.2~6.3	
→Finish turning 精车	IT7~8	Ra0.8~1.6	
→Polishing (rolling) 抛光(滚压)	IT6~7	Ra0.08~0.20	
Rough turning 粗车 →semi-finish turning 半精车→grinding 磨削	IT6~7	Ra0.40~0.80	Used for machining hardening steel, not for non-ferrous metals 用于淬火钢,不适用于非铁(有色)金属
→Rough grinding 粗磨			
→finish grinding 精磨	IT5~7	Ra0.10~0.40	
→Supergrinding 超精磨	IT5	Ra0.012~0.10	

(续)

Machining scheme 加工方案	Economic accuracy 经济精度	Roughness 表面粗糙度 /μm	Application 应用
Rough turning 粗车→semi-finish turning 半精车→finish turning 精车→diamond turning 金刚石车	IT5~6	Ra0.025~0.40	Used for machining non-ferrous metal 用于加工非铁（有色）金属
Rough turning 粗车→semi-finish turning 半精车→rough grinding 粗磨→finish grinding 精磨→mirror grinding 镜面磨	IT5 above IT5 以上	Ra 0.025~0.20	Used for machining parts with high-precision surfaces 主要用于加工表面精度要求高的零件
Rough turning 粗车→semi-finish turning 半精车→finish turning 精车→finish grinding 精磨→lapping 抛光	IT5 above IT5 以上	Ra0.05~0.10	
Rough turning 粗车→semi-finish turning 半精车→finish turning 精车→rough lapping 粗研→polishing 抛光	IT5 above IT5 以上	Ra0.025~0.40	

Table 3-4 Machining scheme and its economical machining accuracy of holes
孔的加工方案及其经济加工精度

Machining scheme 加工方案	Economic accuracy 经济精度	Roughness 表面粗糙度 /μm	Application 应用
Drilling 钻孔	IT11~13	Ra≥50	Used for machining small holes in unhardened metal parts 用于加工未淬火金属零件上的小孔
└→Enlarging 扩孔	IT10~11	Ra25~50	
└→Reaming 铰孔	IT8~9	Ra1.6~3.2	
└→Rough reaming 粗铰→finish reaming 精铰	IT7~8	Ra0.8~1.6	
Drilling 钻孔→reaming 铰孔	IT8~9	Ra1.6~3.2	
Drilling 钻孔→rough reaming 粗铰→finish reaming 精铰	IT7~8	Ra0.8~1.6	
Drilling 钻孔→(enlarging 扩孔)→broaching 拉孔	IT7~8	Ra0.80~1.60	Mass production 大批量生产
Rough boring (or enlarging) 粗镗（扩）	IT11~13	Ra 25~50	Used for machining existing holes in unhardened parts 用于加工已经有孔的未淬火零件
└→Semi-finish boring (finish enlarging) 半精镗（精扩）	IT8~9	Ra 1.6~3.2	
└→Finish boring (reaming) 精镗（铰）	IT7~8	Ra 0.80~1.60	
└→Float boring 浮动镗	IT6~7	Ra 0.20~0.40	
Rough boring(or enlarging) 粗镗→(或扩孔)→semi-finish boring 半精镗→grinding 磨	IT7~8	Ra 0.20~0.80	Used for machining hardened steel 用于加工淬火钢，不适用于非铁（有色）金属
└→Rough grinding 粗磨→finish grinding 精磨	IT6~7	Ra0.10~0.20	
Rough boring 粗镗→semi-finish boring 半精镗→finish boring 精镗→diamond boring 金刚镗	IT6~7	Ra 0.05~0.20	Used for machining holes with high accuracy 用于加工精度要求高的孔
Drilling 钻→(enlarging 扩)→rough reaming 粗铰→finish reaming 精铰→honing (or lapping) 珩（研磨）	IT6~7	Ra0.01~0.20	Used for machining holes with high precision 用于加工高精度要求的孔
Broaching 拉孔→honing (or lapping) 珩（研磨）	IT6~7	Ra0.01~0.20	
Rough boring 粗镗→semi-finish boring 半精镗→finish boring 精镗→honing (or lapping) 珩（研磨）	IT6~7	Ra0.01~0.20	

2）The material to be machined should be considered.

3）The production type of the workpart should be considered.

4）Available production conditions and technical level in your factory should be considered.

Table 3-5 Machining scheme and its economical machining accuracy of plane surfaces
平面的加工方案和及其经济加工精度

Machining scheme 加工方案	Economic accuracy 经济精度	Roughness 表面粗糙度 /μm	Application 应用
Rough turning 粗车 →Semi-finish turning 半精车 　→Finish turning 精车 　　→Grinding 研磨	IT11~13 IT8~9 IT7~8 IT6~7	$Ra \geqslant 50$ $Ra3.2~6.3$ $Ra0.80~1.60$ $Ra0.20~0.80$	Used for machining end faces of workpiece 用于加工零件端面
Rough milling 粗铣→broaching 拉	IT6~9	$Ra0.20~0.80$	Used for machining small planes in mass production 用于大量生产中加工小平面
Rough planing (or rough milling) 粗刨（粗铣） →finish planing (finish milling) 精刨（精铣） 　→scraping 刮研	IT11~13 IT7~9 IT5~6	$Ra \geqslant 50$ $Ra1.6~6.3$ $Ra0.10~0.80$	Used for machining unhardened plane 用于加工未淬火的平面
Rough planing (rough milling) 粗刨（粗铣）→ finish planing (finish milling) 精刨（精铣）→ →grinding 磨 Rough planing (rough milling) 粗刨（粗铣）→ finish planing (finish milling) 精刨（精铣）→ rough grinding 粗磨→finish grinding 精磨	IT6~7 IT5~6	$Ra\ 0.20~0.80$ $Ra\ 0.025~0.40$	Used for machining planes with high accuracy 用于加工高精度平面
Rough planing (rough milling) 粗刨（粗铣）→ finish planing (finish milling) 精刨（精铣）→ finish planing with wide planer tool 宽刀精刨	IT6~7	$Ra0.20~0.80$	Used for machining large planes in mass production 用于大量生产中加工大平面
Rough milling 粗铣→finish milling 精铣→ grinding 磨→lapping 研磨 　　→polishing 抛光	IT5~6 IT5 above IT5 以上	$Ra0.025~0.20$ $Ra0.025~0.10$	Used for machining planes with high quality 用于加工高质量平面

3.4.2　The division of machining phases

In order to ensure the machining quality of the workpiece, and to acquire high productivity and good economy, the process route is usually divided into several machining phases.

(1) Rough machining phase　The main task of this phase is to remove most part of the surplus material from the blank efficiently, and to prepare the finish location surface for successive operations. The main problem to be solved in this machining phase is how to raise the production efficiency.

(2) Semi-finish machining phase　The main task of this phase is to make the major surfaces get to a certain accuracy level, and to make preparation for finishing these surfaces. The minor surfaces should be machined to meet the design requirements.

(3) Finish machining phase　Except for the surfaces required for further polishing, all surfaces should be finished to meet the technical requirements stated on the part drawing.

(4) Burnish machining phase　The main goal of this phase is to decrease the surface roughness, and to improve the machining accuracy at the same time.

The reasons for dividing machining phases are as follows:

1) To avoid unnecessary waste. In rough machining, as large amounts of metal are removed from most surfaces of the part, the faults within the raw material such as slag, a crack or an air vent can be found as early as possible so that the part can be disposed without wasting much machining time.

2) To guarantee the machining quality of the surface. Generally, as the metal removal allowance in rough machining phase is large, the deformation caused by cutting force, cutting heat and residual stress in the workpiece is large. To separate rough machining from finish machining will help to eliminate the internal stress of the work piece, and to ensure the surface quality.

3) Reasonable selection of machine tools. During the rough machining, high power but low accurate machine tools can be used, while during the finish machining, low power but high accurate machine tools can be used. This can not only make full use of the performance of machine tools, but also prolong the service life of precision machine tools.

4) Easy to arrange the necessary heat treatment.

Though dividing machining phases has many advantages, not all the workpices are required to do so. Whether it is necessary to divide the machining process into several phases depends mainly on the effect of the workpiece deformation on machining accuracy. For example, it is not necessary to divide the machining phases when machining the workpieces with simple shape, low machining accuracy and good rigidity. When machining a large and heavy part, it is better not to divide the machining phases so as to avoid the handling and setup times.

3.4.3　Process concentration and dispersal

A process plan is often divided into several operations in order to make the production organization and scheduling more convenient, and to balance the loads of the machine tools. There are two different methods to schedule the process planning, which are process concentration and process dispersal.

1. Process concentration

Process concentration means there are many machining contents in one working operation. Its extreme situation is that all the jobs could be fulfilled in one operation. The process concentration is characterized as follows:

2）确定加工方法时要考虑被加工材料的性质。
3）选择加工方法时应考虑零件的生产类型。
4）选择加工方法时还要考虑本厂现有技术水平、生产条件等。

3.4.2 加工阶段的划分

为了保证零件的加工质量、获得高的生产率和较好的经济性，通常将工艺路线分为以下几个阶段：

（1）粗加工阶段　该阶段的主要任务是高效去除毛坯上的大部分余量，为后续加工准备定位精基准。需要解决的主要问题是如何最大限度地提高生产率。

（2）半精加工阶段　该阶段的任务是使各主要加工表面达到一定的精度并为主要表面的精加工做准备，使各次要表面达到最终要求。

（3）精加工阶段　除少数需要进行光整加工的表面外，其余所有表面均要达到图样规定的要求。

（4）光整加工阶段　该阶段的主要目的是降低表面粗糙度，同时提高零件的尺寸精度。

划分加工阶段的原因如下：

1）避免不必要的浪费。在粗加工阶段，将表面先进行粗加工，去掉加工表面的大部分余量，可以及早发现毛坯材料内的缺陷（夹渣、气孔、裂纹），及时处理，避免损失。

2）保证表面的加工质量。一般情况下，粗加工时加工余量大，切削力、切削热以及内应力引起工件的变形就较大。粗、精加工分开，有利于消除工件的内应力，有利于保证加工表面的质量。

3）合理选择机床。粗加工时可以选择功率大，精度较低的机床；精加工时则选精度较高的机床，这样既有利于发挥机床的性能，又有利于延长精加工机床的使用寿命。

4）合理地安排必要的热处理工序。

尽管划分零件加工阶段有很多好处，但并非所有零件加工都要这样去做。工艺过程是否需要划分阶段，主要取决于工件的变形对加工精度的影响程度。

例如，对于形状简单、加工精度低而刚性又较好的零件，可以不必划分加工阶段。对于大型及重型零件，则应尽量不划分加工阶段，避免多次安装和运输。

3.4.3 工序集中与工序分散

为了便于组织生产，平衡设备的负荷，常将工艺路线划分为若干个工序。有两种划分工序的方法，即采用工序集中和工序分散的原则。

1. 工序集中

工序集中是指在每道工序中安排很多的加工内容。其极端情况是加工一个零件只需要一道工序。工序集中的特点是：

1）减少了工序数目，有利于生产组织和管理。
2）减少了机床数量，从而减少了操作工人和生产面积。
3）减少了零件的安装次数，有利于保证加工质量、提高劳动生产率。
4）有利于采用高生产率的设备。
5）机床结构复杂，调整维护都不方便，生产准备工作量大。

1) Reduction of the operation numbers is helpful to the production organization and management.

2) Reduction of the number of machine tools is helpful to the reduction of the workers and production area.

3) Reduction of the mounting times is helpful to ensuring the machining quality and improving the productivity.

4) The machine tools with high productivity can be used.

5) The structure of machine tools is complex. Adjustment and maintenance are not convenient. The workload of the production preparation is quite heavy.

2. Process dispersal

Process dispersal means a large number of operations and long process routes in a technical process, but there are a fewer machining contents in one operation. Its extreme situation is that there is only one working step in one operation. The process of dispersal is characterized as follows:

1) Both machine tools and process equipment have a simple structure. It is easy to make adjustment and maintenance.

2) It has good production adaptability and is easy to change products.

3) It is helpful to choosing reasonable cutting variables.

4) The work in production organization is more complicated.

Whether process concentrated or process dispersed is adopted depends on the geometrical characteristics and the technical requirements of the part, the production program and available production conditions in the factory. For instance, process concentrated is used for heavy parts in order to reduce the clamping and handing times.

3.4.4 Arrangement of operation sequence

1. Arrangement of machining operations

(1) To machine the finish datum surfaces in advance   That is to say, the finish datum surfaces should always be machined at the beginning of the machining process, and then other surfaces can be machined by taking the finish datum surfaces as the location datum.

(2) To perform the rough machining firstly, then the finish machining   In order to avoid the effect of the forced deformation on machining accuracy and the damage of the machined surface, the rough machining should be done firstly, then the semi-finish machining, the finish machining and the burnish machining in the whole machining process.

(3) To arrange major surfaces, then minor surfaces   That is, the machining of major surfaces should be arranged firstly, then the machining of the minor surfaces is arranged in light of specific conditions.

(4) Machining plane is in advance of hole making   This is a special case principle (1). Taking the machined plane as datum to machine hole is helpful to ensuring the position accuracy of hole.

2. Arrangement of heat treatment operations

The main purpose of the heat treatment is to increase the mechanical properties, improve the cutting performance and release the internal stress. The arrangement of the heat treatment in the process route of the workpiece mainly depends on the workpiece material and the purpose of the heat treatment.

(1) Preparatory heat treatments   Such heat treatments as annealing and normalizing are usually arranged before rough machining. Their purposes are to improve the machinability of the work-

2. 工序分散

工序分散指的是整个工艺过程的工序数目多、每道工序的加工内容较少。工序分散的特点是：

1）机床和工艺装备简单，调整、维修方便。
2）生产适应性好，产品变换容易。
3）有利于选择合适的切削用量。
4）生产组织工作较复杂。

在确定工序集中或分散的问题时，应考虑零件的结构和技术要求、零件的生产纲领、工厂实际的生产条件等因素。例如，对于重型零件的加工，为减少安装和搬运次数，多采用工序集中原则。

3.4.4 工序顺序的安排

1. 切削加工工序的安排

（1）基面先行 即加工一开始，总是先加工精基面，然后用此精基准定位加工其他表面。

（2）先粗后精 整个零件的加工工序应是粗加工在前，然后半精加工、精加工及光整加工。这样可以避免由于工件受力变形而影响加工质量，也避免了精加工表面受到损伤等。

（3）先主后次 先安排主要表面的加工，然后根据具体情况再安排次要表面的加工。

（4）先面后孔 这是基面先行原则的特例。先加工平面，再以该平面定位加工零件上的孔，这样有利于保证孔和平面的相互位置精度。

2. 热处理工序的安排

热处理的目的主要是提高材料的力学性能、改善材料的加工性和消除内应力。热处理在零件加工工艺路线中的安排主要取决于零件材料以及热处理的目的。

（1）预备热处理 如退火、正火，一般安排在粗加工之前，主要以改善材料切削加工性能、消除毛坯制造时产生的内应力为目的。退火有时也安排在粗加工之后，以消除粗加工引起的内应力。

（2）最终热处理 常用的方法有调质、淬火、渗碳淬火、渗氮等。调质一般安排在粗加工和半精加工之间为宜。至于淬火、渗碳淬火，一般均安排在精加工之前。渗氮一般安排在粗磨之后、精磨之前进行。

（3）去应力热处理 如退火、时效，一般应安排在粗加工的前后。对于大而复杂的铸件，为了尽量减少内应力引起的变形，粗加工后进行人工时效处理。粗加工前采用自然时效。

表面处理可以提高零件的耐腐蚀能力，增加耐磨性，使表面美观。一般安排在工艺过程的最后进行。

3. 辅助工序的安排

辅助工序包括检验、清洗、去毛刺、退磁等工序。其中检验工序是主要的辅助工序。在工艺过程中，除了各工序操作工人自行检验外，还必须在下列情况下单独安排检验工序：①粗加工阶段结束之后；②车间转换前后，尤其是热处理前后；③关键工序的前后；④特种性能（如磁力探伤、密封性等）检验之前；⑤零件全部加工结束之后。

piece and release the internal stress produced in blank manufacturing. Sometimes annealing is also arranged after rough machining so as to release the internal stress caused by rough machining.

(2) Final heat treatments   Common used methods are structure improvement (quenching and high-temperature tempering), quenching, carbonizing, nitriding and so on. Quenching and high-temperature tempering is usually arranged in between rough machining and semi-finish machining. Quenching and carbonizing are generally arranged before finish machining. Nitriding is often arranged before finish grinding.

(3) Stress-removing heat treatments   Such heat treatments as annealing and ageing are generally arranged before or after rough machining. For some large and complex castings, artificial ageing is often arranged after rough machining in order to reduce the deformation caused by internal stress. Natural ageing is generally arranged before rough machining.

Surface treatment can raise the corrosion resistance and wear resistance, and improve the appearance. It is usually arranged in the last operation.

3. Arrangement of auxiliary operations

Auxiliary operations include inspection, washing, burring, demagnetizing and so on. Inspection is an important auxiliary operation. In addition to the self-inspection done by the operator in machining process, inspection operations have to be arranged under the following conditions: ① after rough machining phase; ②around the conversion of workpiece in different workshops, especially around heat treatment; ③around the key operations; ④before the inspection of special performance (such as magnetic testing, tightness etc.); ⑤after the whole machining operations.

4. Selection of machine tools

The precision of machine tools has vital influence on operation quality. The selection of machine tools should take into account the following principles:

1) The precision of machine tool should be able to meet the precision requirement for the workpiece.

2) The machining range of machine tool should adapt the contour size of workpiece to be machined.

3) The power, stiffness and working parameters of machine tools should meet the requirements for the most rational machining variables.

4) The production efficiency of machine tools should adapt the production type of the workpiece.

5) The existing machine tools in your factory and domestic machine tools should be used as far as possible.

3.5   Determination of machining allowance

3.5.1   Concept of machining allowance

In order to achieve desired machining quality, redundant materials of a workpiece or a blank should be removed. The thickness of a removed material from a surface of a workpiece or blank is called the machining allowance. The machining allowance can be divided into the total machining allowance and the operation allowance. In the machining process from the blank to the finished prod-

4. 机床的选择

机床的精度对工序的加工质量有很大的影响，在选择时要遵循以下原则：
1) 机床精度应与工序的加工精度相适应。
2) 机床的规格应与工件的轮廓尺寸相适应。
3) 机床的功率、刚度和工作参数应能满足最合理切削用量的要求。
4) 机床的生产率应与零件的生产类型相适应。
5) 选择机床时应充分利用现有设备，并尽量采用国产机床。

3.5 加工余量的确定

3.5.1 加工余量的概念

为了获得所需要的加工质量，要去除零件和毛坯上多余的材料。从零件和毛坯表面上去除的材料层厚度称为加工余量。加工余量有加工总余量和工序余量之分。在毛坯变为成品的过程中，从某被加工表面上切除材料的总厚度称为该表面的加工总余量，即该表面毛坯尺寸与零件图的设计尺寸之差。在某一工序所切除的材料厚度称为工序余量，即相邻两工序的尺寸差。

某个表面的加工总余量 ΔZ_0 与加工该表面各工序余量 ΔZ_i 之间有下列的关系

$$\Delta Z_0 = \Delta Z_1 + \Delta Z_2 + \Delta Z_3 + \cdots = \sum_{i=1}^{n} \Delta Z_i \tag{3-2}$$

式中　n——加工该表面的工序数；
　　　ΔZ_0——加工总余量；
　　　ΔZ_i——第 i 道工序的工序余量。

工序余量分为单边余量和双边余量。

加工非对称表面如平面时，工序余量为单边余量。图 3-15a 所示为加工平面时的单边余量：$Z_b = l_a - l_b$。

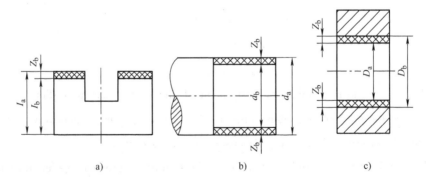

Fig. 3-15　Unilateral allowance and bilateral allowance 单边余量和双边余量

加工内孔或外圆等回转表面时，工序余量为双边余量。图 3-15b 所示为加工外圆时的双边余量：$2Z_b = d_a - d_b$；图 3-15c 所示为加工内孔时的双边余量：$2Z_b = D_b - D_a$。

uct, the total material thickness removed from a surface is called the total machining allowance of the surface. It equals the difference between the blank dimension and the design dimension of the part drawing. The material thickness removed from a surface in one operation is called the operation allowance. It equals the difference between the dimensions of two adjacent operations.

The relationship between the total machining allowance ΔZ_0 and the operation allowance ΔZ_i of a surface is:

$$\Delta Z_0 = \Delta Z_1 + \Delta Z_2 + \Delta Z_3 + \cdots = \sum_{i=1}^{n} \Delta Z_i \qquad (3\text{-}2)$$

where n—the numbers of operations for machining the surface;

ΔZ_0—total machining allowance;

ΔZ_i—i^{th} operation allowance.

The operation allowance can be divided into the unilateral allowance and the bilateral allowance.

For nonsymmetrical surfaces, such as the plane, the unilateral allowance is often used. Fig. 3-15a shows the unilateral allowance in machining plane: $Z_b = l_a - l_b$

The bilateral allowance is used when the rotary surfaces, such as internal or external cylindrical surfaces, are machined. Fig. 3-15b shows the bilateral allowance for external cylindrical surface: $2Z_b = d_a - d_b$; Fig. 3-15c shows the bilateral allowance for internal cylindrical surface: $2Z_b = D_b - D_a$.

Since operation dimension has its tolerance in practical machining, the actual machining allowance would vary in a specific range. The machining allowances can also be nominal machining allowance Z, or maximum machining allowance Z_{max}, or minimum machining allowance Z_{min}.

For the external surface as shown in Fig. 3-16a, there are:

$$Z = L_a - L_b$$
$$Z_{min} = L_{amin} - L_{bmax}$$
$$Z_{max} = L_{amax} - L_{bmin}$$

where L_a, L_{amax}, and L_{amin} are respectively nominal dimension, maximum dimension and minimum dimension of the previous operation; while L_b, L_{bmax}, and L_{bmin} are respectively nominal dimension, maximum dimension and minimum dimension of this operation.

The tolerance T_z of the machining allowance is the sum of the previous operation dimension tolerance T_1 and this operation dimension tolerance T_2.

$$T_z = Z_{max} - Z_{min} = T_1 + T_2 \qquad (3\text{-}3)$$

The relationship between each operation allowance and corresponding operation dimension is shown in Fig. 3-17.

在实际加工中,工序尺寸是有公差变化的,因而实际的切除余量也是有变化的。所以加工余量又有基本余量 Z、最大加工余量 Z_{max} 和最小加工余量 Z_{min} 之分。

对于图 3-16a 所示的外表面:

$$Z = L_a - L_b$$
$$Z_{min} = L_{amin} - L_{bmax}$$
$$Z_{max} = L_{amax} - L_{bmin}$$

其中,L_a、L_{amax}、L_{amin} 分别是上道工序的公称尺寸、上极限尺寸和下极限尺寸;L_b、L_{bmax}、L_{bmin} 分别是本工序的公称尺寸、上极限尺寸和下极限尺寸。

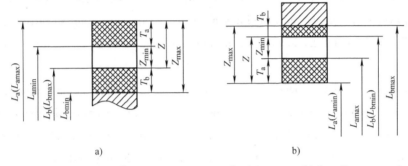

Fig. 3-16 Nominal machining allowance, maximum and minimum machining allowance and tolerance
基本余量、最大和最小加工余量和公差

加工余量的公差 T_z 等于上道工序的工序尺寸公差 T_1 与本工序的工序尺寸公差 T_2 之和。

$$T_z = Z_{max} - Z_{min} = T_1 + T_2 \tag{3-3}$$

各工序加工余量与相应加工尺寸的关系如图 3-17 所示。

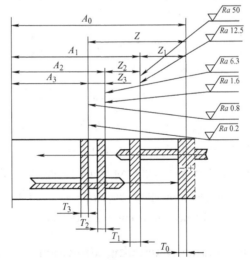

Fig. 3-17 Machining allowance and corresponding dimensions 加工余量与相应加工尺寸
A_0—Dimension of blank 毛坯尺寸　A_1—Dimension of rough machining 粗加工尺寸
A_2—Dimension of finish machining 精加工尺寸　A_3—Dimension of burnish machining 光整加工尺寸
Z—Total machining allowance 总加工余量　Z_1—Rough machining allowance 粗加工余量
Z_2—Finish machining allowance 精加工余量　Z_3—Burnish machining allowance 光整加工余量
T_0—Lower deviation of blank 毛坯下极限偏差　T_1—Tolerance of rough machining 粗加工公差
T_2—Tolerance of finish machining 精加工公差　T_3—Tolerance of burnish machining 光整加工公差

3.5.2 Factors affecting machining allowance

Generally, under the premise of guaranteeing the machining quality, the machining allowance should be as small as possible. But how large the minimum machining allowance should be? It is necessary to analyze the factors which affect the machining allowance.

(1) Surface quality of previous operation   The surface roughness Ra and defective layer Ha left on the surface by previous operation have to be removed by the current operation.

(2) Dimensional tolerance T_a of previous operation   In order to correct the machining errors produced in previous operation, the machining allowance of this operation should include T_a. The magnitude of T_a can be found by looking up the relevant handbooks based on the economic machining accuracy.

(3) Separately considered space error ρ_a left by previous operation   Space error ρ_a means some form and positional errors which are not included in the limit deviation of dimension, such as straightness, centricity and parallelism and so on. These kinds of errors should be removed in the current operation. The value of ρ_a is related to the previous machining method.

(4) Installation error ε_b of current operation ε_b consists of location error and clamping error. It will directly affect relative position between the surface to be machined and the cutting tool.

The space error and installation error are both vectors in space, the resultant of them should be the sum of vectors.

To sum up, machining allowance can be calculated as follows:

For unilateral allowance, there is

$$Z_{\min} = T_a + Ra + Ha + |\vec{\rho_a} + \vec{\varepsilon_b}| \qquad (3\text{-}4)$$

For bilateral allowance, there is

$$Z_{\min} = T_a/2 + Ra + Ha + |\vec{\rho_a} + \vec{\varepsilon_b}| \qquad (3\text{-}5)$$

$$2Z_{\min} = T_a + 2(Ra + Ha) + 2|\vec{\rho_a} + \vec{\varepsilon_b}| \qquad (3\text{-}6)$$

These formulas could be simplified in practical application. For instance, when a shaft is machined by centerless grinding, the clamping error could be ignored. Also, when a hole is machined by floating reamer or broaching tool, the space error ρ_a and clamping error ε_b could be neglected because both of them have less influence on machining allowance.

3.5.3 Methods to determine machining allowance

(1) By calculation   Theoretically, scientific and rational machining allowance can be determined by means of this method. But there must be reliable experimental data. It is not widely used at present. Sometimes, this method is used in key operations in large volume and mass production.

(2) By looking up tables   Combining with specific machining method, the machining al-

3.5.2 影响加工余量的因素

一般情况下，在保证加工质量的前提下尽量减少加工余量。但是，最小加工余量应是多少为合适呢？这就需要分析影响加工余量的因素。

(1) 上道工序加工表面的表面质量　上道工序的表面粗糙度高度 Ra 和表面缺陷层深度 Ha 应在本工序中去除掉。

(2) 上道工序的尺寸公差 T_a　为了纠正上道工序产生的误差，本工序的加工余量必须包含 T_a。T_a 的大小应根据加工方法的经济加工精度从有关手册中查得。

(3) 上道工序留下的需要单独考虑的空间误差 ρ_a　空间误差 ρ_a 指工件上有些不包括在尺寸极限偏差范围内的几何误差，如直线度、同轴度、平行度等。这些误差也应在本工序中去除掉。ρ_a 的大小与上道工序的加工方法有关。

(4) 本工序的安装误差 ε_b　安装误差包括定位误差和夹紧误差，它会影响被加工表面和刀具的相对位置。

空间误差和安装误差在空间都是矢量，因此它们的合成应为矢量和。

综上所述，加工余量的计算公式为

对于单边余量：
$$Z_{\min} = T_a + Ra + Ha + |\vec{\rho_a} + \vec{\varepsilon_b}| \tag{3-4}$$

对于双边余量：
$$Z_{\min} = T_a/2 + Ra + Ha + |\vec{\rho_a} + \vec{\varepsilon_b}| \tag{3-5}$$

$$2Z_{\min} = T_a + 2(Ra + Ha) + 2|\vec{\rho_a} + \vec{\varepsilon_b}| \tag{3-6}$$

以上是基本计算式，实际应用时可根据具体加工条件加以简化。如用无心磨削磨轴时，安装误差可忽略不计；用浮动铰刀或拉刀加工孔时，空间误差对加工余量无影响，且无安装误差。

3.5.3 确定加工余量的方法

(1) 计算法　理论上，该方法能确定比较科学合理的加工余量，但必须有可靠的实验数据资料。目前应用很少，有时在大批量生产中的重要工序中应用。

(2) 查表法　加工余量是结合具体加工方法，通过查阅现有的工艺手册得到的。必要时，查得的数据还需要进行修正。这是一种最简单常用的方法。

(3) 经验法　加工余量是由一些有经验的工程技术人员或工人根据经验确定的。多用于单件小批生产类型中。

lowance is obtained by looking up the handbooks for machining process. If necessary, the data got from handbook should be modified. This is the simplest and most common-used method.

(3) By experience Machining allowance is determined by experienced engineers, technicians or workers. This method is mainly used in job or small batch production.

3.6 Determination of operation dimensions and their tolerances

3.6.1 Introduction

Generally, when operation datum or location datum is coincident with design datum of a workpiece, the final operation dimensions and tolerances for a certain machining surface are determined directly by the parts drawing requirements. The inter-operation dimensions are calculated based on the part drawing dimensions plus or minus the machining allowances of the related operations. Here, dimensions and their tolerances of each operation can be deduced from the last operation. First, machining allowances should be decided. Soon after, basic dimensions including the blank dimension should be calculated one by one from the last operation. Then tolerances of each operation could be identified according to their economical accuracy. The limit deviation of the inter-operation dimensions could be marked based on the "body-in" principle. Fig. 3-18 shows the relationship among operation dimensions when machining a shaft.

As shown in Fig. 3-18, for the external surface-shaft, the dimension of the previous operation is the sum of the current operation dimension and its machining allowance.

$$L_2 = L_1 + Z_3$$
$$L_3 = L_2 + Z_2 = L_1 + Z_3 + Z_2$$
$$L_4 = L_3 + Z_1 = L_1 + Z_3 + Z_2 + Z_1 = L_1 + Z_\Sigma$$

The inter-operation dimensions of a workpiece can be calculated according to its final operation dimensions and the related machining allowances. When calculating the corresponding dimensions, the internal or external machining surfaces, and unilateral or bilateral allowance should be distinguished. The dimension and tolerance of the blank can be determined by looking up the relevant handbooks. If the operation datum or location datum is not coincident with design datum of a part, the operation dimension and its tolerance should be calculated by means of dimension chain method.

【Example 3-1】 A bearing hole in a box body is to be machined. Given that the design requirements for the hole are $\phi 95 Js6$ in diameter, and $Ra0.8 \mu m$. The process route: rough boring—semifinish boring—finish boring—float boring. Try to determine the dimension and tolerance of each operation.

Solution: The machining allowance, operation dimension and tolerance of each operation are listed in Tab. 3-6. The detail steps are as follows:

1) To determine the nominal machining allowance of each operation by looking up related machining handbooks or according to the practices in factory.

3.6 工序尺寸及其公差的确定

3.6.1 概述

一般情况下，当工序基准与定位基准重合时，加工某表面的最终工序的尺寸及公差可直接按零件图的要求来确定。中间各工序的工序尺寸则可根据零件图的尺寸，加上或减去工序的加工余量而得到。中间工序的尺寸和公差可以从最后工序向前推到。首先，确定各个工序的余量，然后从最后工序尺寸逐步向前计算得到工序公称尺寸和毛坯的尺寸。中间工序尺寸的公差根据加工经济精度确定，其极限偏差按"入体"原则标注。图 3-18 所示为加工轴时的各工序尺寸之间的关系。

由图 3-18 可知，对于轴这样的外表面加工，本工序的尺寸加上本工序的余量即为前工序的尺寸。具体计算公式如下

$$L_2 = L_1 + Z_3$$
$$L_3 = L_2 + Z_2 = L_1 + Z_3 + Z_2$$
$$L_4 = L_3 + Z_1 = L_1 + Z_3 + Z_2 + Z_1 = L_1 + Z_\Sigma$$

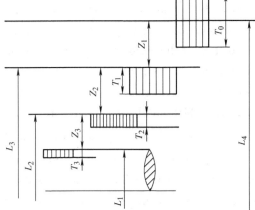

Fig. 3-18　Calculation of operation dimension 工序尺寸的计算

L_1—Nominal dimension of the workpiece 零件公称尺寸　　L_2—Nominal dimension of 2nd operation 工序 2 的公称尺寸
L_3—Nominal dimension of 1st operation 工序 1 的公称尺寸　　L_4—Nominal dimension of the blank 零件的毛坯尺寸
Z_1—Machining allowances of 1st operation 工序 1 的加工余量　　Z_2—Machining allowances of 2nd operation 工序 2 的加工余量
Z_3—Machining allowances of the final operation 最终工序的加工余量
T_0—Dimension tolerance of the blank 毛坯的尺寸公差　　T_1—Dimension tolerance of 1st operation 工序 1 的尺寸公差
T_2—Dimension tolerance of 2nd operation 工序 2 的尺寸公差　　T_3—Dimension tolerance of the workpiece 零件的尺寸公差

零件各个中间工序的公称尺寸根据零件的最终工序尺寸及加工余量计算得到。在计算时应注意区分内、外表面加工类型和单边、双边余量。零件毛坯的尺寸及公差需查阅有关手册来确定。如果加工过程中，工序基准和定位基准不重合，工序尺寸及公差的确定还需利用另一种方法——尺寸链来确定。

【例 3-1】 欲加工某箱体上的轴承孔。已知孔的设计要求为 $\phi 95 Js6$，$Ra = 0.8 \mu m$。加工

2) To calculate the nominal dimension of each operation and blank dimension in turn from the final operation to previous operations.

3) To determine the economic machining accuracy and dimension tolerance of each operation according to the corresponding machining method.

3.6.2 Dimension chain

In the machining process of a workpiece, workpiece dimensions are gradually changed from the blank dimension to each operation dimension, and finally to the design dimension. To reveal the relationships of the related dimensions by means of the dimension chain theory is very important for determining each operation dimension and its tolerance reasonably and ensuring both machining and assembly quality. The dimension chain method is an effective tool used to analyze and calculate the operation dimension in the process planning.

1. Concept of the dimension chain

In product design and manufacturing process, some related and dependent dimensions are usually concerned. A dimension chain is a closed dimensional diagram which is composed of a group of related dimensions connected in end-to-end sequence way. Fig. 3-19 shows an instance of a dimensional chain.

A dimension chain is composed of a closed link and several component links. The link means the dimension of the dimension chain. The last formed dimension in assembling process or machining process is called the closed link (a dimension that is guaranteed indirectly), denoted by A_0. There is only one closed link in a dimension chain.

All of the dimensions which have an effect on the closed link in a dimensional chain are called component links (obtained directly) which are denoted by A_1, A_2, $\cdots$, A_{n-1}. In addition, component links can be further divided into increasing links ($\vec{A_i}$) and decreasing links ($\overleftarrow{A_i}$). Suppose other component links are constant, if the variation of a component link would lead to the variation of a closed link in the same direction, the component link is named as the increasing link; if not, the component link is named as the decreasing link. As shown in Fig. 3-19, if A_2 is the closed link, then A_1 is a increasing link, A_3 is a decreasing link.

2. Classification of the dimension chain

(1) Classification according to the function of the dimension chain

1) Technological dimension chain. The dimension chain formed by some related dimensions of the single component in machining process is called the technological dimension chain, as shown in Fig. 3-19.

2) Assembly dimension chain. It is formed by some related dimensions of different components or parts in assembling process, as shown in Fig. 3-20.

3) Design dimension chain. It is formed by some related dimensions in a part design drawing.

(2) Classification according to the geometrical characteristics of the dimension chain

1) Length dimensional chain. Each dimension in a dimension chain is a measure of the length.

2) Angular dimensional chain. Each dimension in the dimension chain is a measure of the an-

工序为：粗镗—半精镗—精镗—浮动镗。试确定加工该孔的各个工序的工序尺寸与公差。

解：加工箱体零件孔的各个工序的加工余量、工序尺寸与公差见表3-6。具体步骤如下：

1）根据有关手册及工厂实际确定各个工序的基本加工余量。
2）由后工序向前工序逐个计算工序尺寸。
3）根据各种加工方法的经济加工精度确定各个工序尺寸的公差。

Table 3-6 Nominal dimension and tolerance of each operation 各工序的工序尺寸及其公差

Operation name 工序名称	Operation nominal allowance 工序基本余量/mm	Economic machining accuracy 经济加工精度/mm	Operation dimension 工序尺寸/mm	Operation dimension and tolerance 工序尺寸与公差/mm
blank 毛坯	7.9	±1.3	92.1−5=87.1	$\Phi 87.1 \pm 1.3$
rough boring 粗镗	5	H13（ES=+0.44 EI=0）	94.4−2.3=92.1	$\Phi 92.1^{+0.44}_{0}$ $Ra=6.4\mu m$
semi-finish boring 半精镗	2.3	H10（ES=+0.14 EI=0）	94.9−0.5=94.4	$\Phi 94.4^{+0.14}_{0}$ $Ra=3.2\mu m$
finish boring 精镗	0.5	H7（ES=+0.035 EI=0）	95−0.1=94.9	$\Phi 94.9^{+0.035}_{0}$ $Ra=1.6\mu m$
floating boring 浮动镗	0.1	Js6（±0.011）	95	$\Phi 95 \pm 0.011$ $Ra=0.8\mu m$

3.6.2 尺寸链

在零件的加工工艺过程中，零件的尺寸不断地变化，从毛坯尺寸到各个工序尺寸，最后达到设计要求尺寸。应用尺寸链理论揭示各尺寸之间的联系，对于合理确定工序尺寸及其公差、保证加工和装配质量非常重要。尺寸链是在编制工艺规程中分析和计算工序尺寸的有效工具。

1. 尺寸链的概念

在产品设计及制造过程中，常涉及一些互相联系，相互依赖的若干尺寸的组合。通常把互相联系且按一定顺序排列的封闭尺寸组合称为尺寸链。图3-19给出了一个尺寸链的示意图。

尺寸链由一个封闭环和若干个组成环构成。尺寸链中的尺寸称为环。在装配或加工过程中最后形成的（间接获得的）尺寸称为封闭环，记为 A_0。一个尺寸链中只能有一个封闭环。

尺寸链中对封闭环有影响的全部环，称为组成环（直接获得的），用 A_1、A_2、$\cdots A_{n-1}$ 来表示。组成环又分为增环 $\vec{A_i}$ 和减环 $\overleftarrow{A_i}$。设其余组成环不变，若某组成环的变动引起封闭环同向变动，则该组成环称为增环；若某组成环的变动引起封闭环反向变动，则该组成环称为减环。如图3-19所示，设 A_2 是封闭环，则 A_1 是增环，A_3 是减环。

Fig. 3-19 Technological dimension chain 工艺尺寸链

2. 尺寸链的分类

（1）按尺寸链的功能来分

1）工艺尺寸链。由单个零件在加工过程中各有关工艺尺寸所组成的尺寸链称为工艺尺寸链，如图3-19所示。

gle, as shown in Fig. 3-21.

(3) Classification according to the space position

1) Linear dimension chain. All component links are parallel to the closed link, as shown in Fig. 3-19, Fig. 3-20.

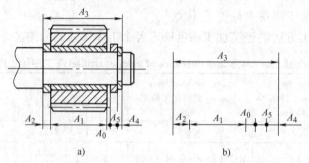

Fig. 3-20 Assembly dimension chain 装配尺寸链

2) Planar dimensional chain. If all the dimensions in a dimensional chain, which comprises both linear and angular dimensions, are located in the same or parallel planes, the dimension chain is called the planar dimensional chain, as shown in Fig. 3-22.

3) Spatial dimension chain. Each dimension in a dimension chain is not in the same plane or mutual parallel planes.

3.6.3 Calculation formulas of dimension chains

The calculation of the dimension chain is to calculate the basic dimensions, tolerances and deviations of the closed link and component links. There are two methods to calculate the dimension chain.

1. Extreme value method

The method is used to analyze and calculate the relation of the closed link and component links when each component link is in the state of extreme value (maximum or minimum). This method is simple and reliable, but it has too strict requirements for the component links. Basically, almost all of the operation dimension chains are calculated by using this method.

A m-link dimension chain with n increasing links can be expressed in Fig. 3-23. Based on the relationship of the dimension chain, the fundamental formulas to calculate dimension chain can be given as follows.

(1) Basic dimension of a closed link   The basic dimension of a closed link equals the difference between the sum of the basic dimensions of the increasing links and the sum of the basic dimensions of the decreasing links, i.e.

$$A_0 = \sum_{i=1}^{n} \overrightarrow{A_i} - \sum_{i=n+1}^{m-1} \overleftarrow{A_i} \tag{3-7}$$

where A_0—basic dimension of the closed link;

$\overrightarrow{A_i}$—basic dimension of the increasing link;

2）装配尺寸链。在机器装配过程中，由相互连接的尺寸形成封闭的尺寸组合称为装配尺寸链，如图 3-20 所示。

3）设计尺寸链。由零件图上的尺寸所组成的尺寸组合称为设计尺寸链。

(2) 按尺寸链的几何特征分

1）长度尺寸链。尺寸链的各环均为长度。

2）角度尺寸链。尺寸链的各环均为角度，如图 3-21 所示。

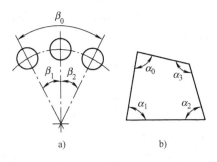

Fig. 3-21　Angular dimension chain　角度尺寸链

(3) 按尺寸链的空间位置可分

1）直线尺寸链。全部组成环都平行于封闭环的尺寸链。如图 3-19 和图 3-20 所示。

2）平面尺寸链。这是既有长度，又有角度尺寸组成的尺寸链，各环均位于一个或几个平行平面内，如图 3-22 所示。

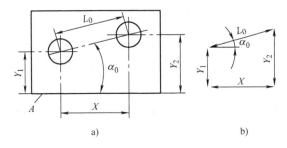

Fig. 3-22　Planar dimension chain　平面尺寸链

3）空间尺寸链。尺寸链中各环不在同一平面或彼此平行的平面内。

3.6.3　尺寸链的基本计算公式

尺寸链计算是指计算封闭环与组成环的公称尺寸、公差及极限偏差。通常尺寸链的计算方法有两种。

1. 极值法

极值法是按各组成环均处于极值条件下来分析计算封闭环与组成环之间的关系。该方法简便、可靠，但对组成环的公差要求过于严格。工艺尺寸链基本上均用极值法计算。

一个具有 n 个增环的 m 环尺寸链可以用图 3-23

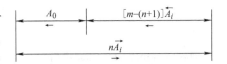

Fig. 3-23　Dimension chain with m links
m 环尺寸链

$\overleftarrow{A_i}$—basic dimension of the decreasing link;

n—number of all the increasing links in the dimension chain;

m—number of all links in the dimension chain.

(2) Limit dimension of a closed link   The maximum dimension of the closed link equals the difference between the sum of the all maximum dimensions of the increasing links and the sum of all minimum dimensions of the decreasing links; the minimum dimension of the a closed link equals the difference between the sum of all minimum dimensions of the increasing links and the sum of all the maximum dimensions of the decreasing links.

$$A_{0\max} = \sum_{i=1}^{n} \overrightarrow{A}_{i\max} - \sum_{i=n+1}^{m-1} \overleftarrow{A}_{i\min} \tag{3-8}$$

$$A_{0\min} = \sum_{i=1}^{n} \overrightarrow{A}_{i\min} - \sum_{i=n+1}^{m-1} \overleftarrow{A}_{i\max} \tag{3-9}$$

(3) Upper and lower deviations of a closed link   The upper deviation of closed link equals the difference between the sum of all upper deviations of the increasing links and the sum of all lower deviations of the decreasing links; the lower deviation of the closed link equals the difference between the sum of all lower deviations of increasing links and the sum of all upper deviations of decreasing links.

$$\mathrm{ES}_0 = \sum_{i=1}^{n} \overrightarrow{\mathrm{ES}}_i - \sum_{i=n+1}^{m-1} \overleftarrow{\mathrm{EI}}_i \tag{3-10}$$

$$\mathrm{EI}_0 = \sum_{i=1}^{n} \overrightarrow{\mathrm{EI}}_i - \sum_{i=n+1}^{m-1} \overleftarrow{\mathrm{ES}}_i \tag{3-11}$$

where, ES_0 and EI_0 are the upper deviation and lower deviation of the closed link; $\overrightarrow{\mathrm{ES}}_i, \overrightarrow{\mathrm{EI}}_i$ are the upper deviation and lower deviation of the increasing link; $\overleftarrow{\mathrm{ES}}_i, \overleftarrow{\mathrm{EI}}_i$ are the upper deviation and lower deviation of the decreasing link.

(4) Tolerance of a closed link   The tolerance of a closed link equals the sum of the tolerances of all component links.

$$T_0 = \sum_{i=1}^{m-1} T_i \tag{3-12}$$

where, T_0 and T_i are the tolerance of the closed link and tolerance of the component links respectively.

2. Probability method

Probability method is the method used to analyze and calculate the dimension chain based on the probability theory. It is mainly used for the dimension chain with many component links.

The basic dimension of a closed link:

$$A_0 = \sum_{i=1}^{n} \overrightarrow{A}_i - \sum_{i=n+1}^{m-1} \overleftarrow{A}_i \tag{3-13}$$

The tolerance of a closed link:

$$T_0 = \sqrt{\sum_{i=1}^{m-1} T_i^2} \tag{3-14}$$

所示的尺寸链图来表示。根据尺寸链的联系性，可以写出尺寸链的基本计算公式。

（1）封闭环的公称尺寸　封闭环的公称尺寸等于各增环基本尺寸之和减去各减环公称尺寸之和。

$$A_0 = \sum_{i=1}^{n} \vec{A}_i - \sum_{i=n+1}^{m-1} \overleftarrow{A}_i \tag{3-7}$$

式中，A_0——封闭环的公称尺寸；

$\vec{A}_i$——增环的公称尺寸；

$\overleftarrow{A}_i$——减环的公称尺寸；

n——增环的环数；

m——尺寸链的总环数。

（2）封闭环的极限尺寸　封闭环的最大值等于各增环最大值之和减去各减环最小值之和；封闭环的最小值等于各增环最小值之和减去各减环最大值之和。

$$A_{0\max} = \sum_{i=1}^{n} \vec{A}_{i\max} - \sum_{i=n+1}^{m-1} \overleftarrow{A}_{i\min} \tag{3-8}$$

$$A_{0\min} = \sum_{i=1}^{n} \vec{A}_{i\min} - \sum_{i=n+1}^{m-1} \overleftarrow{A}_{i\max} \tag{3-9}$$

（3）封闭环的上、下极限偏差　封闭环的上极限偏差等于各增环上极限偏差之和减去各减环下极限偏差之和；封闭环的下极限偏差等于各增环的下极限偏差之和减去各减环的上极限偏差之和。

$$\mathrm{ES}_0 = \sum_{i=1}^{n} \overrightarrow{\mathrm{ES}}_i - \sum_{i=n+1}^{m-1} \overleftarrow{\mathrm{EI}}_i \tag{3-10}$$

$$\mathrm{EI}_0 = \sum_{i=1}^{n} \overrightarrow{\mathrm{EI}}_i - \sum_{i=n+1}^{m-1} \overleftarrow{\mathrm{ES}}_i \tag{3-11}$$

其中，ES_0 和 EI_0 分别是封闭环的上极限偏差、下极限偏差；$\overrightarrow{\mathrm{ES}}_i$、$\overrightarrow{\mathrm{EI}}_i$ 分别是增环的上极限偏差、下极限偏差；$\overleftarrow{\mathrm{ES}}_i$、$\overleftarrow{\mathrm{EI}}_i$ 分别是减环的上极限偏差、下极限偏差。

（4）封闭环的公差　封闭环的公差等于各组成环公差之和。

$$T_0 = \sum_{i=1}^{m-1} T_i \tag{3-12}$$

其中　T_0——封闭环公差；

T_i——组成环公差。

2. 概率法

概率法解尺寸链是利用概率论原理分析和计算尺寸链的一种方法，主要用于组成环数较多的场合。

封闭环的公称尺寸

$$A_0 = \sum_{i=1}^{n} \vec{A}_i - \sum_{i=n+1}^{m-1} \overleftarrow{A}_i \tag{3-13}$$

封闭环的公差

$$T_0 = \sqrt{\sum_{i=1}^{m-1} T_i^2} \tag{3-14}$$

3. Steps of calculating the dimension chain

1) To draw the dimension chain and find out the closed link, then to determine the increasing links and the decreasing links. After the closed link is determined, an arrow can be drawn on the closed link, the direction of the arrow can be set optionally. Then, based on the principle of "end to end of arrow", the arrows are drawn on each component link in turn along the arrow direction. All the component links whose arrow direction is the same as that of the closed link are called decreasing links; those opposite to the arrow direction of the closed link are called increasing links. As shown in Fig. 3-24, A_0 is the closed link; A_1 is a decreasing link; A_2 and A_3 are increasing links.

2) To calculate basic dimension and upper and lower deviations of each component link.

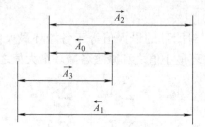

Fig. 3-24 To determine the increasing and decreasing
links of dimension chain 尺寸链增减环判别

3.6.4 Examples of calculating process dimension chains

1. Calculation of process dimensions when the datums are not coincident with each other

(1) Calculation of the operation dimension when the location datum is not coincident with the design datum

【**Example 3-2**】 There is a bush-type component, as shown in Fig. 3-25a. Fig. 3-25b is the operation drawing for milling indentation. Given that the surfaces A, B and C have been machined to the requisite accuracy. It is required that the dimension $10^{+0.18}_{\ 0}$ be ensured after milling the surface D. Try to find the operation dimension L_1 and its deviation..

Solution:

Step 1: To draw the dimension chain. The dimensional chain is shown in Fig. 3-25c.

As 60 ± 0.05 mm and $30^{+0.04}_{\ 0}$ mm are formed in previous operations, L_1 is obtained directly in the current operation, $L_0 = 10^{+0.18}_{\ 0}$ is a closed link. It can be seen from the dimension chain that L_1 and L_2 are increasing link, L_3 is a decreasing link.

Step 2: To calculate the basic dimension and its deviations.

By the formula (3-7), $L_0 = L_1 + L_2 - L_3$
$10\text{mm} = L_1 + 30\text{mm} - 60\text{mm}$
$L_1 = 40\text{mm}$

By the formula (3-10), $ES_0 = ES_1 + ES_2 - EI_3$
$ES_1 = +0.09\text{mm}$

By the formula (3-11), $EI_0 = EI_1 + EI_2 - ES_3$

3. 计算工艺尺寸链的步骤

1) 画出尺寸链图，确定封闭环，判别增、减环。在确定了封闭环以后，可在封闭环上任意设定一个箭头方向。然后按照箭头首尾相接的原则，循此方向依次在各组成环上画箭头。凡是箭头方向与封闭环箭头方向相同者为减环，相反者为增环。图 3-24 中，A_0 为封闭环，A_1 为减环，A_2、A_3 为增环。

2) 计算各环的公称尺寸和上、下极限偏差。

3.6.4 工艺尺寸链计算举例

1. 基准不重合时的工艺尺寸计算

（1）定位基准与设计基准不重合时的工序尺寸计算

【例 3-2】如图 3-25a 所示工件的零件图，图 3-25b 为铣缺口的工序简图。已知表面 A、B、C 均已加工好，即保证了尺寸 (60 ± 0.05) mm 和 $30_{\ 0}^{+0.04}$ mm。求工序尺寸 L_1 及其极限偏差。

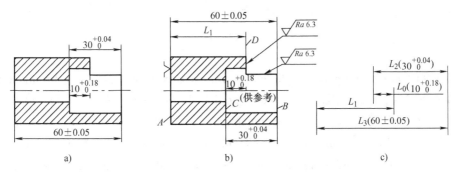

Fig. 3-25 Calculation of dimension chain for milling indentation 铣缺口尺寸链计算

解：1) 根据题意，画出尺寸链，如图 3-25c 所示。

在尺寸链中，因为尺寸 (60 ± 0.05) mm 和 $30_{\ 0}^{+0.04}$ mm 是前工序形成的，尺寸 L_1 是本工序保证的，所以 $L_0 = 10_{\ 0}^{+0.18}$ 是封闭环。由尺寸链可以看出 L_1、L_2 为增环，L_3 为减环。

2) 计算公称尺寸及其极限偏差。

由式（3-7）得

$$L_0 = L_1 + L_2 - L_3$$

$$10\text{mm} = L_1 + 30\text{mm} - 60\text{mm}$$

$$L_1 = 40\text{mm}$$

由式（3-10）得

$$ES_0 = ES_1 + ES_2 - EI_3$$

$$ES_1 = +0.09\text{mm}$$

由式（3-11）得

$$EI_0 = EI_1 + EI_2 - ES_3$$

$$EI_1 = +0.05\text{mm}$$

所以

$$L_1 = 40_{+0.05}^{+0.09}\text{mm}$$

（2）测量基准与设计基准不重合时的工序尺寸计算

【例 3-3】图 3-26a 所示为套类零件，$40_{-0.16}^{\ 0}$ mm 和 $10_{-0.3}^{\ 0}$ mm 为设计尺寸。加工时因尺寸 $10_{-0.3}^{\ 0}$ mm 不便测量，要通过直接测量孔深尺寸 L_1 间接保证。试求工序尺寸 L_1 及其极限

$$EI_1 = +0.05\text{mm}$$

Therefore, $L_1 = 40^{+0.09}_{+0.05}\text{mm}$

(2) Calculation of the operation dimension when the measurement datum is not coincident with the design datum

【Example 3-3】 Fig. 3-26a shows a sleeve-type part. $40^{\ 0}_{-0.16}$mm and $10^{\ 0}_{-0.3}$mm are the design dimensions. As the dimension $10^{\ 0}_{-0.3}$mm is inconvenient for measuring in machining process, it can only be ensured indirectly by measuring the depth L_1 of the hole. Try to find L_1 and its deviations.

Solution:

Step 1: To draw the dimensional chain, and determine the closed link, increasing link. In the dimensional chain shown in Fig. 3-26b, $10^{\ 0}_{-0.3}$ is the closed link L_0; $40^{\ 0}_{-0.16}$ is the increasing link and L_1 is the decreasing link.

Step 2: To calculate the basic dimension and its deviations.

By the formula $L_0 = L_2 - L_1$, $\quad L_1 = 40\text{mm} - 10\text{mm} = 30\text{mm}$

By the formula $ES_0 = ES_2 - EI_1$, $\quad EI_1 = 0\text{mm}$

Then, $\quad ES_1 = +0.14\text{mm}$

Therefore, $\quad L_1 = 30^{+0.14}_{\ 0}\text{mm}$

2. Calculation of the operation dimension marked from the surface to be machined

【Example 3-4】 Fig. 3-27a is a sketch showing the machining of the hole with a keyway. $43.6^{+0.34}_{\ 0}$mm and $\phi 40^{+0.05}_{\ 0}$mm are design dimensions. The machining sequence is as the follows: boring the hole to $\phi 39.6^{+0.10}_{\ 0}$mm; slotting the keyway to the dimension A_1; heat treatment; grinding the hole to $\phi 40^{+0.05}_{\ 0}$mm, and ensure the depth of the keyway $43.6^{+0.34}_{\ 0}$mm at the same time. Find the operation dimension A_1 and its tolerance.

Solution:

Step 1: To draw the dimension chain, and determine the closed link, increasing link. In the dimensional chain shown in Fig. 3-27b, the dimension A_0 is the closed link, A_1 and A_3 are the increasing links, A_2 is a decreasing link.

Step 2: To calculate the basic dimension and its deviations.

Basic dimension: $\quad A_0 = A_1 + A_3 - A_2$

$\quad A_1 = 43.4\text{mm}$

Upper deviation: $\quad ES_0 = ES_1 + ES_3 - EI_2$

$\quad ES_1 = +0.315\text{mm}$

Lower deviation: $\quad EI_0 = EI_1 + EI_3 - ES_2$

$\quad EI_1 = +0.05\text{mm}$

Therefore, $\quad A_1 = 43.4^{+0.315}_{+0.050}\text{mm}$

Based on the "body-in" principle, A_1 can be changed into: $A_1 = 43.45^{+0.265}_{\ 0}\text{mm}$.

偏差。

解：1）画出尺寸链图，确定封闭环和增、减环。在图 3-26b 所示的尺寸链中，$10_{-0.3}^{0}$ 为封闭环 L_0。$40_{-0.16}^{0}$ 为增环、L_1 为减环。

2）计算公称尺寸和上、下极限偏差。

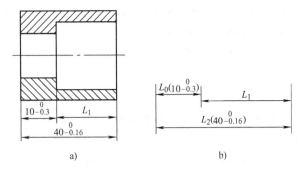

Fig. 3-26　Process dimension chain of a sleeve 套筒零件工艺尺寸链

由 $L_0 = L_2 - L_1$ 得：　　$L_1 = 40\text{mm} - 10\text{mm} = 30\text{mm}$

由 $ES_0 = ES_2 - EI_1$ 得　　$EI_1 = 0\text{mm}$

由 $EI_0 = EI_2 - ES_1$ 得　　$ES_1 = +0.14\text{mm}$

所以　　　　　　　　　　　$L_1 = 30_{0}^{+0.14}\text{mm}$

2. 从尚需继续加工的表面标注的工序尺寸计算

【例 3-4】图 3-27a 为加工一带键槽内孔的简图。设计尺寸为键槽深 $43.6_{0}^{+0.34}\text{mm}$ 及孔径 $\phi40_{0}^{+0.05}\text{mm}$。其加工顺序为：镗内孔至 $\phi39.6_{0}^{+0.10}\text{mm}$；插键槽至工序尺寸 A_1；热处理；磨内孔至 $\phi40_{0}^{+0.05}\text{mm}$，同时保证键槽深度尺寸 $43.6_{0}^{+0.34}\text{mm}$。试计算插键槽的工序尺寸 A_1 及其极限偏差。

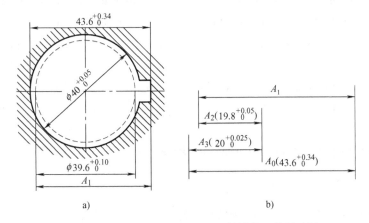

Fig. 3-27　Slotting dimension chain 插键槽工艺尺寸链

解：1）画出尺寸链图，确定封闭环和增、减环。在图 3-27b 所示的尺寸链中，A_0 是封闭环，A_1、A_3 为增环，A_2 为减环。

2）计算公称尺寸及其上、下极限偏差。

公称尺寸　　　　　　　　　　$A_0 = A_1 + A_3 - A_2$

3. Calculation of the operation dimension when several dimensions are ensured simultaneously

(1) Calculation of the dimension chain formed when the main design datum is machined lastly

【Example 3-5】 Fig. 3-28a shows a sleeve – type part, the design requirements of the part are marked on the part drawing. It can be seen that the axial design datum of the part is the surface B. Fig. 3-28b shows the machining operation charts. The surface B is ground in the last operation to ensure the $9_{-0.09}^{\ 0}$ mm, $32_{-0.062}^{\ 0}$ mm and (10 ± 0.18) mm simultaneously. Try to find the operation dimension A_1 and its tolerance.

Solution:

Step 1: To draw the dimension chain, shown as Fig. 3-28c. In the dimension chain, A_0 is the closed link; A_1 and A_3 are the increasing links; A_2 is the decreasing link.

Step 2: To calculate the basic dimension and its deviation.

Basic dimension:
$$A_0 = A_1 + A_3 - A_2$$
$$A_1 = A_0 - A_3 + A_2 = 9.8$$

Upper deviation:
$$ES_0 = ES_1 + ES_3 - EI_2$$
$$ES_1 = ES_0 + EI_2 - ES_3 = +0.09 \text{mm}$$

Lower deviation:
$$EI_0 = EI_1 + EI_3 - ES_2$$
$$EI_1 = -0.09 \text{mm}$$

Therefore,
$$A_1 = 9.8 \pm 0.09 \text{mm}$$

(2) Calculation of the dimension chain in surface treatment

【Example 3-6】 As shown in Fig 3-29a, the silver plating on the surface ϕF is required, and the plating depth is 0.2 – 0.3mm. The final dimension of ϕF is $\phi 63_{\ 0}^{+0.03}$ mm. Try to find the dimension A_1 and its tolerances in order to ensure the plating depth $0.2_{\ 0}^{+0.1}$ mm indirectly.

Solution:

Step 1: To draw the dimension chain shown in Fig. 3-29b, where the plating layer depth $0.2_{\ 0}^{+0.1}$ is the closed link; radius $31.5_{\ 0}^{+0.015}$ (half of $\phi 63_{\ 0}^{+0.03}$) is the decreasing link; A_1 is increasing link.

Step 2: To calculate basic dimension and its deviations.

Basic dimension:
$$A_0 = A_1 - A_2$$
$$A_1 = 31.7$$

Upper deviation:
$$ES_0 = ES_1 - EI_2$$
$$ES_1 = +0.1 \text{mm}$$

Lower deviation:
$$EI_0 = EI_1 - ES_2$$
$$EI_1 = +0.015 \text{mm}$$

Therefore,
$$A_1 = 31.7_{+0.015}^{+0.100} \text{mm} = 31.715_{\ 0}^{+0.085} \text{mm}$$

The operation dimension in the diameter direction is: $2A_1 = 63.43_{\ 0}^{+0.17}$ mm.

$$A_1 = 43.4\text{mm}$$

上极限偏差 $\text{ES}_0 = \text{ES}_1 + \text{ES}_3 - \text{EI}_2$

$$\text{ES}_1 = +0.315\text{mm}$$

同理，下极限偏差为 $\text{EI}_1 = +0.05\text{mm}$

所以 $A_1 = 43.4^{+0.315}_{+0.050}\text{mm}$

按入体原则，A_1 可以转换成 $A_1 = 43.45^{+0.265}_{0}\text{mm}$

3. 多尺寸同时保证的工艺尺寸计算

（1）最后加工主要设计基准的工序尺寸计算

【例 3-5】 图 3-28a 所示为一套筒零件，其设计要求均已标注在图上。由图看出，该零件的轴向设计基准为表面 B。图 3-28b 为零件的加工工序图，其中③为磨端面 B 工序，要求同时保证 $9^{\ 0}_{-0.09}$mm、$32^{\ 0}_{-0.062}$mm 和 (10 ± 0.18)mm。试求钻孔的工序尺寸 A_1。

解：1）画出尺寸链，如图 3-28c 所示。其中，$A_0(10 \pm 0.18)$ 为封闭环，A_1、A_3 为增环，A_2 为减环。

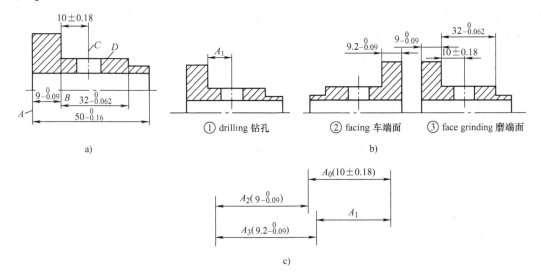

Fig. 3-28 Dimension chain formed when grinding face B lastly 最后磨削 B 面时形成的尺寸链

2）计算公称尺寸和上、下极限偏差。

公称尺寸 $A_0 = A_1 + A_3 - A_2$

$$A_1 = A_0 - A_3 + A_2 = 9.8\text{mm}$$

上极限偏差 $\text{ES}_0 = \text{ES}_1 + \text{ES}_3 - \text{EI}_2$

$$\text{ES}_1 = \text{ES}_0 + \text{EI}_2 - \text{ES}_3 = +0.09\text{mm}$$

下极限偏差 $\text{EI}_0 = \text{EI}_1 + \text{EI}_3 - \text{ES}_2$

$$\text{EI}_1 = -0.09\text{mm}$$

所以 $A_1 = 9.8 \pm 0.09\text{mm}$

（2）表面处理工序的工艺尺寸链计算

【例 3-6】 图 3-29a 所示零件的 ϕF 表面要求镀银，镀层厚度为 $0.2 \sim 0.3$mm，ϕF 的最终尺寸为 $\phi 63^{+0.03}_{0}$mm。为间接保证镀层厚度 $0.2^{+0.1}_{0}$mm，试求镀层工序尺寸 A_1 及其极限偏差。

3.7 Economic analysis of technological processes

3.7.1 Determination of time rating

Time rating is the stipulated time consumed by producing a product or completing an operation under a certain production conditions. It is the basis of arranging production plan, cost checking, determining the number of equipment and personnel, planning production area. It is also the index of the productivity.

The time required for completing an operation of a workpiece is called single-piece time, represented by T_d. T_d consists of the following parts:

(1) Basic time T_j. The time consumed in changing directly the dimension, shape, performance and relative position of the production object is named as basic time. For machining, it refers to the powered time cost in removing the material from the workpiece.

(2) Auxiliary time T_f. The time consumed in carrying out various auxiliary jobs in an operation is named as auxiliary time, such as the time used in loading and unloading the workpiece, starting and stopping the machine tool, changing the cutting variables, measuring operation dimension and so on.

In general, the sum of the basic time and the auxiliary time is called the operation time T_z (the time used directly for producing products or parts and components).

(3) Working area arrangement time T_b. It refers to the time consumed by workers to tend the working area (such as the time used for replacing tools, lubricating machine tools, cleaning up chips and picking up tools). Generally it usually accounts for 2% – 7% of the operation time.

(4) Relax and physiological need time T_x. It refers to the time used by worker to recover his strength or to meet the physiological need on the shift. Generally, it usually accounts for 2% of the operation time.

(5) Preparation and finality time T_z. It refers to the time consumed by workers to make preparation before machining and to do tail-in work after machining in batch production. Suppose the number of a batch of parts is n, the preparation and finality time shared by each part is T_z/n. It can be seen that T_z/n can be ignored in mass production because n is very large.

To sum up, the single-piece time can be expressed as follows:

$$T_d = T_j + T_f + T_b + T_x \tag{3-15}$$

For batch production, the single-piece time can be calculated by following formula:

$$T_{dj} = T_d + T_z/n \tag{3-16}$$

For mass production, the single-piece time can be calculated by Eq. (3-15).

解：1）绘制如图 3-29b 所示的尺寸链。其中，镀层厚度是封闭环。半径 $31.5^{+0.015}_{\ 0}$（$\phi 63^{+0.03}_{\ 0}$ 的一半）是减环，A_1 为增环。

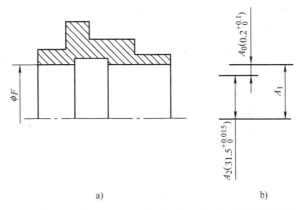

Fig. 3-29 Process dimension chain of silver plating 镀银工艺尺寸链

2）计算公称尺寸和上、下极限偏差。

公称尺寸 $\qquad A_0 = A_1 - A_2$

$\qquad\qquad\qquad A_1 = 31.7 \text{mm}$

上极限偏差 $\qquad \text{ES}_0 = \text{ES}_1 - \text{EI}_2$

$\qquad\qquad\qquad \text{ES}_1 = +0.1 \text{mm}$

下极限偏差 $\qquad \text{EI}_0 = \text{EI}_1 - \text{ES}_2$

$\qquad\qquad\qquad \text{EI}_1 = +0.015 \text{mm}$

所以 $\qquad A_1 = 31.7^{+0.100}_{+0.015} \text{mm} = 31.715^{+0.085}_{\ 0} \text{mm}$

在直径方向上的工序尺寸应为：$2A_1 = 63.43^{+0.17}_{\ 0} \text{mm}$。

3.7 工艺过程的经济分析

3.7.1 时间定额的确定

时间定额是在一定生产条件下，规定生产一件产品或完成一道工序所消耗的时间。它是安排生产作业计划、进行成本核算，确定设备数量和人员编制，规划生产面积的依据，也是说明生产率高低的指标。

完成零件一道工序所需要的时间称为单件时间 T_d，它由下列几部分组成：

（1）基本时间 T_j　直接改变生产对象的尺寸、形状和相对位置关系所消耗的时间称为基本时间。对于机械加工而言，基本时间是指切除金属所耗费的机动时间。

（2）辅助时间 T_f　在一道工序中为完成基本工作所作各种辅助动作而消耗的时间，称为辅助时间；例如，装卸工件、开停机床、改变切削用量、测量加工尺寸、进刀或退刀等动作所花费的时间。辅助时间可以通过实测来确定，也可以用基本时间的百分比来估算。

基本时间和辅助时间之和称为作业时间 T_z（直接用于制造产品或零、部件所消耗的时间）。

（3）布置工作地时间 T_b　为使加工正常进行，工人照管工作地（更换刀具、润滑机床、清理切屑、收拾工具等）所消耗的时间。通常按照作业时间的2%~7%估算。

3.7.2 Economic analysis of process plans

Economic analysis means analyzing and comparing the production costs of different process plans. Production costs consist of two parts: one part is directly related to technological process, the other has not direct relation to technological process (such as the salary of administrative personnel). The former is called process cost, and accounts for 70% - 75% of production costs. Generally, it is required to consider process cost only when making economic analysis of process plan.

1. Elements and calculation of the process cost

Process cost consists of variable cost and constant cost. Variable cost is related to the annual output of the workpart. It includes material cost, wage of machine tool operators, depreciation cost and maintenance cost of general-purpose machine tools and tooling. Constant cost has no relation to the annual output of the workpart. It includes the wage of setters, depreciation cost and maintenance cost of special-purpose machine tools and tooling, etc.

Suppose N stands for the annual output of the workpart, the annual process cost C_n (yuan/year) of the workpart can be expressed as follows:

$$C_n = VN + S \qquad (3\text{-}17)$$

The single-piece process cost C_d (yuan/year) can be expressed as follows:

$$C_d = V + S/N \qquad (3\text{-}18)$$

where V—variable cost of the workpart (yuan/piece);

S—constant cost of a year (yuan/year).

The relation of the annual process cost C_n to the annual output N is shown in Fig. 3-30. The relation of the single-piece process cost C_d to the annual output N is shown in Fig. 3-31.

Fig. 3-30 shows that the variation ΔC_n of annual process cost is proportional to the variation N of the annual output. In Fig. 3-31, the part A is similar to the situation in job and small volume production. A little variation of N would cause C_d to change greatly. But there is something different in the part B. Even though N varies greatly, C_d only has a very little change. The constant cost S has very a little influence on C_d, this is equivalent to the situation in large volume and mass production.

2. Comparison of process plans

Economic analysis for different process plan can be divided into two cases.

1) When the basic investment in one process plan is similar to that in the other, process cost often becomes the basis of economic analysis. Suppose the annual process costs of two different process plans are:

$$C_{n1} = NV_1 + S_1 \qquad (3\text{-}19)$$
$$C_{n2} = NV_2 + S_2 \qquad (3\text{-}20)$$

When the annual output N is given, C_{n1} and C_{n2} can be found. If $C_{n1} > C_{n2}$, then the second process plan is the better one. If N is a variable, the curve of each process plan can be plotted based on the Eq. (3-19) and Eq. (3-20), as shown in Fig. 3-32. It can be seen from Fig. 3-32 that the second process plan is the better one when $N < N_k$. When $N > N_k$, the first process plan is better. When $N = N_k$, $C_{n1} = C_{n2}$, that is

(4) 休息与生理需要时间 T_x 工人在工作班内为恢复体力和满足生理上的需要所消耗的时间。一般按作业时间的 2% 估算。

(5) 准备与终结时间 T_z 在批量生产中,工人在加工前进行准备和结束工作所消耗的时间称为准备与终结时间。设一批工件数为 n,则分摊到每个工件上的准备与终结时间为 T_z/n。由此可见,在大量生产时,由于 n 很大,故 T_z/n 可以忽略不计。

综上所述,单件时间为

$$T_d = T_j + T_f + T_b + T_x \tag{3-15}$$

对于成批生产,单件计算时间为

$$T_{dj} = T_d + T_z/n \tag{3-16}$$

对于大量生产,单件计算时间可用式(3-15)计算。

3.7.2 工艺方案的经济性分析

所谓经济性分析就是分析比较各种不同工艺方案的生产成本。生产成本包括两项费用:一项费用与工艺过程直接有关,另一项费用与工艺过程无直接关系(如行政人员工资等)。与工艺过程直接有关的费用称为工艺成本,占生产成本的 70%~75%。在工艺方案经济分析时,一般只需考虑工艺成本即可。

1. 工艺成本的组成及计算

工艺成本由可变费用与不变费用两部分组成。可变费用与零件的年产量有关。它包括材料费、机床操作工人工资、通用机床和通用工艺装备维护折旧费。不变费用与零件年产量无关,它包括调整工人的工资,专用机床、专用工艺装备的维护折旧费等。

若零件的年产量为 N,则零件全年工艺成本 C_n(元/年)可以表示为

$$C_n = VN + S \tag{3-17}$$

单件工艺成本 C_d(元/年)可以表示为

$$C_d = V + S/N \tag{3-18}$$

式中 V——零件的可变费用(元/件);
　　　S——全年的不变费用(元/年)。

图 3-30、图 3-31 分别给出了全年工艺成本 C_n、单件工艺成本 C_d 与年产量 N 的关系图。

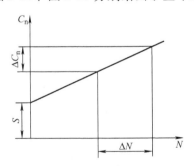

Fig. 3-30 Relation of C_n to N C_n 与 N 的关系

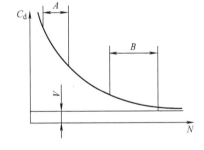

Fig. 3-31 Relation of C_d to N C_d 与 N 的关系

图 3-30 表明全年工艺成本的变化量 ΔC_n 与年产量的变化量 ΔN 成正比。在图 3-31 中,A 区相当于单件小批生产情况,N 略有变化,C_d 值变化很大;而在 B 区,情况则不同,即

$$N_k V_1 + S_1 = N_k V_2 + S_2$$
$$N_k = (S_2 - S_1)/(V_1 - V_2) \tag{3-21}$$

2) If the difference between two basic investments in different process plans is quite large, not only the process cost is considered, the payback period of the difference in investments should also be considered.

For example, the plan 1 adopts advanced special-purpose equipment with a high price. This plan has a larger investment K_1 and a lower process cost C_1, but it has a shorter lead time, and its products can be put on market quickly. The plan 2 adopts the ordinary equipment with low price. It has a smaller investment K_2 and a higher process cost C_2, but it has a longer lead time, and its products are put on market slowly. If we only compare their process cost in this condition, we can't make over-all assessment on their economy. Therefore, the payback period of the difference between two basic investments in different process plans have to be considered. The investment payback period can be determined by the following formula.

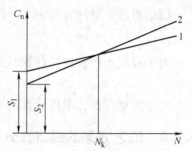

Fig. 3-32　Comparison of process plans 工艺方案比较

$$\tau = \frac{K_1 - K_2}{C_2 - C_1 + \Delta Q} = \frac{\Delta K}{\Delta C + \Delta Q} \tag{3-22}$$

where　ΔK—basic investment difference;

ΔC—savings of the annual process cost;

ΔQ—annual total extra income acquired in product sales.

Investment payback period should satisfy the following requirements:

1) Payback period should be less than the durable years of the special-purpose equipment and technological equipment;

2) Payback period should be less than the market life of the products (year);

3) Payback period should be less than the standard payback period stipulated by government. The standard payback period of special-purpose machine tools is 4 to 6 years, and the standard payback period of special-purpose tooling is 2 to 3 years.

使 N 变化很大，C_d 值变化却不多；不变费用 S 对 C_d 的影响很小，这相当于大批大量生产的情况。

2. 工艺方案的比较

对几种不同工艺方案进行经济分析可分为两种情况。

1）当两种工艺方案基本投资相近或均采用现有设备时，工艺成本往往成为经济分析的依据。若两种方案的全年工艺成本为

$$C_{n1} = NV_1 + S_1 \tag{3-19}$$

$$C_{n2} = NV_2 + S_2 \tag{3-20}$$

当年产量 N 一定时，可由式（3-19）和式（3-20）分别求出 C_{n1} 及 C_{n2}。如果 $C_{n1} > C_{n2}$，则选第 2 方案。当年产量 N 为一变量时，可利用式（3-19）和式（3-20）作图比较，如图 3-32 所示。由图 3-32 可知，当 $N < N_k$ 时，第 2 方案的经济性好；当 $N > N_k$ 时，第 1 方案的经济性好。当 $N = N_k$ 时，$C_{n1} = C_{n2}$，即

$$N_k V_1 + S_1 = N_k V_2 + S_2$$

$$N_k = (S_2 - S_1)/(V_1 - V_2) \tag{3-21}$$

2）当两种工艺方案的基本投资差额较大时，则在考虑工艺成本的同时，还要考虑基本投资差额的回收期限。

例如，方案 1 采用了价格较贵的先进专用设备，基本投资 K_1 大，工艺成本 C_1 较低，但其生产准备周期短，产品上市快；方案 2 采用了价格较低的一般设备，基本投资 K_2 少，工艺成本 C_2 较高，但其生产准备周期长，产品上市慢。这时如果单纯比较其工艺成本，就难以全面评定其经济性，必须同时考虑不同加工方案基本投资差额的回收期限。投资回收期 τ 可用下式求得

$$\tau = \frac{K_1 - K_2}{C_2 - C_1 + \Delta Q} = \frac{\Delta K}{\Delta C + \Delta Q} \tag{3-22}$$

式中　ΔK——基本投资差额；

　　　ΔC——全年工艺成本节约额；

　　　ΔQ——从产品销售中取得的全年增收总额。

投资回收期必须满足以下要求：

1）回收期限应小于专用设备或工艺装备的使用年限。

2）回收期限应小于该产品的市场寿命（年）。

3）回收期限应小于国家所规定的标准回收期。专用机床的标准回收期为 4～6 年，专用工艺装备的标准回收期为 2～3 年。

Chapter 4

Design Principles of Machine Tool Fixtures

第4章

机床夹具设计原理

4.1 Introduction to machine tool fixtures

4.1.1 Introduction to fixtures

Fixtures are the general designation of all technological equipment used to position and clamp the workpiece in machining and other manufacturing operations, such as welding, heat treatment, machining, inspection, and assembly, etc.

To ensure that the workpiece is produced according to the specified shape, dimensions, and tolerances, it is essential that the workpiece should be appropriately located and clamped on the machine tool. Therefore, all fixtures must have location elements, and generally also have clamping device. The configuration of a fixture depends not only on the workpiece characteristics, but also on the sequence of machining operations, magnitude and orientation of the expected cutting forces, capabilities of the machine tool, and cost considerations. The fixtures used in different usage situations have their own particularity. For example, both inspection fixtures and the fixtures of machine tool have higher location accuracy, and other fixtures generally have lower location accuracy. The fixtures of machine tools and its design principles are primarily introduced in this chapter.

4.1.2 Classification of fixtures

The technological equipment used for locating and clamping the workpiece in machine tools is called the fixture of the machine tool. The purpose to discuss the classification of the fixtures is to understand better the characteristics and the scope of application of various fixtures, and further know the universality and particularity of various fixtures in design. The fixtures of machine tools can be classified according to the machine tool, clamping power source, and the application and feature of the fixtures.

1. Classification based on the applications and features of fixtures

(1) General-purpose fixtures The fixtures are generally made by professional manufacturers, and some are supplied with machine tools as its attachments, such as three-jaw chuck, four-jaw chuck, center, face plate, etc. Users need not design and manufacture by themselves. The fixtures have a good adaptability. Without any adjustment, or with a few adjustments, or by exchanging a small number of parts, they can satisfy different applications. Using this kind of fixture can shorten the lead time of production and reduce fixture variety, thus decreasing the production cost of product. But they have lower location accuracy and lower productivity. General purpose fixtures are mainly used in job and small batch production with many varieties and lower machining accuracy.

(2) Special-purpose fixtures This kind of fixture is specially designed and made for an operation of a workpart. Compared with the general-purpose fixture with the same performance, the fixture has a simpler, more compact structure and higher location accuracy and productivity. As it is often designed and made by the users, it has a longer manufacturing cycle. Once the product changes, the fixtures become useless and have to be discarded. Therefore, this kind of fixture is suitable for larger

4.1 机床夹具概述

4.1.1 夹具概述

凡是用来在加工和其他制造工序中对工件进行定位和夹紧的工艺装备统称为夹具。夹具广泛应用于工件的焊接、热处理、机械加工、检测、装配等环节。

为了确保工件加工后达到规定的尺寸、形状和公差要求，有必要使工件在机床上进行合理的定位与夹紧。因此，所有夹具必须具有定位元件，一般也有夹紧装置。夹具的结构形状不仅取决于工件的特征，也取决于加工顺序、切削力的大小和方向、机床的加工能力以及制造成本。但由于使用场合不同，它们在设计中也有各自的特殊性。例如，检验夹具和机床夹具对定位精度要求较高，其他夹具对定位精度要求一般较低。本章主要介绍机床夹具及其设计原理。

4.1.2 夹具的分类

在机床上用来确定工件位置并将其夹紧的工艺装备称为机床夹具（简称夹具）。研究机床夹具分类的目的是为了更好地了解各类夹具的不同特点和应用范围，进而掌握各类夹具设计中的普遍性原理和特殊性问题。机床夹具一般按专门化程度、使用的机床和夹紧动力源进行分类。

1. 按专门化程度分类

（1）通用夹具　指具有较高通用性的夹具，其结构尺寸已经系列化。这类夹具一般由专门厂家生产制造，有些已经作为机床附件随机床一起供应。如自定心卡盘、单动卡盘、台虎钳等。通用夹具均具有适应性强、成本低、可缩短生产准备周期的优点；但其效率较低、定位精度较差也是不容忽视的。因此，通用夹具多用于加工精度要求不高，中、小批和单件生产的场合。

（2）专用夹具　针对某一工件的某一工序的加工精度要求而专门设计、制造的夹具称为专用夹具。由于具有非常高的针对性，所以其效率很高，结构紧凑，定位精度较高；但制造周期较长，成本较高，不具有通用性。同时，高的针对性也决定了专用夹具一般由使用单位自行设计、制造。专用夹具多用于生产批量较大的场合；小批量生产中，当工件加工精度较高或加工困难时也采用专用夹具。

（3）可调夹具　通过更换和调整夹具上的个别元件，就可满足相同或相似类型、但具有不同结构尺寸工件装夹需要的一类夹具称为可调夹具。可调夹具又分为通用可调夹具和成组夹具两种。通用可调夹具是指具有一定通用性的可调夹具，如滑柱钻模；成组夹具是专门应用于成组工艺的夹具，要求夹具在同组工件装夹中能够可调。

（4）组合夹具　由预先制造好的标准元件、合件组装而成的夹具，组成夹具的元件、

volume and mass production of a product. But this kind of fixture is sometimes used in small volume production when the workpiece has higher accuracy or is difficult to machine.

(3) Adaptable fixtures This kind of fixture is specially designed and made for the workpieces with similar structure but different sizes. It can be used for the workpieces with similar structure but different sizes through replacing or adjusting individual location or clamping elements. Adaptable fixtures are generally divided into general adjusting fixtures and group fixtures. The former has larger generality and has not very definite objects to be machined, such as the jig with sliding pillar, the universal vise with various jaws. The latter is designed for machining a group of workpieces in the group process. It can be used for holding the workpieces in the same group by adjusting or replacing a few elements. As this kind of fixture can be used many times, it can avoid the repeat design of fixture and reduce the fixture manufacturing cost. This kind of fixture is suitable for multi-variety and small volume production.

(4) Modular fixtures This kind of fixture is assembled with standard elements and combined members which have been made previously. These elements and combined members can be assembled, disassembled easily and used repeatedly. As the use of modular fixtures can shorten the lead time and reduce the variety of fixture, modular fixtures are suitable for the trial production of new product and the job or small batch production.

(5) Follow-fixtures Contrast to the stationary fixture on the machine tool, the follow-fixture can be moved with the workpiece from one operation to another in the transfer line. The follow-fixture differs from the ordinary fixture in that it has two sets of location datums. One is used for locating the workpiece, the other is for locating itself on the station fixture. The follow-fixture is suitable for the workpieces with complex configuration and without appropriate location surface or conveying surface. When using follow-fixture in machining process, the workpiece to be machined should be mounted in the follow-fixture firstly. Then the follow-fixture carrying the workpiece is moved by means of conveying mechanism from one station to the next in turn to complete different machining contents.

2. Classification according to the placed machine tool

According to the placed machine tool, fixtures can be classified into fixture of lathe, fixture of milling machine, fixture of drilling machine (drilling jig), fixture of boring machine (boring jig), fixture of shaper and so on.

3. Classification according to the clamping power source

According to the clamping power source, fixtures can be classified into manual fixture, electric-drive fixture, pneumatic fixture, hydraulic fixture, electromagnetic fixture, vacuum fixture, etc.

4.1.3 Functions and elements of fixtures

1. Functions of fixtures

1) To ensure the machining accuracy reliably and reduce the influence of human interference on machining. Using a fixture to mount a workpiece, the position accuracy of the workpiece relative to the cutter and the machine tool is ensured by the fixture, and is not affected by the technological level of operators.

2) To shorten the nonproductive time of a workpiece, and increase the productivity, further re-

合件可多次拆装，重复利用。组合夹具的特点就是夹具组装极快；可减少夹具品种，降低夹具的保管、维护费用；可降低工件的加工成本。因此，组合夹具非常适合于新产品的开发试制和单件、小批生产类型。

（5）随行夹具　随行夹具是指在自动线加工中，可随同工件按加工工艺需要一起移动的夹具。随行夹具必须要与固定安装在各加工工位的工位夹具配套使用。随行夹具不同于一般夹具的地方就是具有两套定位基准，一套用于对工件进行定位，另一套用于在工位夹具上对其本身定位。加工时，先将工件装夹在随行夹具上，然后随行夹具带着工件沿自动线依次完成在各工位的装夹和加工。随行夹具适用于被加工工件无可靠定位基准或无可靠输送基面的情况。

2. 按使用的机床分类

根据所使用的机床，夹具可分为车床夹具、铣床夹具、钻床夹具、镗床夹具、拉床夹具、齿轮加工机床夹具等。

3. 按夹紧动力源分类

根据夹具所使用的夹紧动力源，夹具可分为手动夹具、电动夹具、气动夹具、液压夹具、电磁夹具、真空夹具等。

4.1.3　夹具的功用和组成

1. 机床夹具的功用

1）可以稳定保证工件的加工精度。采用夹具装夹工件，工件相对于刀具及机床的位置精度由夹具保证，不受工人技术水平的影响，使一批工件的加工精度趋于一致。

2）可以减少辅助时间，提高劳动生产率。采用夹具后，可以省去对工件的逐个找正和对刀，使辅助时间显著减少；另外，用夹具装夹工件，比较容易实现多件、多工位加工，以及使机动时间与辅助时间重合等；当采用机械化、自动化程度较高的夹具时，可进一步减少辅助时间，从而可以大大提高劳动生产率。

3）可以扩大机床的使用范围，实现一机多能。在机床上配备专用夹具，可以使机床使用范围扩大。例如，在车床床鞍上或在摇臂钻床工作台上安放镗模后，可以进行箱体孔系的镗削加工，使车床、钻床具有镗床的功能。

4）可以改善工人的劳动条件，降低劳动强度。

2. 机床夹具的组成

欲在一批量生产的套筒零件上钻—铰 $\phi6H7$ 的孔。套筒的其他技术要求详见图4-1a。现采用图4-1b所示的钻模加工此孔。工件以内孔 $\phi25mm$ 和左端面在定位心轴6及其台肩面上定位；采用螺母5和开口垫圈4实现工件的快速夹紧和装卸。采用快换钻套1引导钻头，以保证加工孔的位置尺寸要求。

由此可见，被加工孔的尺寸精度（$\phi6H7$）直接由定尺寸刀具（钻头、铰刀）保证，孔的位置尺寸（37.5±0.02）mm以及其他要求均由钻套相对于有关定位元件的位置精度来

ducing the production cost. Once the fixture is used, it is not necessary for the operator to set the cutter or to align each workpiece, thus reducing much nonproductive time. Besides, multi-workpiece or multi-station machining can be realized easily by using some special mechanism.

3) To enlarge the machining range of machine tools. Equipped with a special-purpose fixture, some machine tools can be of more functions. For example, a lathe or a radial drilling machine equipped with boring jig can be used to bore a series of holes in box-type parts.

4) To improve the working condition and decrease the labor intensity.

2. Elements of fixtures

It is desired to drill and ream the hole with ϕ 6H7 in diameter in a batch of sleeves. Other technical requirements for the sleeve are shown in Fig. 4-1a. A drilling jig shown in Fig. 4-1b is used to machine the hole. Taking the hole with ϕ 25mm in diameter and left face as location surfaces, the sleeve is located on the location mandrel 6 and clamped by means of nut 5 and snap washer 4. The quick-change drilling or reaming bush 1 is used to guide the drill or reamer so that the position accuracy of the hole can be ensured.

It can be seen that the dimension accuracy (ϕ6H7) of the hole is ensured by the drill or reamer, and the position accuracy (37.5 ± 0.02) mm as well as other requirements are ensured by the position accuracy of drilling bush relative to the relevant locators.

Though there are many kinds of fixtures with different structural forms, the fixture is generally composed of the elements as follows:

(1) Location elements / Locators The elements directly contact or fit with the location surface in the workpiece and are used to position the workpiece in the fixture. The location mandrel 6 in Fig. 4.1b is the location element.

(2) Clamping device The clamping device is used to clamp the workpiece to prevent the position of workpiece from changing due to all the disturbing forces during the machining. For example, the nut 5 and snap washer 4 in Fig. 4-1b are the components of the clamping device.

(3) Cutter-aligning or cutter-guiding element It is used to specify the position of the cutting tool relative to the fixture. The drilling bush 1 in Fig. 4-1b is the cutter-guiding element.

(4) Connecting components Connecting components are used to specify the position of fixture relative to machine tool and to fasten the fixture on the machine tool. The orienting keys and T-botls in milling fixture belong to connecting components.

(5) Other elements or devices They are used to meet some special requirements during machining, such as auxiliary support, loading/ unloading device, and indexing device, etc.

(6) Body of fixture It is a base of the whole fixture and is used to connect all other parts as mentioned above into an integral fixture. In general, the body of the fixture is the largest component in the fixture, such as the component 7 in Fig. 4-1b.

The elements mentioned above are not always necessary for a practical fixture. Generally, the fixture should have the location element and clamping device. They are the key parts to ensure the machining accuracy of the workpiece. In the following sections, how to position and clamp the workpiece in the fixture will be discussed in detail.

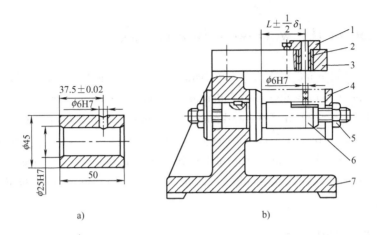

Fig. 4-1　Drilling jig 钻模
1—Drilling bush 钻套　2—Guide bush 衬套　3—Jig plate 钻模板　4—Snap washer 开口垫圈
5—Nut 螺母　6—Location mandrel 定位心轴　7—Body of jig 夹具体

保证。

尽管夹具的种类繁多，夹具结构形式各异，但夹具一般由下列几部分组成。

（1）定位元件　定位元件指与工件定位表面相接触或配合，用以确定工件在夹具中准确位置的元件。图4-1b 中的定位心轴6 就是定位元件。

（2）夹紧装置　夹紧装置用以夹紧工件，防止加工中其他作用力对工件已定好位的破坏。图4-1b 中的螺母5 和开口垫圈4 就属于夹紧装置。

（3）对刀、引导元件　对刀、引导元件是指用来保证刀具相对于夹具或工件之间准确位置的元件。图4-1b 中的钻套1 就是引导元件。

（4）连接元件　连接元件是指用以确定夹具相对于机床之间准确位置，并将夹具紧固在机床上的元件。如夹具与机床工作台之间连接用的 T 形槽螺栓。

（5）其他元件或装置　为了满足工件装卸和加工中其他需要所设置的元件及装置，如辅助支承，装卸工件用的装置以及分度装置等。

（6）夹具体　夹具体是用来连接夹具其他各部分使之成为一个有机整体的基础件。一般情况下，夹具体是夹具中最大的一个元件。如图4-1b 中的7。

以上这些组成部分并不是对每种机床夹具都是缺一不可的。通常，夹具都必须有定位元件和夹紧装置。它们是保证工件加工精度的重要组成部分。以下各节中将详细探讨工件在夹具中的定位和夹紧问题。

4.2 Location of the workpiece in a fixture

The purpose to locate the workpiece in a fixture is to have the batch of workpieces in the same operation occupy the correct position in relation to both the machine tool and the cutting tool. In order to solve the location problem of the workpiece in a fixture, the following questions must be clear firstly: ① How many degrees of freedom does a workpiece have in space? ② How to restrict these degrees of freedom? ③ What is about the relation between the operation accuracy and the degree of freedom of the workpiece? ④ What requirements are there for the restriction of the degree of freedom of the workpiece?

4.2.1 Location principles

1. Six-point location principle

A workpiece can be considered as a free body. Any free body in space has six degrees of freedom (3 linear and 3 rotary) as shown in Fig. 4-2. The three linear degrees of freedom along the x-axis, y-axis, z-axis are expressed by $\vec{x}, \vec{y}, \vec{z}$ respectively; The three rotary degrees of freedom are expressed by $\widehat{x}, \widehat{y}, \widehat{z}$ respectively. Whenever the location is arranged, it is necessary to constrain all these six degrees of freedom to ensure the mechanical stability of the workpiece in the fixture.

For the sake of discussion, the term "locator- location and support point" is introduced. The most common method is 3-2-1, or the six-point location principle: Six locators constrain six degrees of freedom of the workpiece, thus making the workpiece have a definite position in space only. It happened that a locator arrested a degree of freedom.

For the workpiece shown in Fig. 4-2, if the six locators are laid as shown in Fig. 4-3, the three surfaces contacts respectively with six locators, the six degrees of freedom of the workpiece are all arrested.

2. Requirements of the machining accuracy for restricted degrees of freedom

Is it always necessary to restrict all the six degrees of freedom of a workpiece during machining? How many degrees of freedom should be restricted on earth in machining a workpiece? The answer depends on the machining requirement. A plane of the workpiece is to be machined by milling and the dimension A is required, as shown in Fig. 4-4 a. To locate a plane, only three locators are required to restrict three degrees of freedom ($\widehat{x}, \widehat{y}, \vec{z}$). Fig. 4-4b shows that a step surface on a workpiece is to be machined. To ensure the dimension A and B, five degrees of freedom ($\widehat{x}, \widehat{y}, \vec{z}, \widehat{z}, \vec{x}$) must be restricted. When milling the blind slot shown in Fig. 4-4c, all the six degrees of freedom ($\vec{x}, \vec{y}, \vec{z}, \widehat{x}, \widehat{y}, \widehat{z}$) must be restricted.

There are two kinds of degrees of freedom which are required to restrict. The first is the one which has direct influence on the operation accuracy, and must be restricted. The second is the one which should also be constrained to balance the cutting force, clamping force and other disturbing forces. It should be pointed out that not all the six degrees of freedom must be restricted. The first kind of degree of freedom must be restricted, but whether the second kind of degree of freedom is re-

4.2 工件在夹具中的定位

工件在夹具中定位的目的是使同一工序中的一批工件都能在夹具中相对于机床和刀具占据正确的位置。要解决工件在夹具中的定位问题，必须首先搞清楚下列几个问题：①工件在空间有几个自由度？②如何限制这些自由度？③工件的工序加工精度与自由度限制有什么关系？④对工件自由度的限制有什么要求？

4.2.1 工件定位原理

1. 六点定位原理

工件可以视为自由刚体。任何一个自由刚体在空间都有六个自由度（三个直线移动和三个转动），如图4-2所示。三个沿坐标轴 x、y、z 方向移动的自由度分别用 $\vec{x}$、$\vec{y}$、$\vec{z}$ 表示；三个绕坐标轴 x、y、z 转动的自由度分别用 $\hat{x}$、$\hat{y}$、$\hat{z}$ 表示。每当布置定位方案时，有必要限制工件的六个自由度，以保证工件在夹具中的力学稳定性。

为便于讨论，引出了"定位支承点"的概念。最常用的方法是3—2—1，即六点定位原理：用六个定位支承点约束工件的六个自由度，使工件在空间有一个唯一确定的位置，刚好是一个定位支承点约束一个自由度。

对于图4-2所示的工件，如果按图4-3所示那样布置六个支承点，工件的三个面分别与六个支承点保持接触，于是工件的六个自由度就都被限制了。

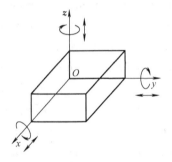

Fig. 4-2 Degrees of freedom of free body 自由刚体的自由度

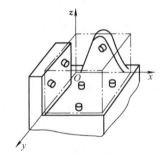

Fig. 4-3 Six-point location 六点定位

2. 加工要求决定必须限制的自由度

难道在加工时工件的6个自由度总是都要限制吗？到底工件在加工时该限制几个自由度？答案是视具体加工要求而定的。欲在图4-4a所示零件上铣削加工一平面，保证尺寸 A。要定位一个平面，只需三个定位支承点就可以限制3个自由度（$\hat{x}$，$\hat{y}$，$\vec{z}$）。如果在图4-4b所示零件上加工台阶面，要保证尺寸 A、B，必须限制5个自由度（$\hat{x}$，$\hat{y}$，$\vec{z}$，$\vec{z}$，$\vec{x}$）。当铣削图4-4c所示零件上的盲槽时，6个自由度（$\vec{x}$，$\vec{y}$，$\vec{z}$，$\hat{x}$，$\hat{y}$，$\hat{z}$）全都要限制。

需要限制的自由度有两类：第一类自由度是指影响工件的工序加工精度要求必须要限制的自由度；第二类自由度是指为了抵消切削力、夹紧力以及其他干扰力需要限制的自由度。需要指出的是，并非工件的六个自由度都要加以限制。第一类自由度必须要限制，至于第二类自由度是否需要限制，要视具体的加工情况而定。

stricted or not depends on the specific machining conditions.

3. The number of the degree of freedom restricted by various location elements

In the design of the fixture, the location and support point is embodied by the specific location element. How many location points a location element is equivalent to is determined by the working manner of the location element and its contact range to the workpiece. Three supporting points under the bottom of the workpiece in Fig. 4-3 may be a large plate, or two long and narrow plates, or just three support tacks in the practical fixture. Therefore, when 6-point location principle is used to analyze the location of the workpiece, the judgement should be made based on the degrees of freedom restricted practically by the location element. When a smaller support plane contacts to a large-size workpiece, it is equivalent to a support point, and can constrain a degree of freedom only. If a long and narrow plate contacts to a workpiece in a larger area in a certain direction, it is equivalent to two supporting points, and can constrain two degrees of freedom. A straight pin which contacts to a cylindrical hole in shorter length is equivalent to two supporting points; a straight pin which contacts to a cylindrical hole in longer length is equivalent to four supporting points. Similarly, there are short V-block and long V-block. The former can constrain two degrees of freedom, and the latter can constrain four degrees of freedom.

4. Several location patterns

(1) Complete location   That all six degrees of freedom are restricted is named as complete location, as shown in Fig. 4-5.

(2) Incomplete location   According to the machining requirement for workpiece, it is unnecessary to restrict all six degrees of freedom. In other word, the machining requirement for a workpiece still can be ensured even if all six degrees of freedom are not restricted. When milling the through slot in the workpiece, as shown in Fig. 4-6, it is enough to restrict five degrees of freedom.

(3) Lack of location   The degree of freedom of the workpiece which must be restricted is not restricted. This location pattern is called lack of location. As shown in Fig. 4-7a, only a large plate is used to restrict the three degrees of freedom of workpiece ($\widehat{x}, \widehat{y}, \vec{z}$) when machining the step surface. Clearly, this pattern can't ensure the machining accuracy of dimension B. In order to ensure the machining accuracy of both A and B, a long and narrow plate is added to contact with the right side of workpiece, as shown in Fig. 4-7b. At the moment, five degrees of freedom of workpiece ($\widehat{x}, \widehat{y}, \vec{z}, \widehat{z}, \vec{x}$) are restricted. This belongs to incomplete location.

(4) Redundant location   A degree of freedom is repeatedly restricted by two or more locators. This kind of location is called the redundant location. Generally, the redundant location would affect the location accuracy of the workpiece, cause uncertainty of location, or make the workpiece/locator produce deformation. Fig. 4-8a shows a location situation of a bush. The fit of the hole with long pin can restrict four degrees of freedom of the workpiece ($\vec{x}, \vec{z}, \widehat{x}, \widehat{z}$), and the large shoulder of the pin can restrict three degrees of freedom of the workpiece ($\widehat{x}, \widehat{z}, \vec{y}$). Clearly, $\widehat{x}, \widehat{z}$ are restricted repeatedly by two locators. If there is a perpendicularity error whether between the central line of the hole and its end face or between the central line of the pin and its shoulder, once a clamping force is applied towards the shoulder of long pin, it would cause either pin or bush distorsion, and finally af-

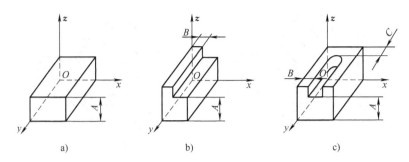

Fig. 4-4　Degrees of freedom constrained by machining requirements　加工要求限制的自由度数

3. 各种定位元件所限制自由度数

在夹具设计中，定位支承点是由具体的定位元件来体现的。一个定位元件究竟相当于几个支承点，要视定位元件的具体工作方式及其与工件接触范围的大小而论。图 4-3 中位于底面的三个支承点在实际的夹具结构中可能是一个大的平板，或是两块狭长平板，或是三个支承钉。因此在运用六点定位原理分析工件的定位问题时，必须从定位元件实际上能够限制的自由度数来分析判断。当一个较小的支承平面与尺寸较大的工件相接触时，只相当于一支承点，因此只能限制一个自由度。一个窄长平面支承在某一方向上并与工件有较大范围的接触，就相当于两个支承点，可以限制两个自由度。一个与工件内孔的轴向接触长度较短的圆柱定位销相当于两个支承点；而一个与工件内孔的轴向接触长度较长的圆柱销则相当于四个支承点。同理，V 形块也有长短之分。固定的短 V 形块限制两个自由度，而长 V 形块则能限制四个自由度。

4. 几种定位方式

（1）完全定位　工件的六个自由度全被限制的定位方式称为完全定位，如图 4-5 所示。

（2）不完全定位　根据工件加工精度要求不需限制的自由度没有被夹具定位元件限制或没有被全部限制的定位。这种定位虽然没有完全限制工件的六个自由度，但仍能保证工件的加工精度要求。当铣削图 4-6 所示工件上的通槽时，只要限制 5 个自由度即可。

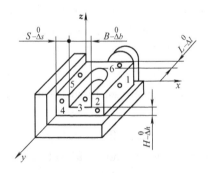

Fig. 4-5　Complete location 完全定位

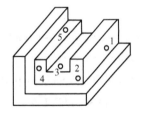

Fig. 4-6　Incomplete location 不完全定位

（3）欠定位　根据工件加工精度要求需要限制的自由度而未加限制的定位称为欠定位。如图 4-7a 所示，当加工台阶面时仅用一块大平板限制工件的三个自由度（$\vec{x}$，$\vec{y}$，$\vec{z}$）。这种定位显然不能保证尺寸 B 的加工精度。为了保证尺寸 A 和 B 的加工精度，在工件的右侧面放置一块狭长平板，如图 4-7b 所示。此时，工件的五个自由度（$\vec{x}$，$\vec{y}$，$\vec{z}$，$\hat{z}$，$\hat{x}$）被限

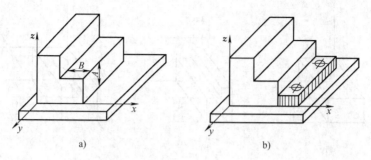

Fig. 4-7 Lack of location and its improvement measure 欠定位及其改进措施

fect the machining accuracy of the workpiece. Therefore, this kind of location pattern should be avoided as far as possible. It is suggested that the location pattern shown in Fig. 4-8b or c should be used.

On the other hand, the redundant location is helpful for improving the location rigidity and stability of the workpiece. Therefore, it is very necessary to apply redundant location to the workpieces with low rigidity.

4.2.2 Locators

1. Locators used when the location surface is a plane

When the plane of a workpiece is used as location surface, the used locators include stationary support (support tack and support plate), adjustable support and self-alignment support. These supports play a role of location and are named as primary supports. There are still other supports, called auxiliary supports, which don't play the role of location.

(1) Primary supports

1) Support tack. The common used structures of support tacks are shown in Fig. 4-9. The support tack with flat head shown in Fig. 4-9a is suitable for finish datum surface. The support tack with round head shown in Fig. 4-9b is suitable for rough datum surface. The support tack with serration pattern shown in Fig. 4-9c can produce larger friction force. But the cutting chips are not easy to clean out, and it may scratch the datum surface of the workpiece. Therefore, it is often used for supporting the rough surface on the side of the workpiece. As the support tack with bush as shown in Fig. 4-9d is convenient for disassembling and replacing, it is often used in such situations as large volume production, easy wear and constant repairs.

2) Support plate. A support plate is often used to locate a larger finish surface. The commonly used structures of A support plate are shown in Fig. 4-10. The support plate with a smooth plane shown in Fig. 4-10a has a simple and compact structure, but the chips in counterbore are not easy to clean out. It is suitable for support the side or top surface of the workpiece. The support plate with slanting slots (Fig. 4-10b) is often used to support the bottom surface of the workpiece. When the location surface of the workpiece is very large, two long and narrow plates are used to compose a large support surface.

3) Adjustable supports. The support whose support point can be adjusted is called adjustable

制。这就属于不完全定位。

（4）过定位　工件的同一自由度被两个或两个以上的定位支承点重复限制的定位方式，称为过定位。过定位会影响工件的定位精度，导致定位不确定性，甚至导致工件或定位元件产生变形。图4-8a为一套筒的定位情况。孔与长销配合可限制工件的四个自由度（$\vec{x}$, $\vec{z}$, $\hat{x}$, $\hat{z}$），而销的大端面限制工件三个自由度（$\hat{x}$, $\hat{z}$, $\vec{y}$）。显然，$\hat{x}$, $\hat{z}$ 两个自由度被两个定位元件重复限制了。如果工件孔与其端面间、长销与其台肩面间存在垂直度误差，则在轴向夹紧力作用下，将导致定位销或工件产生变形，最终影响工件的加工精度。因此在定位设计中应该尽量避免过定位。建议采用图4-8b或c所示的定位方案。

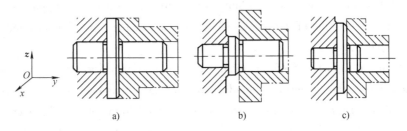

Fig. 4-8　Redundant location and its improvement measures　过定位及其改进措施

但另一方面，过定位可以提高工件的局部刚度和工件定位的稳定性，所以当加工刚性差的工件时，过定位又是非常必要的。

4.2.2　定位元件

1. 平面定位所用定位元件

以平面作定位基准所用的定位元件主要包括固定支承（支承钉、支承板）、可调支承和自位支承。这些支承起定位作用，因此称为基本支承。另外还有不起定位作用的辅助支承。

（1）基本支承

1）支承钉。常用支承钉的结构形式如图4-9所示。平头支承钉（图4-9a）用于支承精基准面；球头支承钉（图4-9b）用于支承粗基准面；网纹顶面支承钉（图4-9c）能产生较大的摩擦力，但网槽中的切屑不易清除，常用在工件以粗基准定位且要求产生较大摩擦力的侧面定位场合。带衬套支承钉（图4-9d）由于便于拆卸和更换，一般用于批量大、磨损快、需要经常修理的场合。

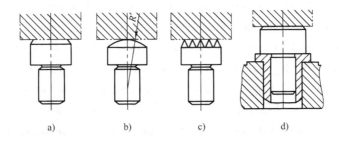

Fig. 4-9　Support tacks　支承钉

2）支承板。一般较大的精基准平面定位多用支承板作为定位元件。常用的支承板结构

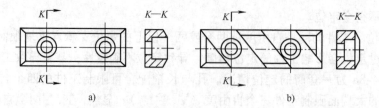

Fig. 4-10 Support plates 支承板

support. Its structures are shown in Fig. 4-11. Adjustable supports are often used to support the rough surface. When the dimensions of a blank vary greatly, the adjustable support is required to adapt to the dimension variation of the blank. Generally, adjustable support is adjusted once every a batch of the workpiece.

4) Self-alignment supports. Self-alignment support is also named as floating support. In the location of the workpiece, the support can be adjusted automatically by itself so as to adapt to the positional variation of the datum in the workpiece. The structural forms of self-alignment supports are shown in Fig. 4-12, where the structure shown in Fig. 4-12a is used to support the plane with angular error; the structure shown in Fig. 4-12b is used to support the blank surface or intermittent surface; and the structure shown in Fig. 4-12c is used to support the step surfaces. As the self-alignment supports are movable or floating, no matter whether it is two points or three points contacting with the workpiece, self-alignment support is only equivalent to one support point, and hence just restricts one degree of freedom.

(2) Auxiliary supports Usually, a auxiliary support doesn't support a workpiece until the workpiece has been located. Therefore, the auxiliary support doesn't play the role of location. It is used primarily to increase the stiffness and stability of the workpiece. Fig. 4-13 shows several auxiliary supports. All auxiliary supports have to be loosed whenever the workpiece is unloaded, and are adjusted to the support the surface and locked again whenever the workpiece is loaded.

2. Locators used when the location surface is an external cylindrical surface

When such workpieces as shafts, sleeves, circular discs are machined, the external cylindrical surface of them is often taken as the location surface. The commonly used locators are V-blocks, location bushes and so on.

(1) V-block With good centering feature, a V-block is one of the most widely used locators. Several V-block structures are shown in Fig. 4-14. The integral long V-block (Fig. 4-14a) is used for the location of long shaft with a finish datum. The other long V-blocks composed two short V-block (Fig. 4-14b) are used for the location of a long shaft with a rough datum or of a step shaft. When two finish datums in a long shaft have a larger distance, the support structure shown in Fig. 4-14c is used. The typical structure of a V-block is shown in Fig. 4-15. The included angle between two inclined surfaces on V-block may be 60°, 90°, or 120°, where the V-block with 90° has been standardized.

(2) Location bushes Several structures of location bushes are shown in Fig. 4-16. Fig. 4-16a and Fig. 4-16b show the long and short bushes separately. Fig. 4-16c and Fig. 4-16d show the struc-

形式如图 4-10 所示。平面型支承板（图 4-10a）结构简单，但沉头螺钉处清理切屑比较困难，适于作侧面和顶面定位；带斜槽型支承板（图 4-10b），在带有螺钉孔的斜槽中允许容纳少许切屑，适于作底面定位。当工件定位平面较大时，可用两块支承板组合成一个平面。

3）可调支承。支承点位置可以调整的支承称为可调支承。常用可调支承结构形式如图 4-11 所示。可调支承多用于未加工过的平面定位，支承高度可以根据需要进行调整，以补偿各批毛坯尺寸误差。一般每批工件（毛坯）调整一次。

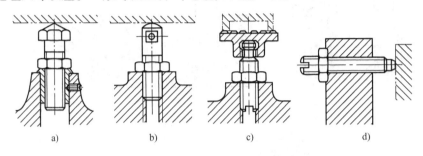

Fig. 4-11　Adjustable supports 可调支承

4）自位支承。自位支承又称浮动支承，在定位过程中，支承本身可以随工件定位基准面的变化而自动调整并与之相适应。常用自位支承的结构形式如图 4-12 所示。其中图 4-12a 所示结构用于有基准角度误差的平面定位；图 4-12b 所示结构用于毛坯平面或断续表面；图 4-12c 所示结构用于阶梯表面。由于自位支承是活动或浮动的，无论结构上是两点或三点支承，其实质只起一个支承点的作用，所以自位支承只限制一个自由度。

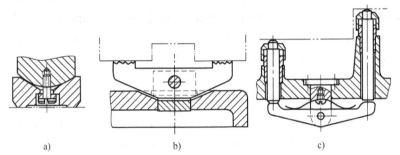

Fig. 4-12　Self-alignment supports 自位支承

（2）辅助支承　辅助支承是在工件定位后参与支承的元件，它不起定位作用，不能限制工件的自由度，它只用以增加工件在加工过程中的刚性。图 4-13 所示为几种辅助支承结构。各种辅助支承在每次卸下工件后必须松开，装上工件后再调整到支承表面并锁紧。

2. 外圆柱面定位所用定位元件

轴、套、盘类零件加工时，常以外圆表面定位。常用的定位元件有 V 形块、定位套等。

（1）V 形块　V 形块具有良好的对中性，是最常用的外圆定位元件。图 4-14 所示为几种 V 形块结构。整体长 V 形块（图 4-14a）用于较长的精基面定位，另一种长 V 形块（图 4-14b）用于较长的粗基面（或阶梯轴）定位。图 4-14c 所示结构则用于两段精基面相距较远的场合。典型的 V 形块结构如图 4-15 所示。V 形块两斜面之间的夹角一般取 60°、90°或 120°，其中 90°夹角的 V 形块已经标准化。

tures with a semi-circle bush. The lower part 1 is fixed to the fixture and plays a role of location. The upper part 2 is assembled on the cover plate with a hinge and used to clamp the shaft. The structures with a semi-circle bush are often used for locating the workpiece which can't be located by complete circle, such as crankshaft.

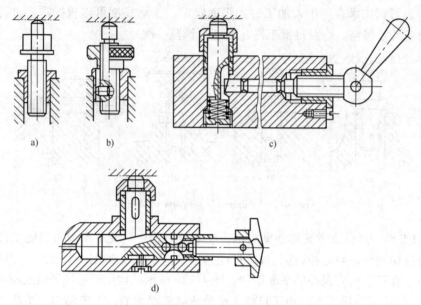

Fig. 4-13　Auxiliary supports 辅助支承

3. Locators used when the location surface is the internal cylindrical surface

Taking the internal cylindrical surface of a workpiece, i. e. hole, as a location datum, the location hole and locator are usually in the fit situation. The commonly used locators are location pins, mandrels and so on.

(1) Cylindrical location pins　Several location pins are shown in Fig. 4-17. Location pins are mainly used for the location of the small hole in the workpiece. Its working diameter d is generally not larger than 50 mm, and machined in light of assembly requirement, such as g5, g6, f6 or f7 etc. Location pins shown in Fig. 4-17a, Fig. 4-17b and Fig. 4-17c are directly fitted into the body of the fixture. Fig. 4-17d shows a replaceable location pin with a bush. For the rhombus pin shown in Fig. 4-17e, it also has four installation forms similar to Fig. 4-17a, Fig. 4-17b, Fig. 4-17c and Fig. 4-17d.

(2) Conical location pins　As shown in Fig. 4-18, conical location pins are often used as the location elements of sleeves, hollow shafts and so on. The conical pin shown in Fig. 4-18a is applied for the rough datum, and Fig. 4-18b is applied for the finish datum. As it is easy to tilt for the workpiece to be located in a single conical pin, the combined location of a conical pin with other location elements is often used. As shown in Fig. 4-18c, the bottom surface is taken as the primary location datum of the workpiece, and the conical pin is inserted into the hole of the workpiece by the spring, thus preventing the axis of the workpiece from the tilt.

(3) Location mandrels　Location mandrels are mainly used as the location elements of a disk-

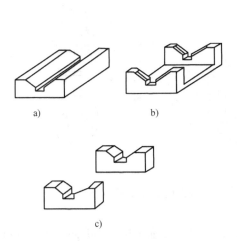
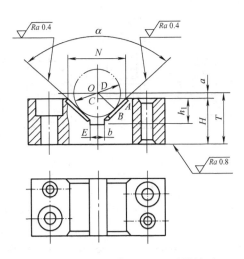

Fig. 4-14　Commonly used V-blocks　常用V形块

Fig. 4-15　Typical structure of V-block 典型的V形块结构

（2）定位套　图4-16所示为几种定位套结构。其中图4-16a、b所示结构为短定位套和长定位套；图4-16c、d所示结构为半圆定位套。这种定位套的下半部1固定在夹具上起定位作用，上半部2装在铰链盖板上，起夹紧作用。常用于曲轴等不宜以整圆定位的轴类零件的定位。

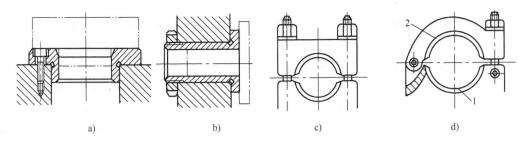

Fig. 4-16　Several structures of location bushes 几种定位套结构

3. 圆孔定位所用定位元件

工件以圆孔定位时，定位孔与定位元件之间处于配合状态。工件以圆孔定位常用定位元件有定位销、定位心轴等。

（1）圆柱定位销　图4-17所示为几种定位销结构。定位销主要用于零件上的小孔定位，一般直径不大于50 mm。其工作部分直径d通常根据加工和装配要求，按g5、g6、f6或f7制造。图4-17中的a、b、c为固定式定位销，图4-17d为带衬套的可换式定位销。图4-17e所示的菱形销也有上述四种结构。

（2）圆锥销　圆锥销常用于套筒、空心轴等零件的定位，如图4-18所示。其中图4-18a用于粗基准，图4-18b用于精基准。由于工件在单个圆锥销上定位容易倾斜，所以常和其他元件组合定位。如图4-18c所示，工件以底面为主要定位面，圆锥销依靠弹簧力插入定位孔中，起到较好的定心作用，从而避免了工件轴线倾斜。

（3）定位心轴　定位心轴主要用于盘类或套类零件的车、铣、磨及齿轮加工。常用的

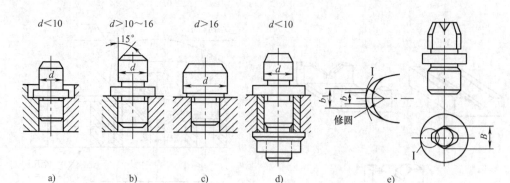

Fig. 4-17 Several structures of location pins 几种定位销结构

type or a sleeve-type workpiece to carry out the turning, milling, grinding and gear machining. Commonly used mandrels include cylindrical and conical mandrels, as shown in Fig. 4-19. Fig. 4-19a shows the mandrel with an interference fit. Fig. 4-19b shows the mandrel with a clearance fit. Fig. 4-19c shows a kind of the taper mandrel with the small conicity (1:5000 – 1:1000). The workpiece is centered and clamped by the conical surface of the taper mandrel. The mandrel with small conicity is generally used in the grinding or finish turning operation.

4.2.3 Analysis and calculation of location errors

According to the workpiece location principle, the location problem about a workpiece in a fixture can be solved by choosing proper locators. But for the location of a lot of workpieces, even though a single location pair is used, as there are differences in the location dimension of different workpieces or locators, the spacial geometric position occupied by different workpieces relative to the fixture is different when the locator is used to position a lot of workpieces. The variation of the workpiece position would lead to the variation of the operation dimension when using adjusting method to machine a lot of workpieces. How does the location error arise? How to calculate the location error? How to assess the rationality of a location scheme based on the location error? All the issues will be discussed in this section.

1. Location error and its causes

When the adjusting method is used to machine a batch of workpieces, the position of the cutter should be set well in advance. That means the position of the surface to be machined in relation to the cutter adjusting datum has been determined before machining. Location error is the maximum positional variation of the operation datum relative to the cutter adjusting datum in the operation dimension direction. In order to control the location error effectively and farthest reduce the effect of the location error on machining accuracy, we must be clear about what would cause the location error. Therefore, the causes for generating location error are analyzed as follows:

(1) Datum non-superposing error Δ_B When using the adjusting method is used to machine a batch of workpieces, if the operation datum is not selected as the location datum, the maximum variation of the operational datum relative to the location datum in the operation dimensional direction is

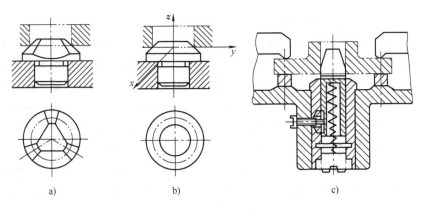

Fig. 4-18　Conical location pins 圆锥定位销

有圆柱心轴和圆锥心轴等，如图 4-19 所示。其中图 4-19a 为过盈配合心轴，图 4-19b 为间隙配合心轴；图 4-19c 为小锥度心轴（1:5000 ~ 1:1000），装夹工件时依靠心轴锥面使工件对中和涨紧。小锥度心轴一般用于磨削或精车。

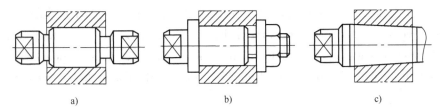

Fig. 4-19　Location mandrels 定位心轴

4.2.3　定位误差的分析与计算

根据工件定位原理，工件在夹具中的定位问题可以通过选择合适的定位元件来解决。但对大批工件定位而言，即使单一定位副，由于不同工件的定位尺寸或不同的定位元件尺寸存在差异，因此用定位元件对一批工件定位时，不同的工件相对于夹具所占有的空间几何位置是不一样的。这种位置的变化就导致了调整法加工工件时工序尺寸和位置精度的变化。定位误差如何产生、如何计算、如何用定位误差评定定位方案的合理性是本节要重点解决的问题。

1. 定位误差及其产生

当用调整法加工一批工件时，刀具的位置应当事先调整好。这就意味着待加工表面相对于调刀基准的位置在加工前就已调好。定位误差是指工序基准相对于调刀基准在工序尺寸方向上的最大变化量。为了有效地控制定位误差和进一步减小定位误差对加工精度的影响，必须搞清楚到底是什么原因引起了定位误差。为此，对定位误差产生的原因进行分析。

（1）基准不重合误差 Δ_B　当调整法加工一批工件时，如果没有把工序基准作为定位基准，工序基准相对于定位基准在工序尺寸方向上产生的最大位置变化量称为基准不重合误差，用符号"Δ_B"表示。

欲在一工件上加工阶梯面。工序尺寸，即加工要求为 $A_{-\delta_a}^{0}$。已知在加工该阶梯面之前其他表面均已加工好。本道工序的定位方案如图 4-20 所示。刀具以支承钉的工作面为调刀基

called the datum non-superposing error, represented by Δ_B.

Suppose a step surface of a workpiece is to be machined. The operation dimension, i.e. machining requirement, is $A_{-\delta_a}^{0}$. Given that all the surfaces of the workpiece have been machined before machining the step surface. The location plan of this operation is shown in Fig. 4-20. Taking the working surface of the support tack 3 as the tool setting datum, the tool setting dimension T is set well. As the operation datum is the left side D, and not the location datum E, the dimension error $\pm \delta_c$ produced in last operation would cause the positional variation of the operational datum D. The positional variation of the surface D will cause the operation dimension A to produce machining error with $2\delta_c$. The location error is caused just by the datum non-superposing error Δ_B. That is:

$$\Delta_{dw} = \Delta_B = 2\delta_c$$

It can be seen that the magnitude of datum non-superposing error equals to the tolerance of the distance from the operational datum to the location datum. Clearly, the Δ_B is caused by the improper selection of the operational datum. Therefore, it can be eliminated by marking the operation dimension correctly.

(2) Datum displacement error Δ_Y When using the adjusting method is used to machine a batch of workpieces, the manufacturing error of the location pairs would cause the positional variation of the location datum relative to the tool setting datum. The maximum positional variation of the location datum in the operation dimensional direction is called the datum displacement error, represented by Δ_Y.

A sleeve-type workpiece is located on the cylindrical pin to machine a slot, as shown in Fig. 4-21. The dimensions of both the hole and the pin and machining requirements for the slot are marked in Fig. 4-21. Because of the inevitable manufacturing error of the hole and the pin, the position of the location hole (location datum) would move relative to the axis of the location pin (tool setting datum). If the location pin is placed vertically, the generatrix of the hole would contact with that of the pin at any point. Then, the moving range of the location datum (the center of the location hole) is a circle, and its diameter is the maximum possible displacement of the location datum, named as the datum displacement error Δ_Y.

$$\Delta_Y = X_{max} = (D + \delta_D) - (d - \delta_d) = \delta_D + \delta_d + X_{min}$$

where, X_{max}—Maximum fit clearance between the hole and the pin;

X_{min}—Minimum fit clearance between the hole and the pin.

Clearly, when Δ_Y occurred along the direction of the operation dimension H, it would cause H to produce the machining error with X_{max}. As the location datum is just the operational datum, $\Delta_B = 0$. Then, the location error is

$$\Delta_{dw} = \Delta_Y = \delta_D + \delta_d + X_{min} \tag{4-1}$$

It is known from the above analysis that both the datum non-superposition and the datum displacement are the cause of the location error. Both the datum non-superposition and the datum displacement can lead to the variation of the operational datum, and further cause the machining error. Therefore, it can be said that the location error is just caused by the variation of the operational datum.

准，以调刀尺寸 T 一次调整好刀具位置。由于工序基准为 D 面而不是定位基准 E，于是上道工序的尺寸误差 $\pm\delta_c$ 就会引起工序基准 D 的位置变化。D 面的位置变化会引起工序尺寸 A 产生 $2\delta_c$ 的加工误差。这一定位误差 Δ_{dw} 就是由基准不重合误差 Δ_B 引起的，即

$$\Delta_{dw} = \Delta_B = 2\delta_c$$

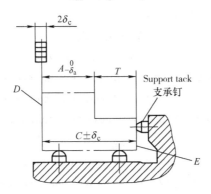

Fig. 4-20 Reason for datum non-superposing error 基准不重合误差的成因

由此可见，基准不重合误差的大小就等于工件上从工序基准到定位基准之间的距离公差。显然，基准不重合误差是由于工序基准选择不当引起的，可以通过不同的工序尺寸标注加以消除。

（2）基准位移误差 Δ_Y　调整法加工一批工件中，定位副的制造误差会引起定位基准相对于调刀基准的位置变动。定位基准在工序尺寸方向上的最大位置变动量称为基准位移误差，用符号"Δ_Y"表示。

图 4-21 所示为一套筒类零件在圆柱销上定位铣槽。图上已标出了孔和销的尺寸以及加工要求。由于孔和销（即定位副）的制造误差，孔的轴线（定位基准）就会相对于销的轴线（即调刀基准）发生移动。如果定位销垂直放置，则孔和销的母线可能在任意方向上接触。那么，定位基准（孔的中心）的移动范围是一个圆，圆的直径就是其可能产生的最大移动量，即基准位移误差 Δ_Y。

$$\Delta_Y = X_{max} = (D + \delta_D) - (d - \delta_d) = \delta_D + \delta_d + X_{min}$$

式中　X_{max}——定位销与定位孔的最大配合间隙；
　　　X_{min}——定位销与定位孔的最小配合间隙。

当 Δ_Y 在工序尺寸 H 方向上发生时，就导致工序尺寸 H 产生 X_{max} 的加工误差。由于定位基准也是工序基准，故 $\Delta_B = 0$。那么，定位误差为

$$\Delta_{dw} = \Delta_Y = \delta_D + \delta_d + X_{min} \tag{4-1}$$

从上述分析可知，基准不重合和基准位移是定位误差产生的原因。但基准不重合和基准位移均通过工序基准发生位置变动，进而使工序尺寸产生加工误差。因此可以说，定位误差产生的根本原因是工序基准的位置变化，即定位误差均是由于工序基准位置变化引起的。

由此可见，基准位移误差是由于定位副制造误差引起的，而基准不重合误差是由于工序基准选择不当产生的。因此，定位误差可以视为两项误差共同作用的结果。由于 Δ_Y 和 Δ_B 具有方向性，那么定位误差的一般计算公式应写成

It can be seen that the datum displacement error is caused by the manufacturing error of the location pairs, and the datum non-superposing error is caused by the improper selection of the operational datum. Therefore, the location error can be regarded as the consequence of the combined actions of two kinds of errors. Since the Δ_Y and Δ_B have the directional nature, the general formula of the location error can be expressed as follows:

$$\overrightarrow{\Delta_Y} + \overrightarrow{\Delta_B} = \overrightarrow{\Delta_{dw}} \tag{4-2}$$

$$\Delta_{dw} = \Delta_B \cos\alpha \pm \Delta_Y \cos\beta \tag{4-3}$$

where, α—angle included between Δ_B and the direction of the operation dimension;

β—angle included between Δ_Y and the direction of the operation dimension.

Using the Eq. (4-3) to calculate the location error is called the error resultant method.

【Discussion】

1) In the calculation of the Δ_{dw}, the first step is to find the Δ_B and Δ_Y respectively, and then to project Δ_B and Δ_Y towards the direction of the operation dimension respectively.

2) On the interrelation between Δ_B and Δ_Y. If both Δ_B and Δ_Y are caused by the same error factor, Δ_B and Δ_Y are thought to be interrelated; otherwise, Δ_B and Δ_Y are unrelated.

3) On the use of " + " or " − " in the formula.

①When Δ_B and Δ_Y are interrelated, take the " + " if the direction of $\Delta_B \cos\alpha$ is the same as that of $\Delta_Y \cos\beta$; take the " − " if the direction of $\Delta_B \cos\alpha$ is opposite to that of $\Delta_Y \cos\beta$.

② If Δ_B and Δ_Y are unrelated, take the " + " constantly.

In conclusion, the premise of producing the location error is adopting the adjusting method to machine a batch of workpieces. That is to say, only when adopting the adjusting method to machine a batch of workpieces can the location error theory be used to analyze and calculate the location error. The meaning of the tool setting datum should be clear in the calculation of the location error. The location datum is generally selected to be the tool setting datum.

When analyzing and calculating the location error, we must clearly understand that there is a one-to-one correspondence between the location error and the operation dimension (or positional accuracy), that is, a location error must be the one of an operation dimension (or positional accuracy), and an operation dimension (or positional accuracy) must have its location error.

2. Analysis and calculation of the location error on using the single surface to locate the workpiece

(1) Location error produced by taking the plane (of the workpiece) as the location surface　If a workpiece is located in a fixture by a plane only, the datum displacement error Δ_Y is caused only by the planeness at most. The influence of this error on operation dimension can be ignored. In this case, location error is caused only by the datum non-superposing error Δ_B, that is

$$\Delta_{dw} = \Delta_B \cos\alpha \tag{4-4}$$

(2) Location error produced by taking the hole (of the workpiece) as the location surface　A slot is to be cut in a sleeve to ensure the operation dimension H. Taking its hole as the location datum, the sleeve is fitted onto the mandrel/pin (of the fixture) with a clearance fit, as shown in Fig. 4-22.

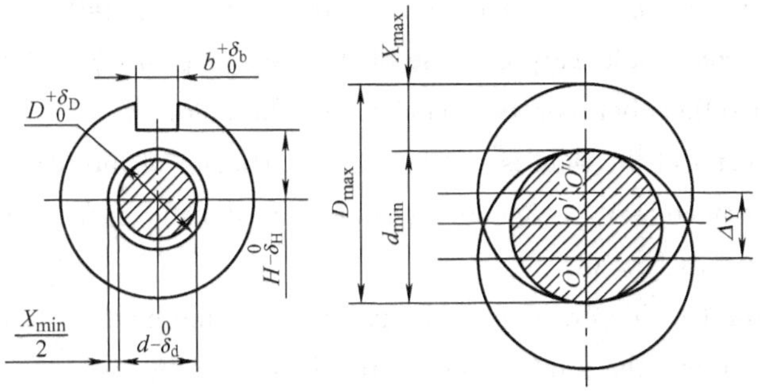

Fig. 4-21　Datum displacement error caused by the manufacturing error of the hole and the pin

孔和销的制造误差引起的基准位移误差

$$\vec{\Delta_Y} + \vec{\Delta_B} = \vec{\Delta_{dw}} \tag{4-2}$$

$$\Delta_{dw} = \Delta_B \cos\alpha \pm \Delta_Y \cos\beta \tag{4-3}$$

式中　α——基准不重合误差 Δ_B 方向与工序尺寸方向间的夹角；

β——基准位移误差 Δ_Y 方向与工序尺寸方向间的夹角。

利用式（4-3）计算定位误差称为误差合成法。

【讨论】

1）计算定位误差 Δ_{dw} 的第一步是分别求出 Δ_B 和 Δ_Y，接着各自向工序尺寸方向投影。

2）关于 Δ_B 和 Δ_Y 的相互关系。如果两者是由同一误差因素产生的，则认为 Δ_B 和 Δ_Y 关联；否则就认为不关联。

3）关于公式中"+"或"-"的使用。

① 当 Δ_B 和 Δ_Y 关联时：如果 $\Delta_B \cos\alpha$ 和 $\Delta_Y \cos\beta$ 方向相同，合成时取"+"号；如果 $\Delta_B \cos\alpha$ 和 $\Delta_Y \cos\beta$ 方向相反，合成时取"-"号。

② 当 Δ_B 和 Δ_Y 不关联时，可恒取"+"号。

综上所述，定位误差产生的前提是调整法加工一批工件。也就是说，只有采用调整法加工一批工件时，才可使用该定位误差理论分析计算。在定位误差计算中，应当清楚调刀基准的含义。调整法加工时，定位基准一般当作调刀基准使用。

在分析计算定位误差时必须清楚，定位误差与工序尺寸（或位置精度）是一一对应的关系，即某一个定位误差一定是某一个工序尺寸（或位置精度）的定位误差，某一个工序尺寸（或位置精度）一定有它自己的定位误差。

2. 工件采用单一面定位时的定位误差分析与计算

（1）用单一平面定位时产生的定位误差　当工件以单一平面定位时，基准位移误差只可能由平面度误差引起，这对工序加工而言，平面度误差的影响一般可以忽略不计。因此，单一平面定位时，定位误差只受基准不重合误差影响。即

$$\Delta_{dw} = \Delta_B \cos\alpha \tag{4-4}$$

（2）工件以孔定位时产生的定位误差　一套筒零件以内孔定位装在心轴上铣槽，保证工序尺寸 H，如图 4-22 所示。

As both the operation datum and the location datum all are the center line of the hole, $\Delta_B = 0$. Because both the hole and the mandrel have manufacturing errors, $\Delta_Y \neq 0$. This means the position of the center line of the hole would vary because of the clearance fit. How large is the Δ_Y? It depends on the contact conditions between the mandrel and the hole.

Given that the diameter of the hole is $(D + \delta_D)$, the diameter of the mandrel/pin is $(d - \delta_d)$, and the minimum fit clearance is Δ_{min}. According to the placed status of the mandrel/pin, two situations should be considered as follows.

1) Calculation of the location error when the generatrix of the hole contacts with that of the pin at any point (equivalent to that the pin is placed vertically) (see Fig. 4-21). The location error can be calculated according to the Eq. (4-1).

2) Calculation of the location error when the hole of the workpiece contacts with the pin in the fixed single side (equivalent to that the pin is placed horizontally) (see Fig. 4-22).

Fig. 4-22a shows the ideal location status between the hole and the pin. Because of the gravity of the workpiece, the location hole always contacts with the pin at the upper generatrix. Fig. 4-22b shows the minimum displacement status of the hole relative to the pin, and Fig. 4-22c shows the maximum displacement status of the hole relative to the pin. The maximum displacement of the centerline of the hole (relative to the centerline of the pin) in the vertical direction is:

$$\Delta_{dw} = \Delta_Y = (\delta_D + \delta_d + X_{min})/2 \quad (4-5)$$

As X_{min} is a constant error and can be eliminated by tool setting, the maximum positional variation of the central line of the hole in the vertical direction is:

$$\Delta_{dw} = \Delta_Y = (\delta_D + \delta_d)/2 \quad (4-6)$$

(3) Location error produced by taking the external cylindrical surface as the location surface

As shown in Fig. 4-23, given that the angle α has no manufacture error, and the diameter error for the external circle is δ_D, then the location datum (axis) of the workpiece always locates on the symmetrical surface of the V-block. Therefore, there is no datum displacement error in horizontal direction. But in the vertical direction, the diameter error δ_D would cause the variation of the location datum. The maximum positional variation of the location datum is:

$$\Delta_Y = \overline{OO_1} = \overline{OC} - \overline{O_1C} = \frac{\overline{OA}}{\sin(\alpha/2)} - \frac{\overline{O_1B}}{\sin(\alpha/2)} = \frac{\delta_D}{2\sin(\alpha/2)} \quad (4-7)$$

The Eq. (4-7) shows that the datum displacement error Δ_Y is reversely proportional to the included angle α of the V-block. The larger the α, the smaller the Δ_Y. When $\alpha = 180°$, $\Delta_Y = \delta_D/2$ (minimum value). But the symmetrical feature becomes worst because it is equivalent to the location of the plane. Therefore, the V-block with $\alpha = 90°$ is generally used for locating the workpiece.

[Example] A shaft with the diameter $(d - \delta_d)$ is located on the V-block with $\alpha = 90°$ to machine a keyseat, as shown in Fig. 4-24. The operation dimensions are marked respectively in H_1, H_2 and H_3. Find the location error to H_1, H_2 and H_3 under this location condition.

Solution:

(1) Calculation of the location error when the operation dimension is H_1 (Fig. 4-24a)

Clearly, $\Delta_B = 0$; and we know from the Eq. (4-7) that

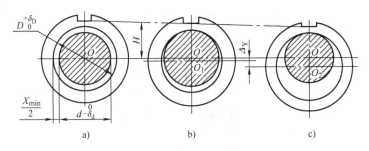

Fig. 4-22　Location error produced by using mandrel/pin as the locator
采用心轴/销定位时产生的定位误差

由于定位基准和工序基准都是孔的轴线，因此 $\Delta_B = 0$。由于工件孔和定位心轴都有制造误差，故 $\Delta_Y \neq 0$。这就意味着孔轴线的位置会由于间隙配合而变化。究竟 Δ_Y 有多大，要视孔与心轴的接触情况而定。

设孔的直径为 $(D + \delta_D)$，心轴或销的直径为 $(d - \delta_d)$，两者的最小配合间隙为 $\Delta_{\min}$，根据心轴的安放状态，需要考虑两种情况。

1）孔和销的母线以任意边接触时的定位误差计算（相当于定位销垂直放置，如图 4-21 所示）。此时定位误差可按式（4-1）计算。

2）孔和销的母线以固定单边接触时的定位误差计算（相当于定位销水平放置，如图 4-22 所示）。

图 4-22a 为理想定位状态。由于工件的重力作用，定位孔和销总在销的上母线处接触，孔轴线相对于销轴线将总是下移。图 4-22b 是孔轴线相对于销轴线的最小下移状态，而图 4-22c 是最大下移状态，孔轴线在垂直方向上的最大位置变动量为

$$\Delta_{dw} = \Delta_Y = (\delta_D + \delta_d + X_{\min})/2 \tag{4-5}$$

由于 $X_{\min}$ 是一个常值误差，可以通过调刀加以消除，因此，孔轴线在垂直方向上的最大位置变动量为

$$\Delta_{dw} = \Delta_Y = (\delta_D + \delta_d)/2 \tag{4-6}$$

（3）工件以外圆柱面定位时产生的定位误差　如图 4-23 所示，若不考虑 V 形块的制造误差，则工件定位基准（工件轴线）总是处于 V 形块的对称面上。因此，在水平方向上，工件定位基准不会产生基准位移误差。但在垂直方向上，工件定位直径尺寸的加工误差 δ_D 会引起工件定位基准产生位置变化。其可能产生的最大位置变化量为

$$\Delta_Y = \overline{OO_1} = \overline{OC} - \overline{O_1C} = \frac{\overline{OA}}{\sin(\alpha/2)} - \frac{\overline{O_1B}}{\sin(\alpha/2)} = \frac{\delta_D}{2\sin(\alpha/2)} \tag{4-7}$$

式（4-7）表明，基准位移误差 Δ_Y 与 V 形块夹角 α 成反比，即夹角 α 越大，Δ_Y 反而越小。当 $\alpha = 180°$ 时，$\Delta_Y = \delta_D/2$ 为最小，但 V 形块的对中作用也最差（无对中作用）。所以，一般多采用 $\alpha = 90°$ 的 V 形块定位。

【例】图 4-24 所示为一外圆直径为 $(d - \delta_d)$ 的轴类零件在夹角为 $\alpha = 90°$ 的 V 形块上定位铣键槽。求工序尺寸分别为 H_1、H_2、H_3 时的定位误差。

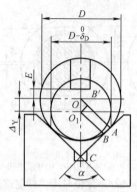

Fig. 4-23　Calculation of location error when using V-block as the locator

V 形块定位误差计算

$$\Delta_Y = \frac{\delta_d}{2\sin(\alpha/2)}$$

Therefore, the location error influencing on the operation dimension H_1 is:

$$\Delta_{dw1} = \Delta_Y = \frac{\delta_d}{2\sin(\alpha/2)}$$

(2) Calculation of the location error when the operation dimension is H_2 (Fig. 4-24 b)

It can be seen from Fig. 4-24 b that the operational datum is the upper generatrix of the shaft (point B), and the location datum is the center of the shaft. Then, $\Delta_B \neq 0$.

$$\Delta_B = \delta_d/2$$

$$\Delta_Y = \frac{\delta_d}{2\sin(\alpha/2)}$$

Because both Δ_B and Δ_Y include the same factor δ_d, i.e. caused by the manufacture error of the workpiece diameter, Δ_B and Δ_Y belong to the interrelated error. Therefore, it is asked to judge whether " + " or " - " is used when calculating the location error. It is found through analyzing the diameter variation from d_{max} to d_{min} that

$$\Delta_{dw2} = \overrightarrow{\Delta_Y} + \overrightarrow{\Delta_B} = \frac{\delta_d}{2\sin(\alpha/2)} + \frac{\delta_d}{2}$$

(3) Calculation of the location error when the operation dimension is H_3 (Fig. 4-24c)

It can be seen from Fig. 4-24c that the operational datum is the lower generatrix of the shaft (point C), and the location datum is the center of the shaft. Then, $\Delta_B \neq 0$.

$$\Delta_B = \delta_d/2$$

$$\Delta_Y = \frac{\delta_d}{2\sin(\alpha/2)}$$

Clearly, Δ_B and Δ_Y belong to the interrelated error. It is found through analyzing the variation of diameter from d_{max} to d_{min} that

$$\Delta_{dw3} = \overrightarrow{\Delta_Y} + \overrightarrow{\Delta_B} = \frac{\delta_d}{2\sin(\alpha/2)} - \frac{\delta_d}{2}$$

3. Location error analysis and calculation when using combined surfaces to locate the workpiece

Using a single surface to locate the workpiece in the fixture is a kind of simple location form. In

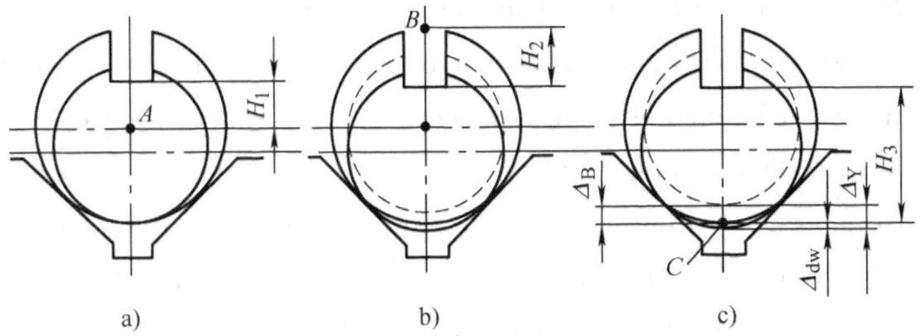

Fig. 4-24　Location error when a shaft is located on V-block

轴在 V 形块上定位时的定位误差

解　(1) 工序尺寸为 H_1 时的定位误差计算（图 4-24a）

显然，$\Delta_B = 0$；由式（4-7）可知

$$\Delta_Y = \frac{\delta_d}{2\sin(\alpha/2)}$$

故影响工序尺寸 H_1 的定位误差为

$$\Delta_{dw1} = \Delta_Y = \frac{\delta_d}{2\sin(\alpha/2)}$$

(2) 工序尺寸为 H_2 时的定位误差计算（图 4-24b）

由于工序基准在外圆的上母线 B 处，而定位基准仍是外圆的中心，两者不重合，故基准不重合误差 $\Delta_B \neq 0$。

$$\Delta_B = \delta_d/2$$

$$\Delta_Y = \frac{\delta_d}{2\sin(\alpha/2)}$$

由于 Δ_B 和 Δ_Y 均含有相同因子 δ_d，即都是由工件直径尺寸制造误差引起的，属于关联性误差，因此采用合成法计算定位误差时需要判断其正负。经过分析得

$$\Delta_{dw2} = \overrightarrow{\Delta_Y} + \overrightarrow{\Delta_B} = \frac{\delta_d}{2\sin(\alpha/2)} + \frac{\delta_d}{2}$$

(3) 工序尺寸为 H_3 时的定位误差计算（图 4-24c）

由于工序基准在外圆的下母线 C 处，与定位基准不重合，故基准不重合误差 $\Delta_B \neq 0$。

$$\Delta_B = \delta_d/2$$

$$\Delta_Y = \frac{\delta_d}{2\sin(\alpha/2)}$$

显然 Δ_B 和 Δ_Y 也属于关联性误差。经过分析得

$$\Delta_{dw3} = \overrightarrow{\Delta_Y} + \overrightarrow{\Delta_B} = \frac{\delta_d}{2\sin(\alpha/2)} - \frac{\delta_d}{2}$$

3. 组合面定位时的定位误差分析与计算

单一表面定位是工件在夹具中定位的一种简单形式，更多情况下需要工件上多个表面共同参与定位来限制工件的自由度。这种定位方式称为组合定位。一面两孔定位是最常用的组

the most cases, several surfaces are required to restrict the freedom degrees of the workpiece. This location is called the combined location. The location layout with a plane and two holes is one of the most often used combined locations. The location error caused by this kind of combined location is analyzed and calculated in this section.

When using two holes as location datums, the commonly used locators are: a cylindrical pin in one hole and a rhombus pin in the other hole, as shown in Fig. 4-25a. The methods of calculating the location error are different in different directions and different positions. There are several cases in calculating the location error.

(1) Datum displacement error $\Delta_{Y(x)}$ in x direction The location of the workpiece in x direction is realized by the left hole 1, and the right hole 2 does not play a role in location. Therefore, the maximum location error produced in x direction is the datum displacement error of the location hole 1 relative to the location pin 1, that is:

$$\Delta_{Y(x)} = \delta_{D1} + \delta_{d1} + X_{1min}$$

(2) Datum displacement error $\Delta_{Y(y)}$ in y direction The datum displacement error in y direction is influenced by two holes together. Its magnitude is different in different positions, and its calculation method is also different in different regions, as shown in Fig. 4-25b.

$\Delta_{Y(y)}$ is just the datum displacement error produced by the location pair composed of the single hole and pin at the center O_1 or O_2. In the region between O_1 and O_2, $\Delta_{Y(y)}$ should be the maximum displacement of two holes in the same direction, such as the datum displacement error $n'n''$ at n point in Fig. 4-25b. In the region outside of O_1 and O_2, $\Delta_{Y(y)}$ should be calculated based on the maximum swing angle error produced by the two holes, such as the datum displacement error $m'm''$ at m point in Fig. 4-25b.

(3) Swing angle error $\pm \Delta\theta$ As shown in Fig. 4-25c, the conditions producing maximum swing angle error are: the diameters of two holes in the workpiece are the largest, i. e. $(D_1 + \delta_{D1})$, $(D_2 + \delta_{D2})$; the diameters of two pines in the fixture are the least, i. e. $(d_1 - \delta_{d1})$、$(d_2 - \delta_{d2})$. As the influence of the distance error of both two holes or two pins on the swing angle error are very small, the basic distance between O_1 and O_2 is only used for the sake of the calculation.

In Fig. 4-25 c, O_1 is the center of the pin 1, and O_2 is the center of the pin 2. When the workpiece swings clockwise, the hole 1 move up to O_1', and the hole 2 move down to O_2', the swing angle error $\Delta\theta$ gets to the maxmum. value. According to the geometric relation shown in Fig. 4-25 c, there is

$$\tan\Delta\theta = \frac{\overline{O_2C}}{L} = \frac{\overline{O_2O_2'} + \overline{O_2'C}}{L} \tag{4-8}$$

where, $\overline{O_2O_2'} = \dfrac{\delta_{d2} + \delta_{D2} + X_{2min}}{2}, \overline{O_2'C} = \overline{O_1O_1'} = \dfrac{\delta_{d1} + \delta_{D1} + X_{1min}}{2}$, then

$$\tan\Delta\theta = \frac{\delta_{d1} + \delta_{D1} + X_{1min} + \delta_{d2} + \delta_{D2} + X_{2min}}{2L}$$

合定位方式之一。本节对这种定位方式的定位误差进行分析与计算。

双孔定位时常采用的定位元件是一个短圆柱销和一个短削边销,如图 4-25a 所示。在不同的方向和不同的位置,其定位误差的计算方法是不同的,定位误差计算有下列几种情况。

(1) x 轴方向上的基准位移误差 $\Delta_{Y(x)}$ 在 x 轴方向上的定位是由定位孔 1 实现的,定位孔 2 不起定位作用。因此,工件所能产生的最大定位误差是定位孔 1 相对于定位销 1 的基准位移误差,即

$$\Delta_{Y(x)} = \delta_{D1} + \delta_{d1} + X_{1min}$$

(2) y 轴方向上的基准位移误差 $\Delta_{Y(y)}$ 在 y 轴方向上,基准位移误差受双孔定位的共同影响,其大小随着位置的不同而不同,且在不同的区域内计算方法也有所不同。如图 4-25b 所示。

在中心 O_1 或 O_2 处,其 $\Delta_{Y(y)}$ 就等于该处单孔、销定位的基准位移误差;在 O_1 和 O_2 的中间区域,应按双孔同向最大位移计算 $\Delta_{Y(y)}$,如图 4-25b 中 n 处的基准位移误差为 $\overline{n'n''}$;在 O_1 和 O_2 的外侧区域,应按双孔的最大转角计算 $\Delta_{Y(y)}$,如图 4-25b 中 m 处的基准位移误差为 $\overline{m'm''}$。

(3) 转角误差 $\pm\Delta\theta$ 如图 4-25c 所示。最大转角发生的条件是:两孔直径最大($D_1 + \delta_{D1}$)、($D_2 + \delta_{D2}$);两销直径最小($d_1 - \delta_{d1}$)、($d_2 - \delta_{d2}$);两销中心距和两孔中心距应取最小相等值,由于其对转角误差影响不大,且考虑计算方便起见,两销中心距和两孔中心距一般取其基本尺寸。

图 4-25c 中 O_1 和 O_2 分别为两销中心。当两孔顺时针转动时,即孔 1 中心上移至 O_1',而孔 2 中心下移至 O_2' 时转角有最大值。根据图 4-25c 中的几何关系得

$$\tan\Delta\theta = \frac{\overline{O_2C}}{L} = \frac{\overline{O_2O_2'} + \overline{O_2'C}}{L} \tag{4-8}$$

其中,$\overline{O_2O_2'} = \frac{\delta_{d2} + \delta_{D2} + X_{2min}}{2}$,$\overline{O_2'C} = \overline{O_1O_1'} = \frac{\delta_{d1} + \delta_{D1} + X_{1min}}{2}$,则

$$\tan\Delta\theta = \frac{\delta_{d1} + \delta_{D1} + X_{1min} + \delta_{d2} + \delta_{D2} + X_{2min}}{2L}$$

$$\Delta\theta = \arctan\left(\frac{\delta_{d1} + \delta_{D1} + X_{1min} + \delta_{d2} + \delta_{D2} + X_{2min}}{2L}\right) \tag{4-9}$$

当两孔逆时针转动时,具有相同的 $\Delta\theta$ 误差,故总的转角误差应为 $\pm\Delta\theta$ 或 $2\Delta\theta$。即

$$2\Delta\theta = 2\arctan\left(\frac{\delta_{d1} + \delta_{D1} + X_{1min} + \delta_{d2} + \delta_{D2} + X_{2min}}{2L}\right) \tag{4-10}$$

4. 对夹具的定位精度要求

影响工件和刀具之间位置关系的因素很多,如工件在夹具中的定位误差 Δ_{dw},夹具在机床上安装时产生的安装误差 Δ_{fi},由对刀、引导元件引起的对刀、引导误差 Δ_{tg},以及其他因素引起的加工误差 Δ_{om}。为了保证工件的加工精度,上述所有误差的合成值不应超出工件的工序公差 T_w。即

$$\Delta_{dw} + (\Delta_{fi} + \Delta_{tg}) + \Delta_{om} \leq T_w \tag{4-11}$$

式 (4-11) 称为误差计算不等式。如果 Δ_{dw}、$(\Delta_{fi} + \Delta_{tg})$ 和 Δ_{om} 各占工序公差 T_w 的三分

$$\Delta\theta = \arctan\left(\frac{\delta_{d1} + \delta_{D1} + X_{1min} + \delta_{d2} + \delta_{D2} + X_{2min}}{2L}\right) \tag{4-9}$$

Similarly, the same $\Delta\theta$ can be obtained when the workpiece swings counter clockwise. Therefore, the total swing angle error is $\pm\Delta\theta$ or $2\Delta\theta$. That is:

$$2\Delta\theta = 2\arctan\left(\frac{\delta_{d1} + \delta_{D1} + X_{1min} + \delta_{d2} + \delta_{D2} + X_{2min}}{2L}\right) \tag{4-10}$$

4. Requirements for the location accuracy of the fixture

There are many factors which have an influence on the positional relation between the workpiece and the cutter, such as the location error Δ_{dw} of the workpiece in the fixture, installation error Δ_{fi} of the fixture on the machine tool, the tool setting and guiding error Δ_{tg} and other machining errors Δ_{om} cause by other factors. In order to ensure the machining accuracy of the workpiece, the sum of above errors would not larger than the operation tolerance T_w of the workpiece, i. e.

$$\Delta_{dw} + (\Delta_{fi} + \Delta_{tg}) + \Delta_{om} \leq T_w \tag{4-11}$$

The Eq. (4-11) is called error calculation equation. If each item, i. e. Δ_{dw}, $(\Delta_{fi} + \Delta_{tg})$ and Δ_{om}, takes up 1/3 of the operational tolerance T_w, the location error Δ_{dw} should satisfies the equation as follows:

$$\Delta_{dw} \leq \frac{1}{3}T_w \tag{4-12}$$

The Eq. (4-12) can be used to judge whether the location plan of the workpiece is rational.

4.3 Clamping of the workpiece

Generally, the workpiece being located should be fixed by applying force on it in the machining process so that the determinate position would not be changed by the cutting force, gravity, centrifugal force and other disturbing forces. This operation is called the clamping. The device used for clamping the workpiece is called the clamping device. The clamping device design is restricted by many factors, such as the location plan, magnitude of cutting force, productivity, machining method, rigidity of the workpiece, machining accuracy, and so on.

4.3.1 Elements and requirements of the clamping device

1. Elements of the clamping device

Clamping devices are classified into manual clamping devices and powered clamping devices in light of the power source used in the clamping device. A clamping device usually consists of three parts.

(1) Power source device Power source device is a necessary part of powered clamping device. Powered clamping devices use a hydraulic cylinder, or pneumatic cylinder, an electric set, a magnetic-actuated component to produce the clamping force. The pneumatic cylinder 1 in Fig. 4-26 is a power source device.

(2) Clamping element The clamping element is the one which contacts directly with the

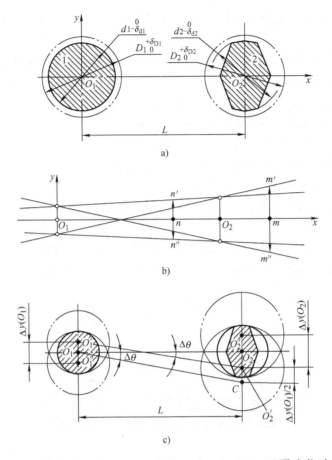

Fig. 4-25　Location error produced by using two holes as location datum　两孔定位时产生的定位误差
a）Using two holes to locate the workpiece　工件以两孔定位
b）Datum displacement error in y direction　y 方向的基准位移误差
c）Calculation of swing angle error　转角误差计算

之一，那么，定位误差 Δ_{dw} 应当满足以下不等式

$$\Delta_{dw} \leq \frac{1}{3} T_w \tag{4-12}$$

式（4-12）可用于判断所采用的工件定位方案是否合理。

4.3　工件的夹紧

通常工件定位后必须进行夹紧，才能保证工件在加工中不会因为切削力、重力、离心力等外力作用而破坏定位。这种操作称为夹紧。对工件进行夹紧的装置称为夹紧装置。夹紧装置的设计受到定位方案、切削力大小、生产率、加工方法、工件刚性、加工精度要求等因素的制约。

4.3.1　夹紧装置的组成和要求

1. 夹紧装置的组成

workpiece and used for holding down the workpiece. The pressing plate 4 in Fig. 4-26 is a clamping element.

(3) Intermediate force-transmission mechanism   The mechanism located in between the power source device and the clamping element is called the intermediate force-transmission mechanism. Its function is to transmit the power from the power source to the clamping element. The intermediate force-transmission mechanism ought to have three features:

1) Be able to change the direction of the clamping force;
2) Be able to change the magnitude of the clamping force;
3) Be provided with self-locking mechanism.

The wedge and other relevant elements belong to the intermediate force-transmission mechanism.

Different clamping devices would have different components. The block diagram shown in Fig. 4-27 illustrates the elements in manual and powered clamping devices.

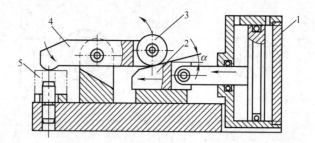

Fig. 4-26   Elements of clamping device 夹紧装置的组成
1—Pneumatic cylinder 气缸   2—Wedge 斜楔   3—Wheel 滚轮   4—Pressing plate 压板   5—Workpiece 工件

2. Requirements for clamping devices

1) Clamping action should be accurate and rapid.
2) Clamping operation should be convenient and labor-saving.
3) Clamping should be secure and reliable.
4) The clamping device should be simple in structure and easy to manufacture.

4.3.2   Principles of determining the clamping force

Whether the design of a clamping device is good or not depends on whether the design of the clamping force is reasonable to large extent. As a force, the clamping force also has three elements: magnitude, direction and action point. To determine the clamping force is just to determine the direction of the clamping force, the clamping position and magnitude of the clamping force. Because of the differences in the fixture structures and specific machining conditions, the first is to consider the total layout of the fixture, the next is to consider the requirements of the workpiece configuration, machining method, machining accuracy, cutting force, etc. for the clamping force.

1. Determination of the direction of the clamping force

The direction of the clamping force depends on the way to clamp the workpiece, the direction of

按照夹紧动力源的不同一般把夹紧机构划分为手动夹紧装置和机动夹紧装置。夹紧装置一般由三部分组成。

（1）力源装置　力源装置指产生夹紧力的装置，它是机动夹紧的必有装置，如气动、电动、液压、电磁等夹紧的动力装置。图4-26中的气缸1就是力源装置。

（2）夹紧元件　夹紧元件是指与工件直接接触并把工件夹紧的元件。图4-26中的压板4即为夹紧元件。

（3）中间递力机构　介于力源装置和夹紧元件之间的机构称为中间递力机构。它把力源产生的力传递给夹紧元件以实施对工件的夹紧。中间递力机构应具有三个特性：

1）能改变夹紧力的方向。
2）能改变夹紧力的大小。
3）具有自锁功能。

图4-26中的斜楔2及相关元件部分属于中间递力机构。不同的夹紧装置会有不同的构成。图4-27所示为机动和手动夹紧装置的不同构成。

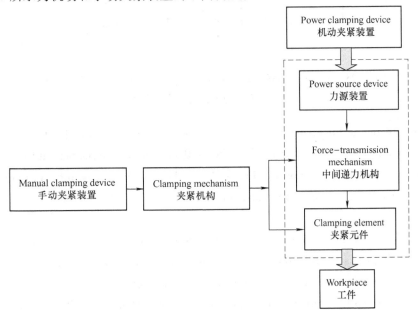

Fig. 4-27　Elements of clamping devices 夹紧装置的构成

2. 对夹紧装置的基本要求

1）夹紧动作要准确、迅速。
2）夹紧操作要方便、省力。
3）夹紧应安全可靠。
4）夹紧装置应结构简单，易于制造。

4.3.2　夹紧力的确定原则

夹紧装置设计的是否好，很大程度上取决于夹紧力的设计是否合理。夹紧力也有三要素：大小、方向和作用点。确定夹紧力就是要确定其大小、方向和作用点的位置。由于夹具

other force acting on the workpiece, the rigidity of the workpiece, and so on. Determination of the direction of the clamping force must adhere to the principles as follows.

1) The main clamping force should point and be perpendicular to the main location surface, making the workpiece location stable and reliable. As shown in Fig. 4-28, it is required that the centerline of the hole be perpendicular to the surface A in boring hole. Therefore, the direction of the clamping force should be perpendicular to the surface A. The clamping plan shown in Fig. 4-28a is correct, and Fig. 4-28b is incorrect.

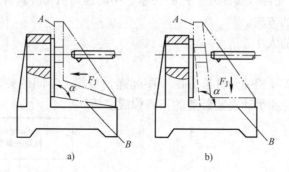

Fig. 4-28 Relation between the clamping direction and the location surface
夹紧力方向与工件定位面的关系

2) The direction of the clamping force should avoid directing towards the direction with low rigidity of the workpiece so as to eliminate the effect of the clamping deformation on the machining accuracy as much as possible. As shown in Fig. 4-29, as the thin sleeve has different rigidities in radius and axial directions, the clamping manner shown in Fig. 4-29b should be used instead of Fig. 4-29a.

3) The direction of the clamping force should do good to reduce the clamping force as far as possible. Fig. 4-30 shows the relations among the clamping force, gravity of the workpiece and cutting force. From the view of reducing the clamping force, it is clear that the clamping force required in Fig. 4-30a is the least. Therefore, the clamping plan shown in Fig. 4-30a is the best, and the next is shown in Fig. 4-30b.

2. Determination of the acting position of the clamping force

The determination of the acting position of the clamping force is involved in determining the position, number, distribution of the acting point. It has a direct influence on the stability of the location and deformation of the workpiece. The following principles should be followed in determining the acting position of the clamping force.

1) The clamping force should act against the locators or within the location area composed by several locators so as to avoid damaging the location of the workpiece. The arrangement of clamping elements in Fig. 4-31 is incorrect. The arrows shown in Fig. 4-31 show the correct acting position of the clamping force.

2) The clamping force should act on the part with good rigidity to reduce the clamping deform-

结构以及具体的加工情况不同,在确定夹紧力时首先要考虑夹具的整体布局问题,其次要考虑加工方法、加工精度、工件结构、切削力等方面对夹紧力的不同需求。

1. 夹紧力方向的确定

夹紧力作用方向取决于工件的夹紧方式、作用在工件上其他力的方向、工件的刚性等因素。确定夹紧力方向时必须遵循以下原则:

1) 主要夹紧力的作用方向应指向并垂直于工件的主要定位面,使工件定位稳定、可靠。图 4-28 所示镗孔时要求孔轴线与 A 面垂直,夹紧力方向应与 A 面垂直。故图 4-28a 方案正确,图 4-28b 方案不正确。

2) 夹紧力的作用方向应尽量避开工件刚性比较薄弱的方向,以减小工件的夹紧变形对加工精度的影响。如图 4-29 所示,由于套筒的径向刚度与轴向刚度不同,故应采用图 4-29b 的夹紧方式,而不要采用图 4-29a 的夹紧方式。

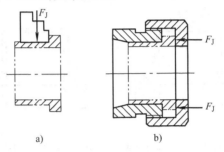

Fig. 4-29　Influence of clamping direction on workpiece deformation
夹紧力方向对工件变形的影响

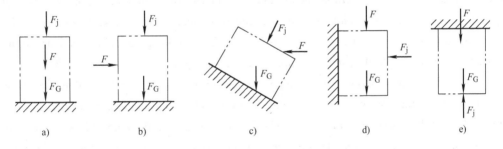

Fig. 4-30　Relation of the clamping force to the gravity and cutting force
夹紧力与切削力、工件重力的关系

3) 夹紧力的作用方向应尽可能有利于减小夹紧力。假设机械加工中工件只受夹紧力 F_j、切削力 F 和工件重力 F_G 的作用,这几种力的可能分布如图 4-30 所示。为保证工件加工中定位可靠,显然只有采用图 4-30a 受力分布时夹紧力 F_j 最小。

2. 夹紧力作用点的确定

夹紧力作用点的选择包括作用点的位置、数量、布局和作用方式。作用点的选择影响工件的定位稳定性和变形。确定夹紧力作用点时应遵循下列原则:

1) 夹紧力作用点应正对定位元件定位面或落在多个定位元件所组成的支承面内,以免破坏工件的定位。图 4-31 所示夹紧元件的布置是不正确的。图 4-31 中用箭头指出了夹紧力作用点的正确位置。

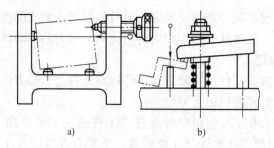

Fig. 4-31 Influence of clamping point on the workpiece location
夹紧力作用点对工件定位的影响

ation of the workpiece. F_j in Fig. 4-32a should be replaced by F_{j1} and F_{j2} in Fig. 4-32 b.

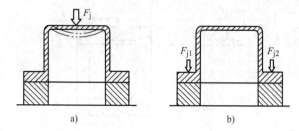

Fig. 4-32 Selection of the acting point of the clamping force 夹紧力作用点的选择

3) The acting point of clamping force should be near to the surface to be machined as far as possible so as to balance the cutting forces and to decrease the vibration caused by the cutting forces. An auxiliary support is added to the part with poor rigidity, if necessary. In Fig. 4-33, the auxiliary support should be near to the part to be machined as far as possible, and an additional clamping force is applied to the auxiliary support at the same time.

4) The proper contact form of the clamping point can reduce the clamping deformation of the workpiece, improve the contact reliability, and avoid the clamping element damaging the location and the surface of the workpiece. Fig. 4-34 a is suitable for clamping the unmachined surface. In order to prevent the thin sleeve from clamping deformation, three special jaws are used to increase the contact area, as shown in Fig. 4-34 b. Fig. 4-34 c adopts a special pressing block to increase the friction coefficient, which is suitable for clamping the machined surface.

3. Determination of the magnitude of the clamping force

The magnitude of the clamping force depends on that of the cutting force. Firstly, the cutting force F_c is calculated under the worst machining conditions based on the theoretical formula. Then, by establishing and solving the static equilibrium equation including the cutting force, gravity of the workpiece, and even inertial force and centrifugal force, the theoretical clamping force F_L can be found. Lastly, a safe coefficient K must be taken into account in determining the practical clamping force F_j for the sake of safty. Therefore, the practical clamping force should be

$$F_j = KF_L \tag{4-13}$$

Usually, $K = 1.5$-3. $K = 2.5$-3 for rough machining, and $K = 1.5$-2 for finish machining.

2）夹紧力作用点应作用在工件刚性较好的部位上，以尽量减小工件的夹紧变形。图 4-32a 中的 F_j 应被图 4-32b 中的 F_{j1} 和 F_{j2} 所取代。

3）夹紧力作用点应尽量靠近加工部位，以抵消切削力，减小切削力引起的振动。必要时应在工件刚性差的部位增加辅助支承。如图 4-33 所示，辅助支承应尽量靠近加工部位，同时给予附加夹紧力 F_{j2}。

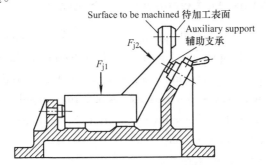

Fig. 4-33　Auxiliary support and additional clamping force 辅助支承和附加夹紧力

4）选择合适的夹紧力作用点的作用形式，可有效减小工件的夹紧变形、改善接触可靠性、提高摩擦因数、增大接触面积、防止夹紧元件破坏工件的定位和损伤工件表面等。图 4-34a 适合于对毛坯面夹紧；图 4-34b 的工件是薄壁套筒，为了减小夹紧变形，应增大夹压面积以使工件受力均匀；图 4-34c 采用特殊压块以增大摩擦因数，适用于对工件已加工面的夹紧。

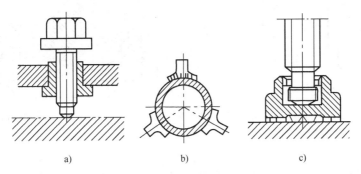

Fig. 4-34　Contact form of clamping point and workpiece 夹紧力作用点与工件的接触形式

3. 夹紧力大小的确定

夹紧力的大小取决于切削力的大小。首先，在最不利的加工条件下，根据理论公式计算切削力。然后通过建立和求解包含切削力、工件重力，甚至离心力在内的静态平衡方程得出理论夹紧力 F_L。最后，在确定工件加工所需要的实际夹紧力 F_j 时，为安全起见，还要考虑一个安全系数 K。因此，实际夹紧力为

$$F_j = KF_L \tag{4-13}$$

通常，$K = 1.5 \sim 3$。对于粗加工，取 $K = 2.5 \sim 3$；对于精加工，则取 $K = 1.5 \sim 2$。

实际上，由于切削力会随如刀具磨损、工件材料、加工余量等诸多因素变化而变化，上述方法很少使用，而常采用经验类比法估计夹紧力。对于在重要工序使用的夹具，其夹紧力大小通常用实验法确定。

In practice, above method is seldom used since the cutting force varies with several factors, such as the tool wear, workpiece materials and machining allowance, etc. Therefore, the experience-analogy method is often used to estimate the clamping force. For the important fixture which is used in the critical operation, the clamping force is usually determined by experiment.

4.3.3 Several commonly used clamping mechanisms

A clamping device is sometimes made up of a simple clamping mechanism, but in most cases, it is made up of a compound clamping mechanism. In the selection of the clamping mechanism, the requirements of machining methods, magnitude of the clamping force, workpiece structure, and productivity should be considered. The working characters of various simple clamping mechanisms have to be learnt firstly in the design of the clamping mechanism. Several commonly used clamping mechanisms are mainly introduced in this section.

1. Clamping device with a wedge

The operation principle of the clamping device with a wedge is shown in Fig. 4-35. Under the action of the original force F_Q from the power source, the wedge moves forward a distance L. The inclined plane of the wedge would produce a clamping journey S, thus clamping the workpiece. The application illustration of the wedge clamping mechanism is shown in Fig. 4-36.

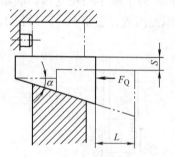

Fig. 4-35 Wedge clamping principle
斜楔夹紧原理

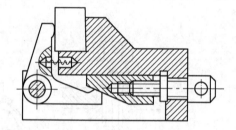

Fig. 4-36 Illustration of wedge clamping mechanism
斜楔夹紧机构实例

(1) Calculation of the clamping force generated by the wedge clamping device Based on Fig. 4-35, the forces acting on the wedge in clamping status are illustrated in Fig. 4-37.

When the wedge is in the equilibrium situation, according to the principle of the statical equilibrium, we know that

$$F_1 + F_{Rx} = F_Q$$
$$F_1 = F_W \tan\varphi_1$$
$$F_{Rx} = F_W \tan(\alpha + \varphi_2)$$

Then the clamping force F_W generated by the wedge can be found by solving above equations.

$$F_W = \frac{F_Q}{\tan\varphi_1 + \tan(\alpha + \varphi_2)} \tag{4-14}$$

where, F_Q—original driving force acting on the wedge;

F_W—counter force of the clamping force produced by the wedge;

4.3.3 几种常用的夹紧机构

夹紧装置有时由简单夹紧机构构成,而多数情况下则是由复合夹紧机构构成。夹紧机构的选择需要满足加工方法、工件所需夹紧力大小、工件结构、生产率等方面的要求。因此,在设计夹紧机构时,首先需要了解各种简单夹紧机构的工作特点。本节主要介绍几种常用的夹紧机构。

1. 斜楔夹紧机构

图 4-35 所示为斜楔夹紧机构的工作原理图。在夹紧源动力 F_Q 的作用下,斜楔向左移动 L 的位移,由于斜楔斜面的作用,将导致斜楔在垂直方向上产生 S 的夹紧行程,从而实现对工件的夹紧。图 4-36 所示为斜楔夹紧机构的应用实例简图。

(1) 斜楔夹紧机构所能产生的夹紧力计算 以图 4-35 为例,夹紧时斜楔的受力分析如图 4-37 所示。当斜楔处于平衡状态时,根据静力平衡可以得出

$$F_1 + F_{Rx} = F_Q$$
$$F_1 = F_W \tan\varphi_1$$
$$F_{Rx} = F_W \tan(\alpha + \varphi_2)$$

解上述方程组可得斜楔夹紧所能产生的夹紧力 F_W

$$F_W = \frac{F_Q}{\tan\varphi_1 + \tan(\alpha + \varphi_2)} \tag{4-14}$$

式中 F_Q——斜楔所受的源动力(N);

F_W——斜楔所能产生的夹紧力的反力(N);

φ_1、φ_2——斜楔与工件、斜楔与夹具体间的摩擦角;

α——斜楔的楔角。

由于 α、φ_1、φ_2 均很小,设 $\varphi_1 = \varphi_2 = \varphi$,式(4-14)可简化为

$$F_W = \frac{F_Q}{\tan(\alpha + 2\varphi)} \tag{4-15}$$

(2) 斜楔夹紧机构的自锁条件 手动夹紧机构必须具有自锁功能。自锁是指对工件夹紧后,即使源动力不存在,夹紧机构依靠静摩擦力仍能保持对工件的夹紧状态。当源动力 F_Q 从图 4-37 中去掉后,应当对斜楔能否退出进行分析。斜楔受力分析如图 4-38 所示。

由图 4-38 看出,摩擦力的方向与斜楔松退方向相反。要使斜楔能够保证自锁,必须满足下列条件

$$F_1 \geqslant F_{Rx}$$
$$F_W \tan\varphi_1 \geqslant F_W \tan(\alpha - \varphi_2)$$

由于 α、φ_1、φ_2 的值均很小,所以上式可近似写成

$$\alpha \leqslant \varphi_1 + \varphi_2 \tag{4-16}$$

式(4-16)说明了斜楔夹紧的自锁条件:斜楔的楔角不得大于斜楔分别与工件和夹具体的摩擦角之和。

对于钢铁接触表面,斜楔夹紧机构满足自锁的条件是 $\alpha \leqslant 11.5° \sim 17°$。为自锁可靠起见,一般取 $\alpha = 6° \sim 8°$。由于气动、液压系统本身具有自锁功能,所以采用气动、液压夹紧的斜楔楔角可以选取较大的值,一般取 $\alpha = 15° \sim 30°$。

φ_1, φ_2—frictional angle included between the wedge and the workpiece, as well as the wedge and the fixture body respectively;

α—ascending angle of the wedge.

As α, φ_1 and φ_2 are very small, set $\varphi_1 = \varphi_2 = \varphi$, then the Eq. (4-14) can be simplified as follows:

$$F_W = \frac{F_Q}{\tan(\alpha + 2\varphi)} \qquad (4\text{-}15)$$

(2) Self-lock conditions for wedge clamping mechanisms   Manual clamping devices must have self-lock function. The self-lock means that after the workpiece is clamped, even though the original driving force is removed, the clamping mechanism can still keep the workpiece in clamping status by means of static friction force. After the original driving force F_Q is removed from Fig. 4-37, whether the wedge would go back or not should be analyzed. The force analysis of the wedge is illustrated in Fig. 4-38.

It can be seen from Fig. 4-38 that the direction of the frictional force F_1 is contrary to the disengaged tendency of the wedge. In order to enssure the wedge in self-locking status, the following conditions should be met:

$$F_1 \geqslant F_{Rx}$$
$$F_W \tan\varphi_1 \geqslant F_W \tan(\alpha - \varphi_2)$$

As α, φ_1 and φ_2 are very small, the above equation can be written approximately as follows:

$$\alpha \leqslant \varphi_1 + \varphi_2 \qquad (4\text{-}16)$$

The Eq. (4-16) shows the self-locking conditions of wedge clamping mechanisms: the wedge angle α should not be larger than the sum of two frictional angles included between the wedge and the workpiece and between the wedge and the fixture body.

For the contact surfaces made of iron and steel, the self-locking condition of wedge clamping mechanisms is $\alpha \leqslant 11.5°\text{-}17°$. From the viewpoint of safety, $\alpha = 6°\text{-}8°$ is generally used. As the pneumatic or hydraulic clamping devices have self-locking function, the wedge angle can be larger, usually $\alpha = 15°\text{-}30°$.

(3) Force transmission coefficient and clamping journey of the wedge   The ratio of the clamping force F_W to the original driving force F_Q is called the force transmission coefficient, represented by i_F. According to the definition, the force transmission coefficient of wedge clamping mechanisms is:

$$i_F = \frac{F_W}{F_Q} = \frac{1}{\tan\varphi_1 + \tan(\alpha + \varphi_2)} \qquad (4\text{-}17)$$

Clearly, the smaller the wedge angle α, the larger the force transmission coefficient i_F. When the original driving force F_Q is definite, the smaller the α, the larger the clamping force F_W. Set $\varphi_1 = \varphi_2 = \alpha = 6°$, then $i_F \approx 3$. Therefore, the wedge clamping mechanisms have the ability to magnify the original driving force.

In Fig. 4-35, the height S is called the clamping journey of the wedge, and the length L is the distance the wedge travels in the clamping workpiece. It can be seen from Fig. 4-35 that

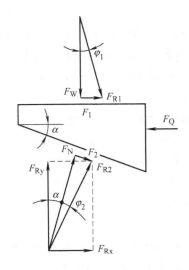
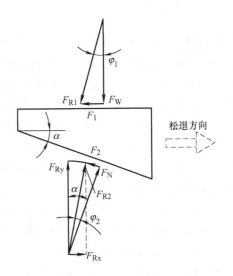

Fig. 4-37　Force analysis of the wedge
斜楔受力分析

Fig. 4-38　Force analysis of the self-locking wedge
斜楔自锁时受力分析

（3）斜楔夹紧的力传递系数和夹紧行程　力传递系数指在夹紧源动力 F_Q 作用下夹紧机构所能产生的夹紧力 F_W 与 F_Q 的比值。根据定义，斜楔夹紧机构的力传递系数为

$$i_F = \frac{F_W}{F_Q} = \frac{1}{\tan\varphi_1 + \tan(\alpha + \varphi_2)} \tag{4-17}$$

显然，楔角 α 越小，力传递系数 i_F 就越大。当夹紧源动力 F_Q 一定时，楔角 α 越小，产生的夹紧力 F_W 就越大。取 $\varphi_1 = \varphi_2 = \alpha = 6°$ 时，$i_F \approx 3$。因此，斜楔夹紧机构具有增力特性。

在图 4-35 中，高度 S 是斜楔的夹紧行程，而长度 L 是在夹紧工件时需要斜楔移动的距离。由图 4-35 可得，

$$S = L\tan\alpha \tag{4-18}$$

由于移动距离 L 受到斜楔长度的限制，因此要想增大斜楔的夹紧行程，就必须增大楔角 α。但楔角 α 过大的斜楔又会失去自锁作用。如果要求斜楔既有较大的夹紧行程，又有良好的自锁作用，建议采用具有两个楔角的斜楔。

斜楔夹紧机构结构简单，自锁性好，并且可以改变夹紧力的方向。斜角 α 越小，扩力比越大，但夹紧行程变小。故一般用于工件毛坯质量高的机动夹紧装置中，且很少单独使用。

2. 螺旋夹紧机构

由螺钉、螺母、垫圈、压板等元件组成的夹紧机构，称为螺旋夹紧机构。螺旋夹紧机构源于斜楔夹紧机构。螺旋面可以看做是绕在圆柱表面上的斜面。

图 4-39 是一个简单的螺旋夹紧装置，但如果使用图 4-39a 所示的夹紧装置，容易压坏工件表面，而且拧动螺钉时容易使工件产生转动，破坏工件的定位。因此，常使用图 4-39b 所示的头部带有浮动压块的螺旋夹紧装置。

（1）单螺旋夹紧机构的夹紧力计算　螺旋可以视为绕在圆柱体上的斜楔，因此可以由斜楔的夹紧力计算公式直接导出螺旋夹紧力的计算公式。如图 4-40 所示，工件处于夹紧状态时，根据力的平衡、力矩的平衡可算得夹紧力 Q。

$$S = L\tan\alpha \tag{4-18}$$

As L is limited by the wedge length, the wedge angle α should be increased in order to augment the clamping journey of the wedge. But the wedge with an overlarge α would lose its self-locking function. When the wedge clamping mechanism with both the self-locking function and the larger clamping journey is required, the wedge with two ascending angles can be used.

The wedge clamping mechanism has a simple structure and a good self-locking function. It can change the direction of the original driving force. It is also a force-magnifying device, and the smaller the ascending angle α, the larger the force magnification. But it has small clamping journey. In practice, the wedge clamping mechanism is rarely used separately, and often used together with the powered clamping device and for clamping the surface with good quality.

2. Clamping device with screw mechanism

The clamping mechanism composed of a screw, a nut, a washer, a pressing plate, etc. is called the screw clamping mechanism. The screw clamping device is originated from the wedge clamping device. The spiral surface of a screw can be seen as an inclined surface wrapped around a cylindrical column.

Fig. 4-39 shows a single screw clamping device. If the screw clamping device shown in Fig. 4-39 a is used, the surface of the workpiece would be damaged. Moreover, turning the screw may bring the workpiece to rotate together and destroy the original location. Therefore, the screw with a floating block assembled to the screw head is often used in practice, as shown in Fig. 4-39 b.

(1) Calculation of the clamping force by the single screw clamping device As the screw can be seen as a wedge wrapped around a cylindrical column, the formula for calculating the screw clamping force can be derived directly from the formula for calculating wedge clamping force. When the workpiece is in the clamping status, as shown in Fig. 4-40, the screw clamping force Q can be calculated based on the principle of the statical equilibrium.

$$Q = \frac{PL}{r_z \tan(\alpha + \varphi_1) + r_1 \tan\varphi_2} \tag{4-19}$$

where α is the ascending angle of the screw ($\alpha = 2°\text{-}4°$); φ_1 is the frictional angle included between the nut and the screw; φ_2 is the frictional angle included between the workpiece and the screw head (or floating bord); r_z is the half of the screw middle diameter; r_1 is the calculation radius of the frictional torque, and its value relates the shape of the screw head or the pressing block.

(2) Force transmission coefficient A single screw clamping device has the function to magnify the clamping force. Its force transmission coefficient is:

$$i_p = \frac{Q}{P} = \frac{L}{r_z \tan(\alpha + \varphi_1) + r_1 \tan\varphi_2} \tag{4-20}$$

As the screw ascending angle is smaller than the wedge ascending angle, and L is usually larger than r_z and r_1, and its force transmission coefficient is larger than that of the wedge clamping device.

(3) Applications of the screw clamping device Screw clamping devices have many advantages, such as simple structure, large clamping distance nearly without any limit, larger force magnification coefficient and good self-locking performance, therefore they are widely used in practice, es-

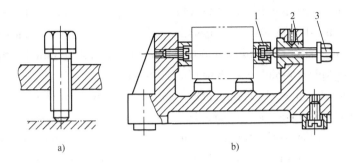

Fig. 4-39 Single screw clamping device 单螺旋夹紧机构
1—Floating block 浮动压块   2—Screw bush 螺母衬套   3—Screw 螺杆

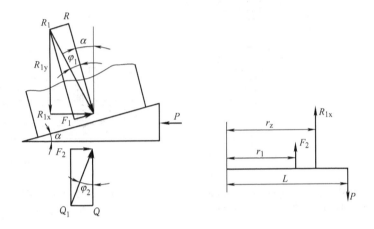

Fig. 4-40 Calculation of screw clamping force 螺旋夹紧力计算

$$Q = \frac{PL}{r_z \tan(\alpha + \varphi_1) + r_1 \tan\varphi_2} \quad (4\text{-}19)$$

式中　α——螺杆的螺旋升角（$\alpha = 2° \sim 4°$）；

　　　φ_1——螺母与螺杆间的摩擦角；

　　　φ_2——工件与螺杆头部（或压块）间的摩擦角；

　　　r_z——螺旋中径的一半；

　　　r_1——摩擦力矩计算半径，其数值与螺杆头部或压块的形状有关。

（2）单螺旋夹紧机构的传力系数　单螺旋夹紧机构具有扩力作用，其传力系数为

$$i_p = \frac{Q}{P} = \frac{L}{r_z \tan(\alpha + \varphi_1) + r_1 \tan\varphi_2} \quad (4\text{-}20)$$

（3）螺旋夹紧机构的应用　螺旋夹紧机构有很多优点，如结构简单、夹紧行程可以很大、传力系数大、自锁性能好等，因此在实际设计中得到广泛应用，尤其适合于手动夹紧装置。当夹紧行程较大时，螺旋夹紧机构的操作就显得比较费时。在实际应用中，可以采用其他方法实现螺旋夹紧机构对工件的快速装卸。图4-41所示为几种实现快速装卸的方法。

pecially used in manual clamping devices. But when the clamping distance is large, the screw clamping device seems to be slow in clamping operation. For this reason, other measures are taken in practical application to realize the loading and unloading of a workpiece rapidly. Several methods to load and unload a workpiece rapidly are shown in Fig. 4-41.

In practice, a single screw clamping mechanism is often combined with a lever and a plate to compose a screw-plate clamping mechanism. The compound forms of common screw-plate clamping mechanisms are shown in Fig. 4-42. Different compound forms have different force transmission coefficients. Which structural form is selected depends on not only the force transmission coefficient, but also the structure of the workpiece.

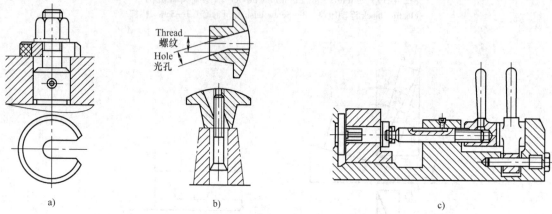

Fig. 4-41 Quick releasing structure with screw clamping 单螺旋夹紧快卸结构
a) Snap washer 开口垫圈 b) Quick releasing nut 快卸螺母 c) Quick releasing plate 快卸垫板

3. Clamping device with an eccentric mechanism

The eccentric clamping mechanism is a kind of clamping mechanism in which an eccentric part is used to clamp the workpiece. It is also originated from the wedge clamping device. This device is often used together with the clamping plate, as shown in Fig. 4-43. The usually used eccentric wheel includes circular eccentric wheel and curvilinear eccentric wheel which involves Archimedean and logarithm curves. The advantages of the two curves are that their ascending angle varies uniformly or keeps constant, and the workpiece can be clamped securely. But as the curvilinear eccentric wheel is difficult to machine, it is rarely used. The circular eccentric wheel is easily manufactured and thereafter used widely. In the following section, the circular eccentric clamping device will be discussed in detail.

(1) Calculation of the clamping force for the circular eccentric clamping device   Actually, the circular eccentric clamping device is the other form of a wedge clamping device—the wedge with variable ascending angle. With the increase of the ascending angle, the clamping force decreases, and the self-locking performance becomes poor. Therefore, the position with maximum ascending angle α_{max} is the important basis for the design of the eccentric wheel. As shown in Fig. 4-44, suppose the ascending angle $\alpha_p \approx \alpha_{max}$ at the point P, then the clamping force at this point is needed to check. For the sake of calculation, provided that the wedge ABC with the ascending angle α (in the

在实际应用中，单螺旋夹紧机构常与杠杆压板构成螺旋压板夹紧机构。常见螺旋压板夹紧机构的组合形式如图 4-42 所示，组合形式不同，其传力系数大小也随之不同。在实际设计中具体采用哪一种组合，除考虑传力系数外，重点还要考虑工件结构的需要。

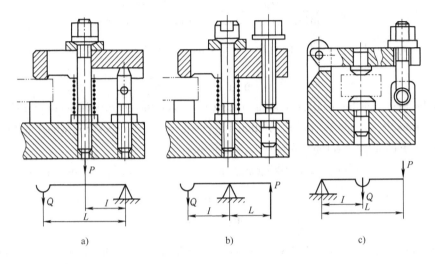

Fig. 4-42　Common screw-plate clamping mechanisms 常见螺旋压板夹紧机构

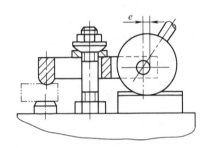

Fig. 4-43　Eccentric clamping mechanism 偏心夹紧机构

3. 偏心夹紧机构

偏心夹紧机构是由偏心件来实现夹紧的一种夹紧机构。偏心夹紧经常与压板联合使用，如图 4-43 所示。偏心件有偏心轮和凸轮两种，其偏心方法分别采用圆偏心和曲线偏心。曲线偏心为阿基米德曲线或对数曲线，这两种曲线的优点是升角变化均匀或不变，可使工件夹紧稳定可靠，但制造困难，故使用较少；圆偏心由于制造容易，因而使用较广。在此主要介绍圆偏心夹紧的原理和方法。

（1）圆偏心夹紧力计算　圆偏心夹紧实际上是斜楔夹紧的另外一种形式——变楔角斜楔。随着楔角增大，斜楔的夹紧力减小，自锁性能变差。因此，最大楔角处是偏心轮设计的重要依据。图 4-44 是偏心轮在 P 点处夹紧时的受力情况。此时，$\alpha_P = \alpha_{\max}$，夹紧力接近最小，一般只需校核该点的夹紧力。为便于计算，在 P 点处将偏心轮看作是一个楔角为 α 的斜楔 ABC，该斜楔处于偏心轮回转轴与工件垫板夹紧面之间。圆偏心夹紧的夹紧力为

clamping position) is put in between the support shaft and the plate, as shown in Fig. 4-44. According to the Eq. (4-14), the clamping force Q of the circular eccentric clamping device can be calculated as follows:

$$Q = \frac{Q_{1x}}{\tan(\alpha + \varphi_1) + \tan\varphi_2} = \frac{Q_1 \cdot \cos\alpha}{\tan(\alpha + \varphi_1) + \tan\varphi_2} \quad (4\text{-}21)$$

As α is very small, $\cos\alpha \approx 1$. Then, the Eq. (4-21) can be expressed as follows:

$$Q = \frac{PL}{\rho[\tan(\alpha + \varphi_1) + \tan\varphi_2]} \quad (4\text{-}22)$$

where L is the length of the handle; ρ is the distance between the center of the support shaft and the clamping position P; φ_1 and φ_2 are the frictional angles between the eccentric wheel and the support shaft, and the eccentric wheel and the workpiece respectively.

As α, φ_1 and φ_2 are very small, taking $\varphi_1 = \varphi_2 = \varphi$, the Eq. (4-22) can be expressed as follows:

$$Q = \frac{PL}{\rho\tan(\alpha + 2\varphi)} \quad (4\text{-}23)$$

(2) Self-locking conditions for circular eccentric wheel According to the wedge self-locking condition, the following relation should be met:

$$\alpha \leqslant \varphi_1 + \varphi_2$$

If the friction between the eccentric wheel and the support shaft is ignored, then $\alpha \leqslant \varphi_2$. We know that $\tan\alpha = 2e/D$, and $\tan\varphi_2 = f_2$, then

$$2e/D \leqslant f_2 \quad (4\text{-}24)$$

where D is the diameter of the circular eccentric wheel, e is the eccentric distance, f_2 is the frictional coefficient between the eccentric wheel and the workpiece (or clamping plate). If $f_2 = 0.1 \sim 0.15$, then

$$D/e \geqslant 14 \sim 20 \quad (4\text{-}25)$$

The Eq. (4-25) is just the self-locking condition for the circular eccentric clamping device. D/e is an important parameter of the circular eccentric wheel.

(3) The force transmission coefficient of the circular eccentric clamping device

$$i_P = \frac{L}{\rho[\tan(\alpha + \varphi_1) + \tan\varphi_2]} \quad (4\text{-}26)$$

(4) Applications of the eccentric clamping device The eccentric clamping device has many advantages, such as compact structure, easy to operation, rapid clamping and so on. But as it has small clamping distance, small clamping force, and poor self-locking performance, eccentric clamping device is usually used in such situations as small cutting force, without vibration, and the workpiece with small dimensional tolerance.

$$Q = \frac{Q_{1x}}{\tan(\alpha+\varphi_1)+\tan\varphi_2} = \frac{Q_1 \cdot \cos\alpha}{\tan(\alpha+\varphi_1)+\tan\varphi_2} \tag{4-21}$$

由于 α 很小，$\cos\alpha \approx 1$。那么式（4-21）可以表示为

$$Q = \frac{PL}{\rho[\tan(\alpha+\varphi_1)+\tan\varphi_2]} \tag{4-22}$$

其中，L 代表手柄长度；φ_1、φ_2 分别为偏心轮与工件、偏心轮与回转轴之间的摩擦角；ρ 代表夹紧点 P 到偏心轮回转轴线的距离。

由于 α、φ_1 和 φ_2 均很小，当取 $\varphi_1 = \varphi_2 = \varphi$ 时，式（4-22）又可写成

$$Q = \frac{PL}{\rho\tan(\alpha+2\varphi)} \tag{4-23}$$

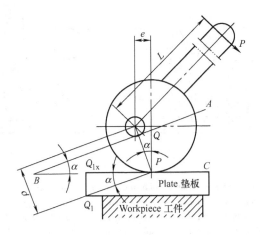

Fig. 4-44　Clamping force of circular eccentric wheel 圆偏心夹紧力

（2）偏心轮夹紧的自锁条件　由于偏心轮夹紧只是斜楔夹紧的另一种形式，因此要保证自锁就必须满足

$$\alpha \leqslant \varphi_1 + \varphi_2$$

如果不计转轴处摩擦，则 $\alpha \leqslant \varphi_2$。因为 $\tan\alpha = 2e/D$，$\tan\varphi_2 = f_2$，于是有

$$2e/D \leqslant f_2 \tag{4-24}$$

式中 D——偏心轮的直径；e——偏心距；f_2——偏心轮与工件（或压板）之间的摩擦因数。

当 $f_2 = 0.1 \sim 0.15$ 时，式（4-24）又可写为

$$D/e \geqslant 14 \sim 20 \tag{4-25}$$

式（4-24）就是取不同摩擦因数时的偏心轮自锁条件。D/e 是偏心轮的重要特性参数。

（3）偏心轮夹紧机构的传力系数

$$i_P = \frac{L}{\rho[\tan(\alpha+\varphi_1)+\tan\varphi_2]} \tag{4-26}$$

4. Other typical clamping devices

(1) **The clamping device with linkage** The clamping device with linkage means the one by which a clamping action can make several clamping elements clamp one or several workpieces in several points or directions simultaneously. The use of the clamping device can reduce the clamping time, raise the productivity, and lighten the labor intensity of workers.

Fig. 4-45 a shows a multi-point clamping device with linkage. Turning the nut 1 clockwise can make two clamping plates clamp a workpiece simultaneously. Fig. 4-45 b shows a multi-piece clamping device. An original force can be used to clamp two workpieces simultaneously. In order to avoid the shafts not being securely clamped, or damaging the clamping device because of the dimensional or geometrical errors of a shaft, two swinging blocks 1 and 2 are often connected with the clamping plate to compensate the diameter tolerances of a shafts through swinging movement.

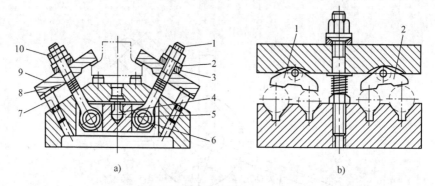

Fig. 4-45　Multi-point and multi-piece clamping devices with linkage 多点与多件夹紧装置
a) Multi-point clamping device with linkage 多点、单向联动夹紧机构
1—Jointed bolt 活节螺栓　2—Spherical nut with shoulder 球面带肩螺母
3—Tapered washer 锥形垫圈　4—spherical pin 球头销　5—Hinge plate 铰链板
6—Cylindrical pin 圆柱销　7—Ball end stud 球头支承　8—Spring 弹簧
9—Rotary clamping plate 转动压板　10—Hexagonal nut 六角扁螺母
b) Multi-piece clamping device 多件夹紧机构
1、2—Floating clamping block 浮动压块

(2) **Centering clamping device** In the location of the workpiece, the centering means the symmetrical center of the workpiece would not vary with the dimension of the workpiece. The locating and clamping device, which can keep the symmetrical center of the workpiece from varying with the dimension of the clamping surface by means of the equal deformation /displacement of clamping elements, or uniform movement of clamping elements in opposite directions, is named as the centering clamping device.

Fig. 4-46 illustrates a centering clamping device. Turning the screw 3, the V-block 1 and the V-block 2 move at equal speed towards the center simultaneously, thus making the workpiece centered and clamped. The furcate block 7 is used to adjust the stud 3 with right-hand and left-hand threads. The correctness of the centering depends on the position of the furcate block 7 to large extent. After the centering position is adjusted well, screw down the screw 6 firstly, then screw down the screw 4

（4）偏心夹紧应用场合　圆偏心夹紧操作方便，动作迅速，结构紧凑。但由于其夹紧力小，自锁性能不是很好，且夹紧行程小，故多用于切削力小、无振动、工件尺寸公差不大的场合。

4. 其他典型夹紧机构

（1）联动夹紧机构　联动夹紧机构是指由一个夹紧动作使多个夹紧元件实现对一个或多个工件的多点、多向同时夹紧的夹紧机构。联动夹紧机构可有效地提高生产率、降低工人的劳动强度，同时还可满足有多点、多向、多件同时夹紧要求的场合。

图 4-45a 所示为多点联动夹紧机构。当向下旋转螺母 2 时，可使两个转动压板 9 同时对工件夹紧。图 4-45b 所示为多件夹紧机构。其特点是：用一个原始力对数个点或数个工件同时进行夹紧。为了避免工件因尺寸或形状误差而出现夹紧不牢或破坏夹紧机构的现象，在压板两边各连接摆动压块，可以通过摆动来补偿各自夹压的两个工件的直径尺寸公差。

（2）定心夹紧机构　保证工件的对称中心不因工件尺寸的变化而变化就称为定心。定心夹紧机构就是利用夹紧元件的等量变形位移或等速相向运动保持工件的对称中心不因夹持面尺寸变化而变化的定位、夹紧装置。

图 4-46 所示为一定心夹紧装置。旋转双头螺柱 3，就可使 V 形块 1、2 作等速相向运动，实现对工件的定心夹紧。当需要调整定心位置时，可拧松固定螺钉 6，通过调节螺钉 5 调整叉座 7，使双头螺柱 3 的轴向位置发生改变，从而就可改变定心的位置。定心位置调整后，需先拧紧固定螺钉 6，再拧紧锁紧螺钉 4 锁死调节螺钉 5 的位置。

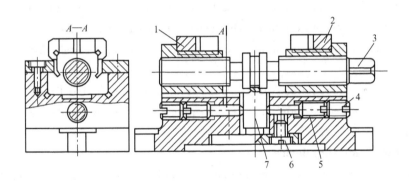

Fig. 4-46　Centering and clamping device with uniform movement 等速运动定心夹紧装置
1、2—V-block V 形块　3—stud 双头螺柱　4—locking screw 锁紧螺钉
5—Adjustable screw 调节螺钉　6—Fixing screw 固定螺钉　7—Furcated seat 叉座

图 4-47 是液性塑料心轴定心夹紧机构，用于定位精度很高的夹具中。在定位夹紧工件时，推动柱塞 4 挤压液性塑料 3，利用液性塑料具有液体的不可压缩性，使薄壁套筒 2 沿径向产生均匀变形，从而实现对工件孔的定心和夹紧。通过调整螺钉 5 限制柱塞 4 的移动位置，以保证夹紧力的大小；当装入或更换液性塑料时，应通过排气螺钉 1 将空气排出。

to lock the screw 5.

Fig. 4-47 illustrates a centering and clamping device with liquid plastic mandrel. It is used in the fixture with a very high location accuracy. In the locating and clamping operation of the workpiece, the plunger 4 is pushed forward to press the liquid plastic 3. As the liquid plastic can not be compressed, the thin-wall sleeve 2 is forced to expand in radial direction, which makes the workpiece centered and clamped. The adjustable screw 5 is used to limit the position of the plunger 4 so as to ensure a suitable clamping force. The screw 1 is used to vent air when encasing or replacing liquid plastic.

4.4 Other devices in machine tool fixtures

Besides location device and clamping device, there are other devices in the fixtures of machine tools. Some of them are the peculiar elements of a certain fixture. Some typical devices or elements in the fixtures of machine tools are introduced simply in this section.

4.4.1 Indexing device

1. Concept of indexing device

In machining, we often see some workpieces with a group of surfaces (holes, slots, or polyhedron), which are required to machine in one-step installation on fixture. The group of surfaces is distributed in a certain angle or distance. It is required that the fixture should have the indexing function. After a surface is machined, some part of the fixture together with the workpiece on it turns a certain angle or moves a certain distance. The device which can change the machining position of the workpiece in one installation by turning a certain angle or moving a certain distance is called the indexing device.

The indexing device can be classified into circle indexing device and linear indexing device. Both have the similar structural form and operation principle. The circle indexing device is widely used in practice.

2. Elements of indexing device

Fig. 4-48 shows a rotary jig used for drilling three radial holes uniformly distributed on a sector workpiece. Taking its hole, keyway and side plane as location surfaces, the workpiece is located by the location pin 6, key 7 and round support plate 3, and is clamped by nut 5 and snap washer 4. Indexing device is composed of an indexing plate 9, an indexing location sleeve 2, an indexing pin 1 and a locking handle 11.

It can be seen from Fig. 4-48 that an indexing device is usually composed of following elements:

(1) Rotary (or movable) part It is used to realize the rotation (or movement) of the workpiece, such as the indexing plate 9 in Fig. 4-48.

(2) Stationary part It is the base of a indexing device, and usually is joined with a fixture body, such as the base 13 in Fig. 4-48.

(3) Indexing location device It can ensure the workpiece in the correct indexing position,

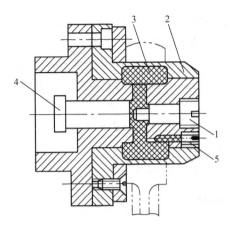

Fig. 4-47 Centering clamping device with liquid plastic mandrel 液性塑料心轴定心夹紧机构
1—Vent screw 排气螺钉　2—Thin-wall sleeve 薄壁套筒　3—Liquid plastic 液性塑料
4—Plunger 柱塞　5—Adjustable screw 调整螺钉

4.4　机床夹具的其他装置

除了定位装置、夹紧装置外，机床夹具上还有一些其他装置，其中有些还是某种夹具的特殊元件。本节就机床夹具上一些典型装置作一简单介绍。

4.4.1　分度装置

1. 分度装置的概念

在机械加工中，往往会遇到一些工件要求在夹具的一次安装中加工一组表面（孔系、槽系或多面体等），而此组表面是按一定角度或一定距离分布的，这样便要求该夹具在工件加工过程中能进行分度。即当工件加工好一个表面后，应使夹具的某些部分连同工件转过一定角度或移动一定距离。使工件在一次装夹中，每加工完一个表面之后，通过夹具上的可动部分连同工件一起转动一定的角度或移动一定的距离，以改变工件加工位置的装置，称为分度装置。

分度装置可分为两类：回转分度装置和直线分度装置。两者的基本结构形式和工作原理都是相似的，而生产中又以回转分度装置应用较多。

2. 分度装置的组成

图 4-48 是用来加工扇形工件上三个等分径向孔的回转式钻模。工件以内孔、键槽和侧平面为定位基面，分别在夹具上的定位销轴 6、键 7 和支承板 3 上定位，限制 6 个自由度。由螺母 5 和开口垫圈 4 夹紧工件。分度装置由分度盘 9、分度定位套 2、分度销 1 和锁紧手柄 11 组成。

由图 4-48 可知，分度装置一般由以下几个部分组成。

（1）转动（或移动）部分　它实现工件的转位（或移位），如图 4-48 中分度盘 9。

（2）固定部分　它是分度装置的基体，常与夹具体连接成一体，如图 4-48 中的底座 13。

（3）对定机构　它保证工件正确的分度位置，并完成插销、拔销动作，如图 4-48 中的

and carry out inserting pin and pulling pin. The indexing plate 9, indexing location sleeve 2 and indexing pin 1 are the parts of the indexing location device.

(4) Locking mechanism It is used for tightening the rotary (or movable) part to the stationary part and plays a part of reducing the vibration and protecting the indexing location device, such as the locking handle 11 and sleeve 10.

Using an indexing device can reduce the installation time of the workpiece, lighten the labor intensity and raise the productivity. Therefore, indexing devices are widely used in drilling, milling, turning, boring and other machining processes.

4.4.2 Tool guiding element

The guiding elements are usually used in fixtures to guide the cutting tool so that the correct position of holes can be ensured. The commonly used guiding elements are drilling bushes in drilling jigs and boring bushes in boring jigs.

1. Drilling bush

A drilling bush is a special element of a drilling jig. Its function is to guide the drill, core drill or reamer to prevent them deflecting in the machining process. There are four types of drilling bushes.

(1) Stationary drilling bush As shown in Fig. 4-49, the stationary drilling bush is directly pressed into the hole on the drilling jig plate with the fit H7/n6 or H7/r6. This drilling bush has higher position accuracy, but is not easy to replace after it is worn. The stationary drilling bush is only suitable for drilling hole in job and small volume production.

The basic diameter D of a guiding hole in the drilling bush equals to the maximum diameter of the drill to be guided. Generally, the tolerance band of the guiding hole is F7 for drilling and core drilling, G7 for rough reaming, G6 for finish reaming.

The height H of the drilling bush is the length of the guiding hole. The larger the H, the better the guidance quality, but the more serious the tool and bush wear. The smaller the H, the worse the guidance quality, and the cutting tool is easy to deflect. To determine the height H of a drilling bush, the factors, such as the diameter, depth, position accuracy of the hole to be drilled, and workpiece material and its surface condition, stiffness of cutting tool and so on, should be considered. Generally, $H = (1 \sim 3)D$; if the hole is located on the slope or curved surface, or has higher position accuracy, then $H = (4 \sim 6)D$.

The distance from the bottom of the bush to the end face of the hole is called the chip removal gap. It has an influence on the chip removal and guidance quality. Generally, $h = (0.3 \sim 0.7)D$ for brittle material, and $h = (0.7 \sim 1.5)D$ for ductile material. When drilling the hole on slope, $h = (0 \sim 0.2)D$ to prevent the drill from deflecting. Sometimes, $h = 0$ when the position accuracy of the hole is quite high, or the diameter of drill is very small.

(2) Renewable drilling bush The renewable drilling bush shown in Fig. 4-50 a is suitable for drilling hole in large volume production. There is a lining bush between the drilling bush and the jig plate to protect the jig plate from wear. The fit in between the drilling bush and the lining bush is

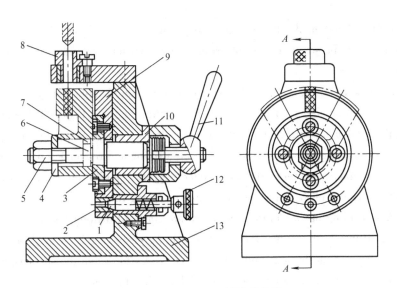

Fig. 4-48　Rotary jig 回转式钻模
1—Indexing pin 分度销　2—Indexing location sleeve 分度定位套　3—Support plate 支承板
4—Snap washer 开口垫圈　5—Nut 螺母　6—Location pin 定位销轴　7—Key 键　8—Drilling bush 钻套
9—Indexing plate 分度盘　10—Sleeve 套筒　11—Locking handle 锁紧手柄　12—Handle 手柄　13—Base 底座

分度盘9、分度定位套2、分度销1等。

（4）锁紧机构　它将转动（或移动）部分与固定部分紧固在一起，起减小加工时的振动和保护对定机构的作用，如图 4-48 中的锁紧手柄 11、套筒 10 等。

使用分度装置可以减少工件的安装次数，减轻劳动强度和提高生产率，因此广泛用于钻、铣、车、镗等加工中。

4.4.2　刀具引导元件

在钻、镗等孔加工夹具中，常用引导元件来保证孔加工的正确位置。常用引导元件主要有钻床夹具中的钻套、镗床夹具中的镗套等。

1. 钻套

钻套是钻床夹具的特殊元件，其作用是引导钻头、扩孔钻或铰刀，以防止其在加工过程中发生偏斜。按钻套的结构和使用情况，可分为以下四种类型。

（1）固定钻套　如图 4-49 所示，固定钻套直接被压装在钻模板上，其配合为 H7/n6 或 H7/r6，其位置精度较高，但磨损后不易更换。适用于只需要钻孔的单一工步及生产批量较少的场合。

钻套导引孔的公称尺寸 D，应等于所导引刀具刀刃部分直径的上极限尺寸。对于钻孔和扩孔，钻套导引孔尺寸公差带一般取 F7；粗铰时，钻套导引孔尺寸公差带取 G7；精铰时，钻套导引孔尺寸公差带取 G6。

钻套高度 H 即导引孔长度，H 较大则导向性好，但刀具和钻套的磨损较大；H 过小则导引作用差，刀具容易倾斜。设计时，应根据加工孔直

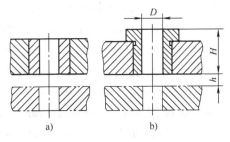

Fig. 4-49　Stationary drilling bush 固定钻套

F7/k6, and the fit in between the lining bush and the jig plate is H7/n6. The renewable drilling bush should be fixed by the screw to prevent drilling bush from rotation with drill or lift by the withdrawing drill.

(3) Rapid-replacing drilling bush The structure of the rapid-replacing drilling bush is shown in Fig. 4-50 b. We can see clearly the difference in structure between Fig. 4-50 b and Fig. 4-50 a. This kind of drilling bush is suitable for drilling, core drilling and reaming in one operation. Because the diameter of cutting tools used in drilling, core drilling and reaming is different, the drilling bushes with different inner diameters (ID) are needed to guide the different cutting tools. The use of the rapid-replacing drilling bush can reduce the bush-replacing time.

Three kinds of drilling bushes introduced above are standard drilling bushes.

(4) Special drilling bush When the standard drilling bushes can't be used because of limitation in workpiece shape or machining condition, it is necessary to design the special-purpose drilling bushes by oneself. Fig. 4-51 shows us some special drilling bushes, where the drilling bush in Fig. 4-51 a is used to drill the hole on slope, Fig. 4-51 b is a lengthening bush used to drill the hole in concave pit, and Fig. 4-51 c is used for drilling two holes with very small central distance.

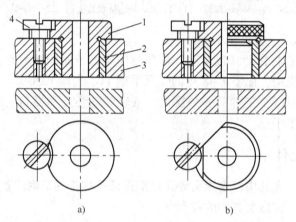

Fig. 4-50 Renewable drilling bush and rapid-replacing drilling bush 可换钻套与快换钻套的结构
1—Drilling bush 钻套 2—Lining bush 衬套 3—Jig plate 钻模板 4—Screw 螺钉

2. Boring bush

The boring bush is used to guide the boring bar in boring operation. According to whether it is movable or not, the boring bush is classified into stationary boring bush and rotary boring bush. The structure of the stationary boring bush is similar to that of the drilling bush, and has higher position accuracy. As the relative motion between the boring bush and the boring bar would cause the wear, the stationary boring bush is applied for the boring at lower speed. The rotary boring bush is used when the velocity of the boring bar is larger than 20 m/min. Fig. 4-52 shows a rotary boring bush. The structure shown at left end a is an inner rolling bush in which the bush 2 is stationary, the boring bar 4 is installed on the rolling bearing inside the sliding bush 3. The boring bar rotates relative to the sliding bush and move along with the sliding bush relative to the boring bush. This kind of boring bush has a good guidance quality but large size. Therefore, it is usually used as rear guid-

径大小、深度、位置精度、工件材料、孔口所在的表面状况及刀具刚度等因素而定。一般情况下取 $H=(1\sim3)D$，若刀具容易偏斜（如在斜面或曲面上钻孔）或位置精度要求高时，钻套的高度应按 $(4\sim8)D$ 选取。

钻套距工件孔端距离（间隙）h 会影响排屑和刀具导向。h 的大小要根据工件材料和加工孔位置精度要求而定，总的原则是引偏量要小又利于排屑。在加工铸铁等脆性材料时，一般取 $h=(0.3\sim0.7)D$；加工塑性材料时，常取 $h=(0.7\sim1.5)D$。当在斜面上钻孔时，为防止引偏可按 $h=(0\sim0.2)D$ 选取。当被加工孔的位置精度要求很高或钻头直径很小时，也可以不留间隙（即 $h=0$）。

(2) 可换钻套　图 4-50a 所示的可换钻套适合于生产批量较大零件的钻孔工序。在可换钻套与钻模板之间装有衬套，以防止钻模板磨损。钻套与衬套孔常用 F7/k6 配合，而衬套与钻模板采用 H7/n6 配合。这类钻套需要压套螺钉固定，以防其随刀具转动或在退刀时被刀具带起。

(3) 快换钻套　图 4-50b 所示为快换钻套的结构。由图可以清晰地看出快换钻套与可换钻套在结构上的不同。当孔需要钻、扩、铰等多工步加工时，由于刀具直径尺寸不同，需要内径不同的钻套来引导刀具。采用快换钻套可减少更换钻套的时间。

以上介绍的三种钻套均为标准钻套。

(4) 特殊钻套　因工件的形状或工序加工条件而不能使用以上三种标准钻套时，需自行设计的钻套称特殊钻套。常见的特殊钻套如图 4-51 所示。其中图 4-51a 为在斜面或圆弧面上钻孔的钻套；图 4-51b 为在凹形表面上钻孔的加长钻套；图 4-51c 为小孔距钻套。

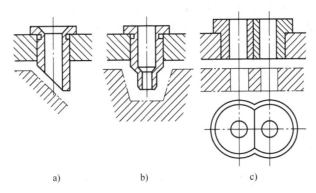

Fig. 4-51　Special drilling bush　特殊钻套

2. 镗套

镗套用于引导镗杆，根据其在加工中是否运动可分为固定式镗套和回转式镗套两类。固定式镗套的结构与钻套相似，且位置精度较高。但由于镗套与镗杆之间的相对运动使之易于磨损，一般用于速度较低的场合。当镗杆的线速度大于 20m/min 时，应采用回转式镗套。图 4-52 所示为回转式镗套，其中左端 a 所示结构为内滚式镗套，镗套 2 固定不动，镗杆 4 装在导向滑套 3 内的滚动轴承上。镗杆相对于导向滑套回转，并连同导向滑套一起相对于镗套移动。这种镗套的精度较好，但尺寸较大，因此多用于后导向。图中右端 b 的结构为外滚式镗套，镗杆 4 与镗套 5 一起回转，两者之间只有相对移动而无相对转动。镗套的整体尺寸小，应用广泛。

ance. The structure shown at right end b is an outside rolling bush in which the boring bar rotates together with the boring bush and moves relative to the boring bush. This kind of boring bush has smaller integral size and is widely used in practice.

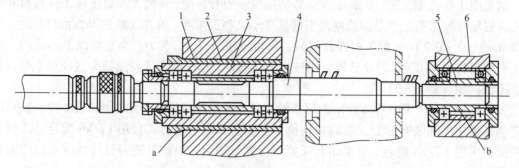

Fig. 4-52　Rotary boring bush 回转式镗套
1、6—Guiding support 导向支架　2、5—Boring bush 镗套　3—Sliding bush 导向滑套　4—Boring bar 镗杆

4.4.3　Tool aligning device

A tool aligning device is composed of a tool aligning block and a feeler gauge, and its structural form depends on the shape of the surface to be machined. The tool aligning block is the element to decide the position of the cutting tool relative to the locator on the fixture. The structure of the tool aligning block has been standardized. Specific structural dimension of tool aligning block can be looked up in "Standards of Fixture" (JB/T 8031.1~4—1995).

In tool aligning operation, the milling cutter is not allowed to contact with the tool aligning block directly. Their mutual position is trued up by means of feeler gauge to avoid damaging the cutting edge or the premature wear of the tool aligning block. Fig. 4-53 illustrates the applications of various tool aligning blocks.

4.4.4　Connecting elements

A fixture has to be located and clamped on the machine tool. The elements used to locate and clamp a fixture on the machine tool are named as the connecting element. Generally, connecting elements have several forms as follows:

1) The fixtures installed on the milling machine, shaper and planer, and boring machine are located by location keys which are fitted with the T-slot on the worktable. Fig. 4-54 illustrates the structure of the location key and its application. The clearance fit (H7/h6) between the location key and the T-slot is usually used. A fixture is generally equipped with two location keys. The location key is not suitably used in the fixtures with high location accuracy or heavy duty. This fixture is located on the machine tool by means of a long and narrow plane which has been machined previously on the fixture body.

2) The fixtures used on both the lathe and the internal/external grinding machine are generally installed on the spindle of the machine tool. The connecting forms of the fixture to machine tools are

4.4.3 对刀装置

对刀装置由对刀块和塞尺组成,其形式视加工表面的情况而定。对刀块是用来确定刀具与夹具相对位置的元件。对刀块结构已经标准化,具体结构尺寸可参阅"夹具标准"(JB/T 8031.1~4—1995)。

对刀时,铣刀不能与对刀块的工作表面直接接触,应通过塞尺来校准它们之间的相对位置,以免损坏刀刃或造成对刀块过早磨损。图4-53所示为各种对刀块的应用情况。

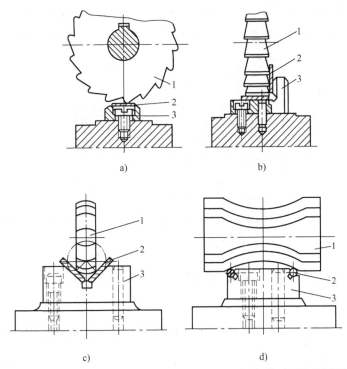

Fig. 4-53 Applications of various tool aligning blocks 各种对刀块使用举例
1—Milling cutter 铣刀 2—Feeler 塞尺 3—Tool aligning block 对刀块

4.4.4 连接元件

夹具在机床上必须定位夹紧。在机床上进行定位夹紧的元件称为连接元件,它一般有以下几种形式。

1) 在铣床、刨床、镗床上工作的夹具通常通过定位键与工作台T形槽的配合来确定其在机床上的位置。图4-54为定位键结构及其应用情况。定位键与夹具体的配合多采用H7/h6。一副夹具一般要配置两个定位键。对于定位精度要求高的夹具和重型夹具,不宜采用定位键,而采用夹具体上精加工过的狭长平面来找正安装夹具。

2) 车床和内外圆磨床的夹具一般安装在机床的主轴上,连接方式如图4-55所示。图4-55a采用长锥柄(莫氏锥度)安装在主轴锥孔内,这种方式定位精度高,但刚度较差,多用于小型机床。图4-55b所示夹具以端面A和圆孔D在主轴上定位,孔与主轴轴颈的配合一般取H7/h6。这种连接方法制造容易,但定位精度不很高。图4-55c所示夹具以端面T和

shown in Fig. 4-55. Fig. 4-55 a shows that a fixture with a long Morse taper shank is installed into the taper hole of the spindle. This installation form has high location accuracy but low stiffness, and is often used in small-size machine tools. Fig. 4-55 b shows that the fixture is located on the spindle with its end face A and hole D. The fit between the hole D and the journal of the spindle is H7/h6. This installation form is easy to manufacture, but its location accuracy is not high. Fig. 4-55 c shows that the fixture is located on the spindle with its end face T and short taper surface K. This installation form has not only high location accuracy but high stiffness. It is noticed that this location form belongs to redundant location. Therefore, this installation form has higher requirement for the manufacturing of the fixture, and the match grinding is needed for the location surfaces.

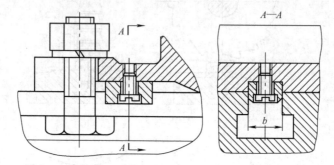

Fig. 4-54　Installation of location key on Fixture 定位键在夹具上的安装

4.5　Methods to design special-purpose fixtures

4.5.1　Basic requirements for the design of special-purpose fixtures

1) To ensure reliably the machining accuracy of the workpiece. The key is to determine correctly the location and clamping plans, and to select or design the locators, clamping device and tool-aligning element.

2) To improve the productivity and the reduce production cost. According to the lot size of the workpiece, rapid and efficient clamping devices with different complexities should be selected so as to shorten the assistant time.

3) Convenient for use. The operation of the fixture should be easy, labor-saving, safe, reliable, and convenient for chip removal.

4) Good manufacturability. The fixture designed should be easy to manufacture, inspect, assemble and maintain.

4.5.2　Approaches and procedures for the design of special-purpose fixtures

1. To be clear about the design task and collect the design information

Firstly, the designers should analyze the structure feature, material, production scale, the technical specifications of the current operation, as well as the relationship between the current oper-

短锥面 K 定位。这种方法不但定心精度高，而且刚度也好。值得注意的是这种定位方法是过定位，因此，要求制造精度很高，夹具上的端面和锥孔需进行配磨加工。除此之外还经常使用过渡盘与机床主轴连接。

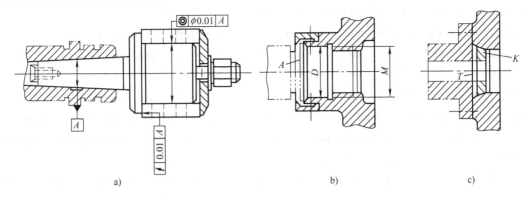

Fig. 4-55　Installation of fixture on spindle　夹具在机床主轴上的安装

4.5　专用夹具的设计方法

4.5.1　专用夹具设计的基本要求

1）可靠保证工件的加工精度。关键在于正确确定定位方案和夹紧方案，合理选用与设计定位元件、夹紧装置以及对刀元件等。

2）提高生产率，降低生产成本。应根据工件生产批量的大小选用不同复杂程度的快速高效夹紧装置，以缩短辅助时间。

3）使用性好。夹具的操作应简便、省力（可采用气动、液压和气液联动等机械化夹紧装置）、安全可靠、排屑方便。

4）工艺性好。所设计的夹具应便于制造、检验、装配、调整和维修等。

4.5.2　专用夹具设计的方法和步骤

1. 明确设计任务，收集设计资料

首先应分析研究工件的结构特点、材料、生产规模和本工序加工的技术要求以及前后工序的联系；然后了解加工所用设备、辅助工具中与设计夹具有关的技术性能和规格；了解工具车间的技术水平等。必要时还要了解同类工件的加工方法和所使用夹具的情况。

2. 拟订夹具结构方案，绘制夹具草图

确定夹具结构方案时，主要解决如下问题：

1）确定工件定位方案。虽然定位基准在工序图中已经确定，仍然要分析研究其合理性，如定位精度和夹具结构实现的可能性。在确定定位方式以后，选择、设计定位元件，进行定位误差分析计算。

2）确定夹紧方案。选择或设计夹紧机构，并对夹紧力进行验算。

3）确定刀具的引导方式。选择或设计引导元件或对刀元件。

ation and the preceding operation etc. Then, the designers should learn about the machining equipment and the auxiliary tools (primarily including the technical performance and standard related to a fixture design) and also the technical level of the tool workshop. They should learn about the machining methods and fixtures applied for the similar workpieces, if necessary.

2. To draw up the structural scheme and draft the sketch of the fixture

In the determination of the structural scheme of the fixture, the designers should solve mainly the problems as follows:

1) To determine the location plan for the workpiece. Though the location datum has been determined in operation drawing, its rationality should still be analyzed, especially in the location accuracy and the probability for realizing the structure of the fixture. After the location plan is determined, to select or design the locator and to calculate the location error can be done.

2) To determine the clamping plan for the workpiece. It involves in the selection or design of the clamping mechanism and the examination of the clamping force.

3) To determine the tool leading manner, and to select and design the tool-guiding element or tool-aligning element.

Then, the structures of other elements or devices, such as the location key, indexing device and so on, should also be determined.

When determining the general structure of the fixture, it is better to consider several options and draft each sketch. Finally, the more rational scheme is selected from them by the analysis and comparison.

3. To draw the general assembly drawing and part drawing

The general assembly drawing of the fixture should be drawn following National Standard. The drawing scale should be in the scale of 1∶1 as far as possible, which can make the drawing view more intuitionistic. The workpiece is regarded as the transparent object and drawn in double dots line. The general assembly drawing of the fixture should be able to reflect its operational principle, such as the location principle, the relative position relation of various elements, the operational principle of the clamping mechanism and so on. It should also show the general layout and primary structure of the fixture at the same time. The sequence to plot the general assembly drawing is: workpiece→locator→tool-guiding/aligning element→clamping device→other devices→fixture body →marking the necessary dimension tolerances and technical requirements→drawing the title bar and parts list of the fixture.

All the non-standard parts in fixtures must have the part drawings. The dimension and its tolerance, surface roughness and necessary form and positional tolerance, material and its heat treatment for each non-standard part should be marked on the part drawing. When determining the dimensions, tolerances and technical specifications for these parts, designers should pay attention to that whether their dimensions and tolerances could meet the needs of the assembly drawing.

4.5.3 A case of special-purpose fixture design

Fig. 4-56 illustrates a hole-making operation drawing of a rocker arm. Given that the rocker arm

确定其他元件或装置的结构形式,如定位键、分度装置等。

对夹具的总体结构,最好考虑几个方案,画出草图,经过分析比较,从中选取较合理的方案。

3. 绘制夹具总装图及零件图

夹具总图应遵循国家标准绘制。夹具总图设计尽量采用1:1的比例,工件用双点画线绘制,并把工件视为透明体。夹具总图须清楚表明其工作原理,如定位原理、各元件的位置关系及夹紧机构工作原理等,同时应尽量清楚地反映夹具的总体及主要结构。夹具总图绘制顺序为:工件→定位元件→对刀、导引元件→夹紧装置→其他装置→夹具体→标注必要尺寸公差及技术要求→编制夹具明细表及标题栏。

夹具总图上所有的非标准零件都必须有零件图。在零件图中须标注出全部尺寸、表面粗糙度及必要的形状和位置公差、材料及热处理要求以及其他技术要求。在确定这些零件的尺寸、公差和技术要求时,设计人员应当注意这些尺寸及公差是否满足装配图的要求。

4.5.3 专用夹具设计案例

图 4-56 所示为摇臂零件孔加工工序简图。零件材料为 45 钢,毛坯为模锻件,成批生产规模,所用机床为 Z525 型立式钻床,现设计加工 $\phi 18H7$ mm 孔的钻床夹具。

(1) 零件加工情况分析与夹具总体方案确定 本工序加工孔尺寸为 7 级精度,孔中心距离精度要求较高,由于零件属批量生产,为可靠保证精度要求,需使用专用夹具进行加工。但考虑到生产批量不是很大,因而夹具结构应尽可能的简单,以采用手动夹具为宜,以减小夹具制造成本。

(2) 确定夹具的结构方案

1) 确定定位方案,设置定位元件。本工序加工要求保证的主要尺寸是 (120 ± 0.08) mm 和 $\phi 18H7(^{+0.018}_{0})$ mm 孔。根据工艺规程给定的定位基准和夹紧力,用 $\phi 36H7(^{+0.025}_{0})$ 孔和端面及小头外圆定位。定位设计如图 4-57 所示。

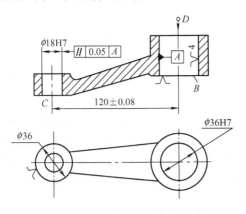

Fig. 4-56　Operation drawing of the rocker arm 摇臂零件工序图

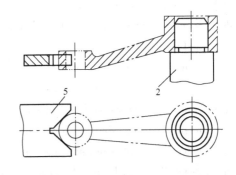

Fig. 4-57　Design of location scheme 定位方案设计

定位孔与定位销的配合尺寸取为 $\phi 36H7/g6$(定位销 $\phi 36^{-0.009}_{-0.025}$ mm)。对于工序尺寸 (120 ± 0.08) mm 而言,定位基准与工序基准重合 $\Delta_B = 0$;定位副制造误差引起的基准位移

is made in batch production, its blank made of 45 steel is manufactured by die forging, an upright drilling machine with model Z525 is selected. It is required to design a drilling jig to realize the machining of the hole $\phi 18H7$.

(1) To perform operation analysis and determine the general scheme of a jig It can be seen from Fig. 4-56 that the machining accuracy of the hole is IT7, and the accuracy requirement for the center distance between $\phi 18$ and $\phi 36$ is also high. As the rocker arm is in batch production, a special-purpose fixture is required in order to ensure the desired machining accuracy. Considering that the production volume is not very large, the fixture configuration should be simple as far as possible. It is better to use the hand-operated fixture so as to reduce its manufacturing cost.

(2) To determine the structural scheme of the fixture

1) To determine the location scheme and arrange locators. The requirement for this operation is to ensure (120 ± 0.08) mm and $\phi 18H7$. According to the location datum and the arrangement of the clamping force given in technical schedule, the hole $\phi 36 H7$ and its end face, and the external cylindrical surface with $\phi 36$ are used as the location surface. Location design is shown in Fig. 4-57.

Suppose the fit of the location hole to the location pin is $\phi 36H7/g6$ (the diameter of the location pin is $\phi 36_{-0.025}^{-0.009}$ mm). For the operation dimension (120 ± 0.08) mm, we know that the operation datum is consistent with the location datum, i.e. $\Delta_B = 0$. The datum displacement error caused by the manufacturing error of the location pair is: $\Delta_Y = T_d + T_D + \Delta_{min} = (0.025 + 0.016 + 0.009)$ mm $= 0.050$ mm. Therefore, the location error of this operation is: $\Delta_D = \Delta_Y = 0.050$ mm. As $\Delta_D < 0.16$ mm $\times 1/3 = 0.0533$ mm, this location scheme is feasible.

2) To design a tool-guiding device. In order to ensure the machining accuracy of hole ($\phi 18H7$), the machining process of drilling→core drilling→reaming is used. Therefore, it is determined to use a rapid-changing drilling bush, as shown in Fig. 4-58.

The drilling bush height $H = 1.5D = 1.5 \times 18$ mm $= 27$ mm; the chip removal gap $h = D = 18$ mm.

3) To design a clamping mechanism. With a view to simplifying the fixture structure and loading-unloading workpiece rapidly, it is determined to use screw clamping mechanism with a snap washer, as shown in Fig. 4-59.

4) To arrange an auxiliary support. An auxiliary support is arranged in the machining position to improve the stiffness of the workpiece.

5) To design a fixture body. The fixture body is a base part to join each component so as to form an integral fixture. Many factors, such as assembling, strength and rigidity, operation, chip removal, connection to the machine tool and so on, should be considered synthetically. As the drilling jig is generally trued by the drilling bush and fixed by the pressing plate, you just set aside the clamping place on the fixture body for the pressing plate.

(3) To draw the general assembly drawing of the fixture The general assembly drawing of the drilling jig is shown in Fig. 4-60.

(4) To mark the dimensions, tolerances and technical requirements Relative handbooks can be taken for reference to mark the dimension and fit. Specific dimension and fit for the drilling jig are shown in Fig. 4-60. Other technical requirements which are not marked in Fig. 4-60 are listed as

误差是 $\Delta_Y = T_d + T_D + \Delta_{\min} = (0.025 + 0.016 + 0.009)\text{mm} = 0.050\text{mm}$，因此本工序的定位误差为 $\Delta_D = \Delta_Y = 0.050\text{mm}$。由于 Δ_D 小于该工序尺寸制造公差 0.16mm 的 1/3，说明上述定位方案可行。

2）设计刀具导引装置。本工序小头孔加工的精度要求较高，需采用钻→扩→铰工艺，才能达到零件设计要求（ϕ18H7mm）。因此，钻套需要采用快换式，如图 4-58 所示。

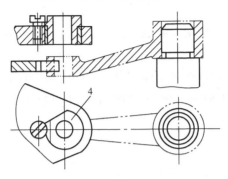

Fig. 4-58　Design of tool-guiding device 刀具导引设计

钻套高度 $H = 1.5D = 1.5 \times 18\text{mm} = 27\text{mm}$，排屑空间 $h = D = 18\text{mm}$。

3）设计夹紧机构。根据工艺规程设计内容，考虑到简化结构和工件快速装卸，确定采用螺旋夹紧机构，用螺母和开口垫圈实现快速装卸工件，如图 4-59 所示。

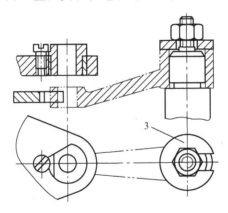

Fig. 4-59　Design of clamping device
夹紧机构设计

4）辅助支承设计。考虑到加工部位的刚性，在加工部位设计一辅助支承以提高刚性。

5）夹具体设计。夹具体是连接夹具各组成部分、使之成为一个有机整体的基础零件，设计时需要综合考虑装配、强度和刚度、操作应用、排屑以及与机床连接等诸多因素。因立式钻床夹具安装一般用钻套找正并用压板固定，故只需在夹具体上留出压板压紧的位置即可。

（3）绘制夹具总图　夹具总图如图 4-60 所示。

（4）在夹具装配图上标注尺寸、配合及技术要求　夹具总图的尺寸与配合标注可参照有关手册，具体标注如图 4-60 所示。其他未在图中标出的技术要求有：

follows:

1) The parallelism between the centerlines of the drilling bush and the location pin is 0.02mm.

2) The symmetry of the symmetric plane of movable V-block relative to the centerline of the drilling bush is 0.05mm.

3) The parallelism between the location plane on the jig and the underside of the jig body is 0.01mm.

4) The perpendicularity of the centerline of the drilling bush to the underside of the jig body is 0.02mm.

(5) To number each component of a jig, fill in the detail list, measure and obtain dimensions of the non-standard parts and complete detail drawings (omitted).

1) 钻套中心线对定位销中心线的平行度公差取为 0.02mm。
2) 活动 V 形块对称平面相对于钻套中心线的对称度公差取为 0.05mm。
3) 夹具上的定位平面对夹具底面的平行度取为 0.01mm。
4) 钻套中心线对夹具底面的垂直度误差取为 0.02mm。
（5）对零件进行编号、填写明细表、测绘零件图（略）

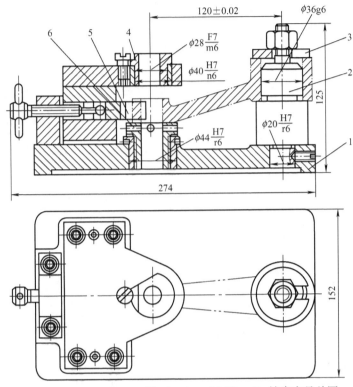

Fig. 4-60　General assembly drawing of drilling jig 钻床夹具总图

Chapter 5

Analysis and Control of Machining Quality

第5章

机械加工质量分析与控制

5.1 Introduction

5.1.1 Introduction to machining quality

All mechanical products are assembled with many interrelated components. The working performance and service life of a machine always relate directly to the machining quality of components and the assembly accuracy of the product, whereas the machining quality of components is the fundamental to ensure the product quality.

Machining quality includes machining accuracy and surface qualities. The machining accuracy consists of the dimension accuracy, form accuracy and mutual position accuracy between surfaces. The machining accuracy is also called geometric accuracy. The surface quality includes three aspects: surface roughness, surface waviness and physical and mechanical performance of the surface. The contents of the machining quality are shown in Fig. 5-1.

5.1.2 Basic concepts of machining accuracy

1. Machining accuracy and machining error

Machining accuracy means the tallying level of the practical geometric parameters of a workpiece machined with that of an ideal workpiece. The higher the tallying level, the higher the machining accuracy.

Machining error means the deviation degree of the practical geometric parameters of a workpiece machined from the ideal geometrical parameters. The machining accuracy and machining error are the indexes used to evaluate the exactness of geometric parameters of a workpiece from different sides. The magnitude of the machining error denotes the grade of the machining accuracy.

It is impossible to manufacture parts to the exact dimensions as prescribed at the design stage because of the inevitable errors at every stage of the manufacturing process. Considering the service performance of a product, it is not necessary to machine every component to the exact dimensions. Under the precondition of ensuring the service performance of a product, a certain amount of the machining error is permitted only that the error is not beyond predetermined range.

2. Economical machining accuracy

As there are many factors which have an influence on the machining accuracy in the machining process, the same machining method can reach different accuracies under different operation conditions. No matter what machining method is used, only serious operation, careful adjustment and choosing proper machining parameters in processing would make the machining accuracy improved. However, this would decrease the productivity and increase the machining cost.

Machining error is always inversely proportional to the machining cost, as shown in Fig. 5-2. It is noticed from Fig. 5-2 that:

1) The higher the machining accuracy, the larger the cost.
2) The improvement of machining accuracy has its limits. After point A (the curve on the left of

5.1 概述

5.1.1 机械加工质量概述

所有机械产品都是由许多相关零件装配而成的。机器的工作性能和使用寿命总是与零件的加工质量、产品的装配精度有着密切的关系，而零件的加工质量是保证产品质量的基础。

机械加工质量包括机械加工精度和加工表面质量。加工精度包括尺寸精度、形状精度和相互位置精度。加工精度也称几何精度。表面质量包含三个方面的内容：表面粗糙度、表面波度和表面的物理力学性能。机械加工质量的内容如图 5-1 所示。

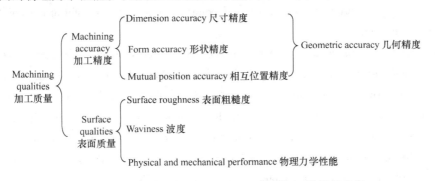

Fig. 5-1　Contents of machining quality 机械加工质量的内容

5.1.2 机械加工精度的基本概念

1. 加工精度与加工误差

加工精度是指零件加工后的实际几何参数与理想几何参数的符合程度。符合程度越高，加工精度就越高。

加工误差是指零件加工后的实际几何参数与理想几何参数的偏离程度。加工精度和加工误差是从不同角度评价零件几何参数精确性的两个指标。加工误差的大小反映了加工精度的高低。

实际加工中由于种种原因，任何加工方法都不可能把零件加工的绝对准确，总会出现这样或那样的加工误差。而且从满足产品使用性能和降低加工成本的角度来看，也没有必要把零件加工的绝对准确。因此，在保证产品使用性能的前提下，零件存在一定的加工误差是允许的，只要把加工误差控制在零件图上所规定的公差范围内即可。

2. 经济加工精度

由于在实际加工中影响加工精度的因素很多，即使是同一种加工方法，在不同的加工条件下所达到的加工精度不相同。无论采用何种加工方法，只要精心操作，细心调整，并选用合适的切削参数进行加工，都能使加工精度得到较大的提高。但这样会降低生产率，增加加工成本。

如图 5-2 所示，加工误差 Δ 总是与加工成本 C 成反比。由图 5-2 可知：

point A), even the cost continues to increase, the improvement of machining accuracy is not apparent.

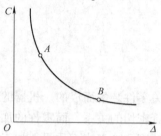

Fig. 5-2 Relation between machining error and cost
加工误差与成本的关系

3) The cost has a certain limit too. After B point (the curve on the right of the point B), cost keeps nearly unchanged.

Economical machining accuracy means the accuracy which can be easily gained by a specific machining method under the normal production conditions. Normal production conditions mean using the equipment conforming to the quality standard and the workers with standard technique grade, and not prolonging the machining time. In fact, the economical accuracy of a machining method has a certain range, and the machining accuracy in this range can be termed as economical.

On the other hand, the economical machining accuracy is not a fixed value, and is changing with the time and the progress of technology, as shown in Fig. 5-3.

3. Methods to study machining accuracy

There are two methods to study the machining accuracy. One is the single factor analysis method. On the basis of understanding the influence law of various original errors on the machining accuracy, this method is trying to determine which original error would cause this kind of error in machining process, and find the relation between the original error and machining error. The machining error is determined firstly by rough calculation, and then verified by experiment or inspection. Another one is the statistic method in which the practical geometric parameters of a batch of workpieces obtained under specific machining conditions are measured, and then processed by means of statistical theory so as to find the law and nature of machining errors and further control the machining quality.

In practice, these two methods are often used together. The statistical analysis method is used firstly to find out the rule of error appearance and estimate the possible reasons for producing the machining error. Then the single factor analysis method is used to make analysis and test so that the key factors affecting the machining accuracy can be found rapidly and effectively.

5.1.3 Basic concepts of machined surface quality

1. Meaning of machined surface quality

The surfaces machined by any machining method cannot be absolutely ideal ones. There are various geometrical errors on the machined surfaces. They are macro-geometry error (or geometrical error), surface waviness and micro-geometry error (or surface roughness); at the same time, the

1）加工精度越高，加工成本就越大。

2）加工精度的提高是有限度的，过 A 点后（A 点左侧曲线），即使再增加成本，加工精度的提高也不明显。

3）加工成本下降也有一定极限，过 B 点后（B 点右侧曲线），成本基本不变。

经济加工精度指在正常加工条件下，某种加工方法可容易达到的加工精度。正常加工条件指的是采用符合质量标准的设备和工艺装备，使用标准技术等级工人，不延长加工时间。事实上，某种加工方法的经济加工精度有一定的精度范围，在此范围内的加工精度被认为是经济的。

另一方面，经济加工精度不是固定不变的，而是随着时间的推移和技术的进步在不断地变化的，如图 5-3 所示。

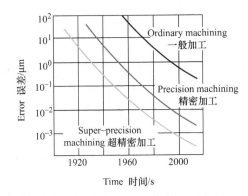

Fig. 5-3　Relation between machining error and time 加工误差与时间的关系

3. 研究加工精度的方法

研究加工精度的方法一般有两种：一是单因素分析法。在了解各种原始误差对加工精度影响规律的基础上，该方法试图确定究竟是哪项原始误差引起了该项加工误差，并找出原始误差和加工误差之间的关系。通常是先粗略计算加工误差的大小，然后通过实验、测试等方法进行验证。二是统计分析法，即运用数理统计方法对生产中一批工件的实测结果进行数据处理，以便确定加工误差的变化规律和性质，从而进一步控制加工质量。

上述两种方法在生产中往往结合起来使用。通常先用统计分析法找出误差的出现规律，初步判断产生加工误差的可能原因，然后运用因素分析法进行分析、实验，以便迅速有效地找出影响加工精度的关键因素。

5.1.3　加工表面质量的基本概念

1. 加工表面质量的含义

任何加工方法加工过的表面都不可能绝对理想。零件加工后的表面上有各种各样的几何形状误差。分别是宏观几何形状误差（或几何形状误差）、表面波度和微观几何形状误差（或表面粗糙度）；同时，表面层金属的物理力学性能也会在加工过程中发生改变，甚至化学性能也会在某种条件下发生改变。

通常，加工表面质量是指零件经过机械加工后表面层的物理力学性能和表面粗糙度。其主要内容如下：

physical and mechanical properties of metallic material on the surface layer may be changed in machining process, and even the chemical nature would be changed under some conditions.

Generally, machined surface quality means the physical and mechanical properties of the surface layer and the surface roughness of the machined parts. The main contents of the surface quality are as follows:

(1) Geometrical errors of machined surface layer

1) Surface roughness. It means the micro-geometry error of the machined surface, and the peak-valley uneven track on the surface formed by the cutter cutting movement. Its characteristic is that the ratio between its wave length L_1 and its wave height H_1 is within $L_1/H_1 < 50$, as shown in Fig. 5-4.

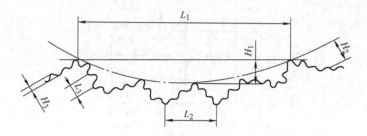

Fig. 5-4　Geometrical shape of surface　表面的几何形状

2) Surface waviness. It is the periodic geometry error in between the geometry error ($L_3/H_3 > 1000$) and the surface roughness. The ratio between its wave length L_1 and wave height H_2 is within $50 \leqslant L_2/H_2 \leqslant 1000$, as shown in Fig. 5-4. It is caused mainly by the low-frequency vibration of the process system.

(2) Physical and mechanical properties of the surface layer

Physical and mechanical properties of the surface layer include the work hardening of the surface layer, residual stress, and metallurgical structure change of the surface layer.

2. Influence of surface qualities on service performances of products

(1) Influence of surface qualities on the wear resistance of components　When the friction pair material, heat treatment status and lubricating condition have been determined, the surface quality of the component plays a decisive role in the wear resistance of the component.

1) Influence of the surface roughness on the wear resistance. It can be seen from Fig. 5-5 that under certain conditions, always there is an optimal roughness value on the surface of the friction pair. That is, a component with such roughness will have the minimum initial wear rate. Over-large or over-small surface roughness would make the initial wear rate increase.

2) The influence of the surface vein direction on the wear resistance. It can be seen from Fig. 5-6 that it has the best wear resistance when the vein direction of two surfaces of a friction pair is parallel to the relative motion direction (the curve 3); and it has the worst wear resistance when the vein direction of two surfaces of a friction pair is perpendicular to the relative motion direction (the curve 1).

（1）加工表面层的几何形状误差

1）表面粗糙度。它指的是加工后表面的微观几何形状误差以及由刀具切削运动在零件表面留下的凹凸不平的痕迹。其特征是波长 L_1 与波高 H_1 之比 $L_1/H_1 < 50$，如图 5-4 所示。

2）表面波度。它是介于宏观几何形状误差（$L_3/H_3 > 1000$）和表面粗糙度之间的周期性的几何形状误差。而波长与波高之比为 $50 \leqslant L_2/H_2 \leqslant 1000$，如图 5-4 所示。它主要由工艺系统的低频振动引起。

（2）表面层的物理力学性能　表面层金属的物理力学性能主要包括表面层的加工硬化、残余应力、金相组织变化等三个方面的内容。

2. 表面质量对产品使用性能的影响

（1）表面质量对零件耐磨性的影响　当摩擦副的材料、热处理状态以及润滑条件确定后，零件的表面质量对其耐磨性起决定性的作用。

1）表面粗糙度对耐磨性的影响。由图 5-5 可见，在一定条件下，在摩擦副表面总是存在一个最佳的粗糙度。即具有这种大小粗糙度的零件将会有最小的起始磨损率。表面粗糙度过大或过小都会使起始磨损量增大。

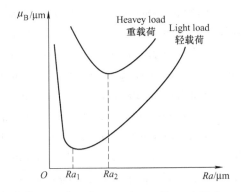

Fig. 5-5　Influence of surface roughness Ra on initial wear rate μ_B
表面粗糙度 Ra 对起始磨损率 μ_B 的影响

2）表面纹理方向对零件耐磨性的影响。在轻载条件下，当一对摩擦副的两个表面的纹理方向与相对运动方向一致时其耐磨性最好（图 5-6 中的曲线 3）；当一对摩擦副的两个表面的纹理方向与相对运动方向垂直时其耐磨性最差（图 5-6 中的曲线 1）。

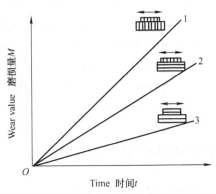

Fig. 5-6　Influences of the surface vein direction on the wear resistance 表面纹理方向对耐磨性影响

3) The influence of physical and mechanical properties of the surface layer on the wear resistance. The work hardening of the surface layer is beneficial to improving the wear resistance. The reason is that the cold hardening increases the micro-hardness of the surface layer. But, the wear resistance does not always increase with the increase of the surface hardness at all, as shown in Fig. 5-7. If the surfaces are over hardened, the hardness difference between the interior and the surface layer of the component is too large and surface peeling would happen, which will aggravate the wear.

If the metallurgical structure of the surface layer changes, the original hardness of the matrix material will change. This will bring a direct influence on the component wear.

(2) Influence of surface qualities on the fatigue strength　The surface roughness has a very large influence on the fatigue strength of the component. The influence of the surface roughness on the fatigue strength is shown in Fig. 5-8. To decrease the surface roughness can increase the fatigue strength. The internal stress of a surface layer has quite a large influence on the fatigue strength of the component. A residual compression stress in the surface layer is beneficial to improving the fatigue strength of the component. In comparison, a residual tension stress in the machined surface facilitates crack forming and extending, so it decreases the fatigue strength of the component. Proper cold hardening of the surface layer can strengthen the surface layer of the component and hamper the formation and expansion of cracks, thus increasing the fatigue strength of the component.

(3) The influence of surface qualities on the corrosion resistance　The corrosion resistance of components depends to a large extent on the surface roughness. If the damp air and corrosive medium contact with the component surface, they will accumulate easily on it and cause corrosion. The larger the surface roughness is, the larger the contact areas between the damp air and the machined surface would be, and the corrosion action would become more serious.

Both cold hardening and metallurgical structure change of the surface would produce internal stress. Working under the internal stress state, the machinery components would suffer stress corrosion. Therefore, internal stress inside the surface layer can decrease the corrosion resistance of the component.

(4) Influence of surface qualities on the fit properties of mating parts　The surface quality, particularly the surface roughness, has a very large influence on the fit properties of mating parts. For clearance fit, the larger the surface roughness is, the more serious the surface wear is, thus leading to the increase of the fit clearance and the decrease of the fit accuracy. For interference fit, the peaks on the parts with larger surface roughness would be pressed down, which reduces the practical interference and decreases the fit strength of mating parts. Therefore, the surfaces with higher fit accuracy should have smaller surface roughness.

5.2　Factors affecting machining accuracy

Part machining happens in the technological system composed of machine tools, fixtures, cutting tools and workpieces. All the factors in the system that may cause machining errors are called

3）表面层金属的物理力学性能对耐磨性的影响。加工表面冷作硬化一般有利于提高耐磨性，其原因是因为冷作硬化提高了表面层的显微硬度，但是，并非硬化程度越高耐磨性越好，如图5-7所示。过度的冷作硬化会使表面层金属变得很硬，零件的内部组织和外表的硬度差过大，容易造成表面层脱落和撕裂，这种情形将会加剧磨损。

如果表面层的金相组织结构发生变化，母体材料的原始硬度将会改变，这将对零件磨损产生直接影响。

（2）表面质量对耐疲劳强度的影响　表面粗糙度对零件的疲劳强度影响很大。图5-8所示为表面粗糙度 Ra 对相对疲劳强度 σ_r 的影响。减小零件的表面粗糙度，可以提高零件的疲劳强度。表面层残余应力对疲劳强度的影响极大。表面层残余压应力有利于提高零件的耐疲劳强度。相比之下，表面层残余拉应力有助于裂纹的形成和扩展，从而降低零件的疲劳强度。适度的冷作硬化可以强化零件的表面层，阻止裂纹的形成和扩展，从而提高零件的耐疲劳强度。

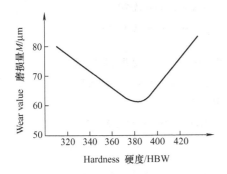

Fig. 5-7　Relation between work hardening and wear resistance 加工硬化与耐磨性的关系

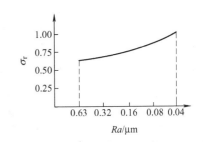

Fig. 5-8　Influence of surface roughness on fatigue strength 表面粗糙度对疲劳强度的影响

（3）表面质量对耐腐蚀性的影响　零件的耐腐蚀性很大程度上取决于零件的表面粗糙度。空气中所含的潮湿空气和腐蚀介质与零件接触时会凝聚在零件表面上使表面腐蚀。表面粗糙度越大，加工表面与气体、液体接触面积越大，腐蚀作用就越强烈。

表面的冷作硬化和残余应力一般都会降低零件表面的耐腐蚀性。表面冷作硬化和金相组织变化都会产生内应力。零件在应力状态下工作时会产生应力腐蚀。因此，表面内应力会降低零件表面的耐腐蚀性。

（4）表面质量对配合件配合性质的影响　表面质量，尤其是表面粗糙度对配合件配合性质有很大的影响。对于间隙配合，表面粗糙度越大，磨损越严重，导致配合间隙增大，配合精度降低；对于过盈配合，装配时表面粗糙度较大部分的凸峰会被挤平，使实际的配合过盈减少，降低配合表面的结合强度。因此配合精度要求较高的表面，应具有较小的表面粗糙度。

5.2　影响加工精度的因素

零件的机械加工是在由机床、夹具、刀具和工件组成的工艺系统中进行的。工艺系统中凡是能直接引起加工误差的因素都称为原始误差。原始误差的存在，使工艺系统各组成部分之间的位置关系或速度关系偏离了理想状态，致使加工后的零件产生了加工误差。若原始误

the original errors. The existence of the original errors makes the positional or motional relationship deviate from its desired state, and brings about the machining error. If an original error has been existed before machining, and is inspected under no cutting load, this kind of original error is named as static error of technological system; if an original error is produced under the action of cutting load, it is named as dynamic error of technological system. Various original errors appeared in machining process can be classified as follow.

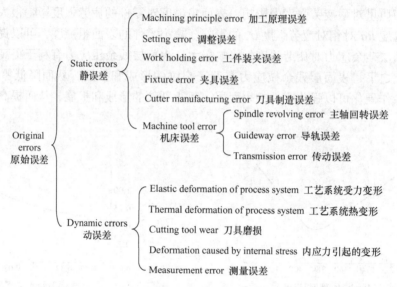

Fig. 5-9 illustrates various original errors appeared in boring the hole of the piston pin. A location error appeared when the location datum is not design datum. Overlarge clamping force would cause a clamping error. Guideway wear would lead to a machine tool error. In addition, there are many other kinds of errors, such as tool setting errors, errors caused by thermal deformation of machine tools, errors caused by tool wear, measurement error caused by measuring method and instrument error, and so on.

5.2.1 Machining principle errors

In order to machine specified surface, the accurate forming motions between the cutting tool and the workpiece must be realized. This is called the machining principle in mechanical machining field. The machining principle error is caused by approximate forming motion, or approximate cutting edge profile. Theoretically, an ideal machining principle or completely exact forming motion should be used so as to obtain the accurate component surface. But it is often very difficult to realize the completely exact machining principle in practices. Sometimes the exact machining principle has a low machining efficiency; sometimes it would make the structure of the machine tool or cutting tool very complex and difficult in manufacturing; sometimes it would cause larger transmission errors as there are more transmission members. Therefore, using the approximate machining principle is helpful for simplifying the design and manufacturing of the machine tool or cutting tool, reducing the manufacturing cost, and increasing the productivity of products.

差是在加工前已存在的，即在无切削负荷的情况下检验的，称为工艺系统静误差；若在有切削负荷情况下产生的则称为工艺系统动误差。加工中可能出现的种种原始误差归纳如下。

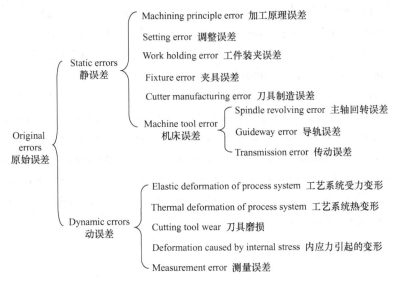

图 5-9 所示为镗活塞销孔中的各种原始误差。当定位基准不是设计基准时产生定位误差。过大的夹紧力会引起夹紧误差。导轨磨损会导致导轨误差。此外，还有许多其他误差，如刀具调整误差，机床热变形引起的误差，刀具磨损引起的误差，由测量方法和仪器误差引起的误差等。

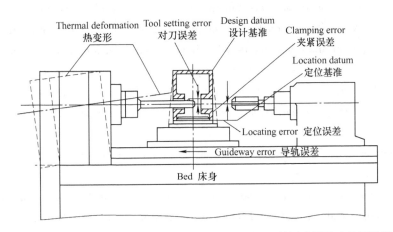

Fig. 5-9　Original errors in boring the hole of the piston pin 镗活塞销孔中的原始误差

5.2.1　加工原理误差

在机械加工中为了获得规定的加工表面，刀具和工件之间必须实现准确的成形运动。在机械加工领域将此称为加工原理。加工原理误差是指采用了近似的成形运动或近似的刀刃轮廓进行加工而产生的误差。理论上应采用理想的加工原理和完全准确的成形运动以获得精确的零件表面。但实际上完全精确的加工原理常常很难实现。即使能采用准确的加工原理，有时会使加工效率很低；有时会使机床或刀具的结构极为复杂，制造困难；有时由于结构环节

Take the gear hobbing for example. There are two kinds of machining principle errors in the gear hobbing. One is the approximate cutting edge profile, that is, because of the difficulties in manufacturing the involute hob, Archimedes' worm or basic worm with a straight profile in normal direction instead of an involute worm is used to fabricate the hob. The other is that because of limit numbers of the cutting edges, the practical tooth face cut by the hob is formed by lots of polygonal line instead of smooth involutes, as shown in Fig. 5-10.

The other example is about gear milling by means of module milling cutter. We know that the tooth parameters are different for the gears with the same module and different tooth numbers. In theory, the gears with the same module and different tooth numbers should be cut by means of different milling cutters corresponding to it. In fact, in order to reduce the cutter numbers and machining cost, a handle of the module milling cutter is often used for cutting the gears within a range of tooth numbers. Because of the approximate cutting edge profile, there is a machining principle error in gear milling, as shown in Fig. 5-11.

5.2.2 Errors of machine tools

Errors of machine tools arise from the manufacture error, installation error and wear without cutting load. There are many kinds of machine tool errors. The guideways error, spindle revolving error and transmission error of the transmission chain, which have a larger influence on the machining accuracy, are mainly analyzed in this section.

1. Errors of machine tool guideways

Guideways are the support members and linear motion datum of a machine tool. The errors of guideways have a direct influence on the machining accuracy. The manufacture error, installation error, fit clearance, wear and other factors would have the guideways produce guiding errors. In the precision standards for machine tools, the guiding precisions of linear guide ways generally include the straightness of the guideway in horizontal/vertical plane (bending of guide way), parallelism between the front and back guideways (twist of guide ways), the parallelism or perpendicularity of the guideway to the spindle revolving axis, and so on.

Take the center lathe for example to illustrate the influences of guideways errors on the machining accuracy.

(1) Straightness of a guideway in horizontal plane The straightness error of the guideway in horizontal plane would make a shaft after longitudinal turning present the "drum shape" or "saddle shape". The straightness error of the guideway in horizontal plane (protruding to the back side) is shown in Fig. 5-12a. Fig. 5-12b illustrates the "saddle shape" error of a shaft caused by the straightness error of the guideway in horizontal plane. In the same way, if the guideway protrudes to the front, the shaft after longitudinal turning would present in the "drum shape". It can be seen from Fig. 5-12b that the straightness of the guide way in horizontal plane is directly reflected on the workpiece.

(2) Straightness of the guideway in vertical plane Similarly, this error would also cause the form error of the workpiece. It will be seen that the effect of this error on machining accuracy is very

多,造成机床传动中的误差增加,或使机床刚度和制造精度很难保证等。因此,采用近似的加工原理有助于简化机床或刀具的设计与制造,降低制造成本以及提高产品的生产率。

以滚齿加工为例。滚齿加工中存在两种加工原理误差:一是采用了近似的刀刃轮廓,即由于制造上的困难,采用阿基米德基本蜗杆或法向直廓基本蜗杆代替渐开线基本蜗杆制造滚刀;二是由于滚刀刀刃数有限,所切出的齿形实际上是一条由微小折线组成的折线面,和理论上的光滑渐开线有差异,这些都会产生加工原理误差,如图 5-10 所示。

另一个例子是关于用模数铣刀铣削齿轮。模数相同而齿数不同的齿轮,齿形参数是不同的。理论上,同一模数、不同齿数的齿轮就要用相应的齿形刀具加工。实际上,为精简刀具数量,常用一把模数铣刀加工某一齿数范围内的齿轮,即采用了近似的刀刃轮廓,同样产生了加工原理误差,如图 5-11 所示。

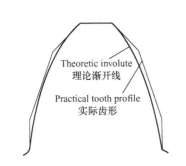

Fig. 5-10　Tooth error in gear hobbing
滚齿的齿形误差

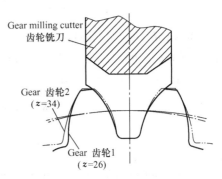

Fig. 5-11　Tooth error in gear milling
铣齿的齿形误差

5.2.2　机床误差

机床的误差是指在无切削负荷下,来自机床本身的制造误差、安装误差和磨损。机床误差的项目很多,这里主要分析对工件加工精度影响较大的主轴回转误差、导轨误差和传动链误差。

1. 机床导轨误差

机床导轨是机床的支承件和直线运动基准。机床导轨误差对加工精度影响很大。由于导轨副的制造误差、安装误差、配合间隙以及磨损等因素的影响,会使导轨产生导向误差。在机床的精度标准中,直线导轨的导向精度一般包括:导轨在水平面/垂直面内的直线度(弯曲)、前后导轨的平行度(扭曲)以及导轨对主轴回转轴线的平行度(或垂直度)等。

下面以车床为例,分析导轨误差对加工精度的影响。

(1) 导轨在水平面内的直线度误差　导轨在水平面内的直线度误差会使被加工的工件产生鼓形或鞍形。由于导轨在水平面内向"后凸"(图 5-12a),纵向车削圆柱面时使工件产生"鞍形"误差,如图 5-12b 所示。同理可知,如果导轨在水平面内向"前凸",纵向车削圆柱面时使工件产生"鼓形"误差。如果导轨在水平面内的直线度误差为 ΔY,则由此引起的工件半径误差 $\Delta R' = \Delta Y$,如图 5-12c 所示。可见导轨在水平面内的直线度误差对加工精度的影响很大。

little, even can be ignored.

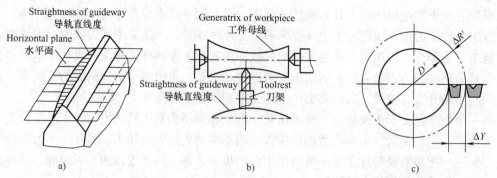

Fig. 5-12　Machining error caused by the straightness of the guideway in horizontal plane
导轨在水平面内的直线度误差引起的加工误差

As shown in Fig. 5-13, when the tool nose moves up or down ΔZ along the tangential to the workpiece surface, the radius of the workpiece increases from R to R', the increment is ΔR.

According to the geometric relation, it can be seen from Fig. 5-13b that

$$R' = \sqrt{R^2 + \Delta Z^2} \approx R + \frac{\Delta Z^2}{2R}$$

$$\Delta R = R' - R = \frac{\Delta Z^2}{2R} = \frac{\Delta Z^2}{D} \tag{5-1}$$

where D stands for the diameter of the workpiece.

It can be seen from Fig. 5-12 and Fig. 5-13 that the original errors in different directions have a different influence on machining error. Suppose $\Delta Z = \Delta Y = 0.1$ mm, $D = 2R = 40$ mm, then

$$\Delta R = \frac{0.1^2}{40} \text{mm} = 0.00025 \text{mm}$$

$$\Delta R' = \Delta Y = 0.1 \text{mm} = 400 \Delta R$$

The example tells us that when the original error presents in the normal direction of the machining surface, its influence on machining accuracy is the largest; and the influence of the original error in the tangential direction of the machining surface is the smallest, and even can be ignored. The direction (normal to machining surface) with the largest machining error is named as error sensitive direction.

(3) Parallelism between two guideways in a lathe (twist)　As shown in Fig. 5-14, this error would cause a worktable to incline crosswise when the worktable moves and change the position of the workpiece relative to the cutting tool, thus producing geometric error on the workpiece. Suppose the height of the center is H, the width of two guideways is B, parallelism between two guideways is Δ, then the radial variation of the tool point is:

$$\Delta R \approx \Delta y = \Delta \cdot H/B \tag{5-2}$$

For a general lathe, $H/B \approx 2/3$; for a grinding machine, $H/B \approx 1$. Therefore, the influence of the parallelism error on machining accuracy can't be ignored.

(4) The position error of guideways relative to the spindle axis　Influence of the position error of guideways relative to the spindle axis on machining accuracy (see Fig. 5-15).

（2）导轨在垂直面内的直线度误差　同样，这种误差也会引起工件出现形状误差。后面将会发现这种误差对加工精度的影响微乎其微，甚至可以忽略不计。

如图 5-13 所示，当刀尖沿着工件表面的切线向上或向下移动 ΔZ 时，工件的半径从 R 增大到 R'，其增量为 ΔR。

根据图 5-13b 中的几何关系，得

$$\Delta R = R' - R = \frac{\Delta Z^2}{2R} = \frac{\Delta Z^2}{D} \tag{5-1}$$

其中，D 表示工件的直径。

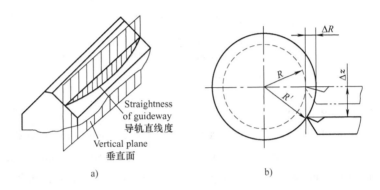

Fig. 5-13　Straightness of guideways in vertical plane
导轨在垂直面内的直线度误差

由图 5-12 和图 5-13 看出，不同方向的原始误差对加工误差的影响程度不同。假设 $\Delta Z = \Delta Y = 0.1\text{mm}$，$D = 2R = 40\text{mm}$，那么

$$\Delta R = \frac{0.1^2}{40}\text{mm} = 0.00025\text{mm}$$

$$\Delta R' = \Delta Y = 0.1\text{mm} = 400\Delta R$$

此例说明，当原始误差出现在加工表面的法线方向时，对加工精度的影响最大；而在加工表面切线方向的原始误差对加工精度的影响最小，甚至可以忽略不计。为此把具有最大加工误差的方向，即加工表面的法线方向称为误差敏感方向。

（3）车床前后导轨之间的扭曲　如图 5-14 所示，这种误差会造成工作台移动时产生横向倾斜，改变工件相对刀具的位置，从而在工件上出现几何形状误差。设车床中心高为 H，导轨宽度为 B，导轨的平行度误差是 Δ，那么刀尖在工件径向的变化量为

$$\Delta R \approx \Delta y = \Delta \cdot H/B \tag{5-2}$$

对于一般车床，$H/B \approx 2/3$，外圆磨床，$H/B = 1$。因此，导轨扭曲对加工精度的影响不能轻视。

（4）导轨相对主轴回转轴线的位置误差　导轨与主轴回转轴线的相互位置误差引起的工件形状误差如图 5-15 所示。

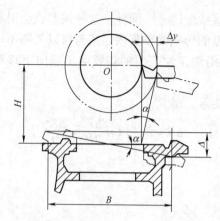

Fig. 5-14 Effect of parallelism between two
guide ways on the workpiece 导轨平行度对工件的影响

2. Spindle revolving error

The spindle of a machine tool is an important part which is used to mount a workpart or cutting tool and to transmit the cutting motion and power. Its revolving accuracy is one of the primary accuracies of a machine tool, and has an effect on geometrical accuracy, mutual position accuracy of machined surfaces, and surface roughness.

Theoretically, when spindle revolves, the space position of its revolving axis is stationary. In practical, various factors in spindle system would change the position of the spindle revolving axis. The maximum variation of the real revolving axis of the spindle relative to its ideal revolving axis (generally replaced by the average revolving axis) is called the spindle revolving error.

(1) Basic forms of spindle revolving errors For the sake of analyzing the influence of the spindle revolving error on machining errors, spindle revolving errors are divided into three basic forms: pure radial runout, pure axial runout, pure angular swing, as shown in Fig. 5-16.

Because the spindle revolving errors are caused by above three error forms, the error moving tracks of spindle centers in different sections of the spindle are neither the same nor similar.

(2) Primary factors influencing on the spindle revolving error Primary factors influencing on the spindle revolving error have to do with the manufacturing accuracy of spindle assembly. One is the manufacturing errors of bearing itself, such as the roundness, waviness and coaxiality of the slide bearing's journal and hole, perpendicularity of end face to the rotating axis of the slide bearing; or roundness, waviness of a ball bearing race, roundness and dimension error of a roller, coaxiality of race and bearing hole; bearing clearance and perpendicularity of the race of thrust ball bearing to rotating axis, etc. The other is the manufacturing accuracy of the spindle and bearing hole, such as the roundness, waviness and coaxiality of the spindle journal and bearing hole, perpendicularity of a spindle shoulder to a spindle axis and so on.

(3) Effect of spindle revolving errors on machining accuracy Different revolving error forms have different effects on machining accuracy. Even the same type of revolving errors also has differ-

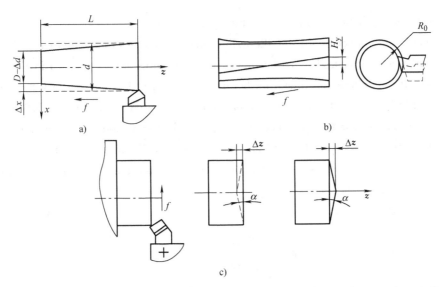

Fig. 5-15 Workpiece form errors caused by the position error of guideways relative to the spindle axis
导轨与主轴回转轴线的相互位置误差引起的工件形状误差

2. 主轴回转误差

机床主轴是用来装夹工件或刀具并传递切削运动和动力的重要零件，它的回转精度是机床主要精度指标之一，对零件加工表面的几何形状精度、相互位置精度和表面粗糙度都有影响。

为了保证加工精度，机床主轴回转时，其回转轴线的空间位置应当固定不变，但实际上由于主轴部件在制造、装配、使用过程中的种种因素影响，主轴在每一瞬间回转轴线的空间位置都是变动的，即存在着回转误差。主轴回转误差是指主轴实际回转轴线相对其理想回转轴线（一般用平均回转轴线来代替）的偏离程度。

（1）主轴回转误差的基本形式　为便于分析，主轴回转误差可以分解为三种基本形式：纯径向圆跳动、纯轴向窜动和纯角度摆动，如图 5-16 所示。

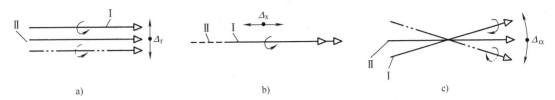

Fig. 5-16 Basic forms of spindle revolving errors 主轴回转误差的基本形式
a）Pure radial runout 纯径向圆跳动　b）Pure axial runout 纯轴向窜动　c）Pure angular swing 纯角度摆动

由于主轴回转误差是上述三项误差综合作用的结果，所以主轴不同截面内轴心的误差运动轨迹既不相同，也不相似。

（2）影响主轴回转误差的主要因素　影响主轴回转误差的主要因素与主轴部件的制造精度有关：一是轴承本身的制造误差，如滑动轴承轴颈和轴承孔的圆度、波度和同轴度，端面与回转轴线的垂直度，或滚动轴承滚道的圆度和波度，滚动体的圆度和尺寸误差，滚道与轴承内孔的同轴度，轴承间隙以及推力滚动轴承的滚道与主轴回转轴线的垂直度等；二是轴

ent effect on the accuracy under different machining conditions.

For example, the influence of a spindle radial runout on the external cylindrical turning is very little, the roundness error caused by radial runout even can be ignored, as shown in Fig. 5-17. But the influence of the spindle radial runout on boring is very large, it would cause a large roundness error, as shown in Fig. 5-18. The axial runout of a spindle has a large influence on facing and threading. Fig. 5-19a shows that the end face is not perpendicular to the axis of a workpiece, and Fig. 5-19b shows that the machined thread presents on the periodic pitch error. The influence of the spindle angular swing on boring is very clear, it would cause the roundness error (ellipse) of the hole, as shown in Fig. 5-20.

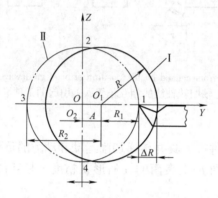

Fig. 5-17　Influence of radial runout on turning
径向跳动对车外圆的影响

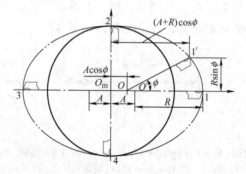

Fig. 5-18　Influence of radial runout on boring
径向跳动对镗孔的影响

In summary, machining errors caused by spindle revolving errors are listed in Tab. 5-1.

承相配件（如主轴、箱体孔等）的精度和装配质量，如主轴颈和轴承内孔各自的圆度、波度和同轴度，轴肩与主轴回转轴线的垂直度等。

（3）主轴回转误差对加工精度的影响　不同的回转误差形式对加工精度的影响不同。即使是同一类型的回转误差在不同的加工条件下对加工精度的影响也不同。

例如，主轴径向跳动对外圆车削的影响很小，由此引起的圆度误差甚至可以忽略不计，如图 5-17 所示。但是，主轴径向跳动对镗孔的影响却很大，造成大的圆度误差，如图 5-18 所示。主轴的轴向窜动对车端面和车螺纹有较大的影响。图 5-19a 所示车削过的端面与工件轴线不垂直，而图 5-19b 表明加工过的螺纹导程存在周期误差。图 5-20 清楚地表明了主轴的角度摆动对镗孔的影响，它将使镗过的孔出现圆度误差，即呈现椭圆形。

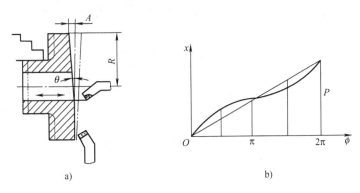

Fig. 5-19　Influence of axial runout on facing and threading 轴向窜动对车端面和车螺纹的影响

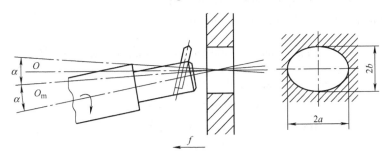

Fig. 5-20　Influence of spindle angular swing on boring 主轴的角度摆动对镗孔的影响

归纳起来，表 5-1 中列出了主轴回转误差所引起的加工误差。

Table. 5-1　Machining errors caused by spindle revolving errors 主轴回转误差引起的加工误差

Error forms 误差形式	Turning 车削			Boring 镗削	
	Straight turning 内、外圆车削	Facing 车端面	Threading 车螺纹	Boring hole 镗孔	Facing 车端面
Radial runout 径向跳动	Very little 影响极小	No 无影响	No 无影响	Roundness 圆度	No 无影响
Axial runout 轴向窜动	No 无影响	Flatness 平面度 Perpendicularity 垂直度	Pitch error 螺距误差	No 无影响	Flatness 平面度 Perpendicularity 垂直度
Angular swing 角度摆动	Very little 影响极小	Very little 影响极小	Pitch error 螺距误差	Cylindricity 圆柱度	Flatness 平面度

3. Errors of transmission chain

When machining some surfaces such as threads, gears, worm gears, the constant motion relationship between the cutting tool and the workpiece is ensured by the transmission system of the machine tool. Transmission chain error of the machine tool refers to the relative motion error of the elements at the two ends of internal connection transmission chain.

Take threading for example to illustrate the influence of the transmission chain error on machining accuracy. Fig. 5-21 shows a sketch of transmission system in a threading lathe.

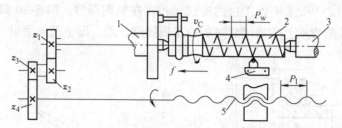

Fig. 5-21 Transmission system of a threading lathe 螺纹车床的传动系统示意图
1—Spindle 主轴　2—Workpiece 工件　3—Tailstock 尾座　4—Toolrest 刀架　5—Leadscrew 丝杠

According to Fig. 5-21, the relation between the pitch P_w of the workpiece and the pitch P_1 of the leadscrew can be expressed by Eq. (5-3).

$$P_w = i \cdot P_1 \tag{5-3}$$

where, P_w——Pitch of the workpiece to be machined;

P_1——Pitch of the leadscrew of machine tool;

i——Transmission ratio from the spindle 1 to the leadscrew 5, $i = \dfrac{z_1}{z_2} \times \dfrac{z_3}{z_4}$.

Differentiating with respect to Eq. (5-3), then

$$\Delta P_w = i \cdot \Delta P_1 + P_1 \cdot \Delta i \tag{5-4}$$

The lead error of the workpiece can be expressed as follows:

$$dP_w = i \cdot dP_1 + P_1 \cdot di \tag{5-5}$$

It is known from Eq. (5-5) that the lead error dP_w of the workpiece is composed of two items. One is $i \cdot dP_1$ which is caused by the transmission error of lead screw pairs. The other, i.e. $P_1 \cdot di$, is caused by total drive ratio error.

If the drive ratio error is considered only, then

$$dP_w = P_1 \cdot di \tag{5-6}$$

The variation of Δi reflects a finally angle error on the workpiece. So Δi can be expressed as:

$$\Delta i = \Delta \theta_s / 2\pi \tag{5-7}$$

where, $\Delta \theta_s$ is total angle error of the lead screw, i.e. accumulating errors of all drive parts to the lead screw.

Set $\Delta \varphi_j$ is the angle error of the j th drive part in one revolution of a workpiece; i_j is the drive ratio from the jth drive part to the lead screw; $\Delta \theta_j$ is the angle error of the lead screw caused by the angle error of the jth drive part, then

3. 传动链误差

当加工如螺纹、齿轮、蜗轮等零件时，机床的传动系统必须保证刀具与工件之间有恒定的运动关系。机床传动链误差是指内联传动链首末两端传动件间相对运动的误差。

下面以车螺纹为例来说明传动链误差对加工精度的影响。图 5-21 所示为某螺纹车床传动系统示意图。

根据图 5-21 所示的传动关系，被加工工件螺距与机床丝杠螺距之间的关系可以用式（5-3）来表示。

$$P_w = i \cdot P_l \tag{5-3}$$

式中　P_w——被加工工件的螺距；

　　　P_l——机床丝杠螺距；

　　　i——从主轴 1 到丝杠 5 之间的传动比，$i = \dfrac{z_1}{z_2} \times \dfrac{z_3}{z_4}$。

对式（5-3）两边及进行微分，得

$$\Delta P_w = i \cdot \Delta P_l + P_l \cdot \Delta i \tag{5-4}$$

被加工工件的导程误差可以表示为

$$dP_w = i dP_l + P_l di \tag{5-5}$$

由式（5-5）可知，工件的导程误差 dP_w 由两项构成。一项是由丝杠传动副传动误差引起的 $i \cdot dP_l$，另一项则是由总传动比误差引起的 $P_l di$。

如果仅考虑传动比误差，则

$$dP_w = P_l di \tag{5-6}$$

Δi 的变化最终反映出工件的转角误差。因此，Δi 可以用下式来表示

$$\Delta i = \Delta \theta_s / 2\pi \tag{5-7}$$

其中，$\Delta \theta_s$ 为丝杠的总转角误差，即各传动元件所引起末端元件（丝杠）转角误差的叠加。

设 $\Delta \varphi_j$ 是工件转一转第 j 个传动元件的转角误差；i_j 是从第 j 个传动元件到丝杠的传动比；$\Delta \theta_j$ 是由第 j 个传动元件的转角误差引起的丝杠转角误差，那么

$$\Delta \theta_j = i_j \cdot \Delta \varphi_j \tag{5-8}$$

丝杠的总转角误差为

$$\Delta \theta_s = \sqrt{\sum_{j=1}^{n} \Delta \theta_j^2} \tag{5-9}$$

因为

$$\Delta \theta_{z1} = \frac{z_1}{z_2} \cdot \frac{z_3}{z_4} \cdot \Delta \varphi_{z1}, \quad \Delta \theta_{z2} = \frac{z_3}{z_4} \cdot \Delta \varphi_{z2}, \quad \Delta \theta_{z3} = \frac{z_3}{z_4} \cdot \Delta \varphi_{z3}, \quad \Delta \theta_{z4} = \Delta \varphi_{z4}$$

把上述各转角误差代入式（5-9），得

$$\Delta\theta_j = i_j \cdot \Delta\varphi_j \tag{5-8}$$

Total angle error of the lead screw is

$$\Delta\theta_s = \sqrt{\sum_{j=1}^{n} \Delta\theta_j^2} \tag{5-9}$$

$\Delta\theta_{z1} = \dfrac{z_1}{z_2} \cdot \dfrac{z_3}{z_4} \cdot \Delta\varphi_{z1}$, $\Delta\theta_{z2} = \dfrac{z_3}{z_4} \cdot \Delta\varphi_{z2}$, $\Delta\theta_{z3} = \dfrac{z_3}{z_4} \cdot \Delta\varphi_{z3}$, $\Delta\theta_{z4} = \Delta\varphi_{z4}$

Substitute each angle error into Eq. (5-9), then

$$\begin{aligned}\Delta\theta_s &= \sqrt{\Delta\theta_{z1}^2 + \Delta\theta_{z2}^2 + \Delta\theta_{z3}^2 + \Delta\theta_{z4}^2} \\ &= \sqrt{\left(\dfrac{z_1}{z_2} \cdot \dfrac{z_3}{z_4}\right)^2 \cdot \Delta\varphi_{z1}^2 + \left(\dfrac{z_3}{z_4}\right)^2 \cdot \Delta\varphi_{z2}^2 + \left(\dfrac{z_3}{z_4}\right)^2 \cdot \Delta\varphi_{z3}^2 + \Delta\varphi_{z4}^2}\end{aligned} \tag{5-10}$$

It can be seen from the error analysis that: ①short transmission chain is beneficial to improving drive accuracy; ②smaller drive ratio i has higher drive accuracy, and using speed-down drive is an important rule to ensure drive accuracy; ③the last drive member should have higher machining and assembling accuracy.

5.2.3 Errors caused by elastic deformation of process systems

In machining process, the process system will present elastic and plastic deformation under the action of external forces such as cutting force, driving force, clamping force, weight and inertial force and so on. The deformation will change the correct position between the cutting tool and the workpiece and cause machining errors. You might as well go to machine shop to observe carefully the machining situations as follows.

1) Turning a long shaft (slender) on a new lathe.
2) Turning a short shaft (strong shaft) on an old lathe.
3) Longitudinal grinding a long shaft without feed-in or "smooth grinding" on an external cylindrical grinding machine.

What have you found? What do these phenomena mean?

A batch of shafts are machined by means of a center lathe, as shown in Fig. 5-22a. After turning, the shafts would have form error, either in the shape of a "drum", or in the shape of a "saddle". Try to explain what the reason is.

1. Rigidity of process system

(1) Basic concept of system rigidity In material mechanics, the static rigidity K of a body means the ratio of the force F exerting on the body to the deformation y, caused by the force F, in the acting direction of the force. That is

$$K = F/y \tag{5-11}$$

Correspondingly, the process system rigidity K_s refers to ratio of the normal cutting force F_p acting on the machining surface to the deformation value y_s of process system in the normal direction. That is

$$K_s = F_p / y_s \tag{5-12}$$

【Discussion】

① Which force causes the radial/normal deformation y_s?

$$\Delta\theta_s = \sqrt{\Delta\theta_{z1}^2 + \Delta\theta_{z2}^2 + \Delta\theta_{z3}^2 + \Delta\theta_{z4}^2}$$

$$= \sqrt{\left(\frac{z_1}{z_2} \cdot \frac{z_3}{z_4}\right)^2 \cdot \Delta\varphi_{z1}^2 + \left(\frac{z_3}{z_4}\right)^2 \cdot \Delta\varphi_{z2}^2 + \left(\frac{z_3}{z_4}\right)^2 \cdot \Delta\varphi_{z3}^2 + \Delta\varphi_{z4}^2} \tag{5-10}$$

由上述传动误差分析可以得出：①缩短传动链有利于提高传动精度；②传动比 i 小的传动链具有较高的传动精度，因此采用降速传动是保证传动精度的一个重要原则；③末端传动元件应具有较高的加工精度和装配精度。

5.2.3 工艺系统弹性变形引起的加工误差

机械加工过程中，工艺系统在切削力、传动力、夹紧力、重力以及惯性力等外力作用下会产生相应的弹性变形和塑性变形。变形会改变刀具和工件之间的正确位置关系，使工件产生加工误差。不妨深入加工车间仔细观察以下几种加工情况：

1）在一台新车床上车削细长轴。
2）在一台旧车床上车削短粗工件。
3）在一台外圆磨床上"无进给光磨"一根长轴。

你都发现什么现象？这些现象的发生说明了什么问题？

如图 5-22a 所示，在卧式车床上加工一批轴类零件。轴在车削后会出现形状误差，要么呈"鼓形"，要么呈"鞍形"，试解释其产生的原因。

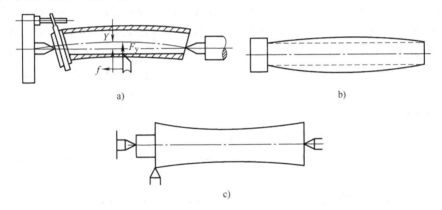

Fig. 5-22 Form errors produced in turning a shaft 车削轴类零件时产生的形状误差

1. 工艺系统刚度

（1）工艺系统刚度的基本概念　在材料力学中，物体的静刚度指的是作用力与由它所引起的在作用力方向上的变形量的比值。即

$$K = F/y \tag{5-11}$$

对应地，工艺系统的刚度 K_s 指的是切削力在加工表面法线方向的分力 F_p 与在总切削力作用下工艺系统产生的沿法向的变形 y_s 的比值，即

$$K_s = F_p / y_s \tag{5-12}$$

【讨论】

① 变形 y_s 是由什么力引起的？

② Why is the F_p used only in formula? Why does not use the resultant force F?

$$F = \sqrt{F_x^2 + F_p^2 + F_z^2}$$

(2) Calculation of process system rigidity Under the action of cutting force, the relevant assemblies of the machine tool, fixture, cutting tool and workpiece in process system would appear different deformation, which would change the relative position between the cutting tool and the workpiece in normal direction and cause machining error. According to the principle of deformation superposition, there is

$$y_s = y_m + y_f + y_t + y_w \tag{5-13}$$

where, y_m, y_f, y_t, y_w represent respectively the deformation of machine tool, fixture, cutting tool and workpiece. According to the definition of rigidity, there is:

$$y_s = F_p / K_s,\ y_m = F_p / K_m,\ y_f = F_p / K_f,\ y_t = F_p / K_t,\ y_w = F_p / K_w$$

By substituting y_s, y_m, y_f, y_t, y_w into Eq. (5-13), then

$$\frac{1}{K_s} = \frac{1}{K_m} + \frac{1}{K_f} + \frac{1}{K_t} + \frac{1}{K_w} \tag{5-14}$$

It can be seen from Eq. (5-14) that the rigidity K_s of whole process system is even less than that of the element with minimum rigidity.

Eq. (5-14) is a general formula for calculation of process system rigidity. Under specific machining conditions, it can be simplified. For the external cylindrical turning, the deformation of the lathe cutter has a very little influence on the machining error and can be neglected. Therefore the $1/K_t$ can be omitted. As for the boring hole on the boring machine, there is something different. The forced deformation of the boring bar has a serious effect on the machining accuracy; but the rigidity of the workpiece, saying a box, is often very large, and its deformation can be ignored.

[Discussion]

① Eq. (5-14) shows that if the rigidity of each component, i.e. K_m, K_f, K_t, K_w is given, the rigidity K_s of the process system can be calculated. But how can you know the rigidity of each component?

② Further, take a turning process system for example. For the shaft held by chuck only, it can be viewed as a cantilever to calculate its rigidity in light of theoretical formula. For the shaft supported by two centers, it can be viewed as a simply supported beam to calculate its rigidity. But how do you determine the rigidity K_m of the machine tool? Do you expect to calculate the rigidity of the headstock, tailstock, and compound rest based on some available theoretical formulae?

2. Influence of the elastic deformation of process system on machining accuracy

(1) Machining error caused by the change of the cutting force acting point (suppose the magnitude of cutting force is a constant)

1) The workpiece is first assumed to be infinitely rigid and the centers are assumed to be simple supports which do not exert any moment on the workpiece. In this case, the deformation of machine tool is thought to be the deformation of process system. When the turning tool moves longitudinally to the position shown in Fig. 5-23, under the action of F_A, F_B, F_P, the deformations produced by

② 为什么使用切削分力 F_p 而不使用总切削力 F？
$$F = \sqrt{F_x^2 + F_p^2 + F_z^2}$$

（2）工艺系统刚度计算　在切削力作用下，工艺系统中机床的有关部件、夹具、刀具和工件都会产生不同程度的变形，导致刀具和工件在法线方向的相对位置发生变化，从而产生加工误差。根据变形叠加性原理，有

$$y_s = y_m + y_f + y_t + y_w \tag{5-13}$$

其中，y_m、y_f、y_t、y_w 分别代表机床、夹具、刀具、工件的受力变形。根据刚度的定义，工艺系统各部分的变形可写成

$$y_s = F_p / K_s,\ y_m = F_p / K_m,\ y_f = F_p / K_f,\ y_t = F_p / K_t,\ y_w = F_p / K_w$$

把 y_m、y_f、y_t、y_w 代入式（5-13），得

$$\frac{1}{K_s} = \frac{1}{K_m} + \frac{1}{K_f} + \frac{1}{K_t} + \frac{1}{K_w} \tag{5-14}$$

由式（5-14）看出，整个工艺系统的刚度 K_s 比工艺系统中任何一个组成部分的刚度都要小。

式（5-14）是计算工艺系统刚度的一般计算公式。在具体的加工条件下，该公式还可以进行简化。如外圆车削时，车刀的受力变形很小，可忽略不计，于是可省去刀具刚度一项。至于在镗床上镗孔时，情况有所不同。镗杆的受力变形严重影响加工精度，而箱体零件的刚度一般较大，其受力变形可忽略不计。

【讨论】

① 式（5-14）表明，如果已知工艺系统各组成部分的刚度，就可以计算出工艺系统的刚度。但如何能够知道各组成部分的刚度 K_m、K_f、K_t、和 K_w？

② 进一步以车削工艺系统为例。对于仅用卡盘夹持的轴来讲，完全可以视为悬臂梁，按理论公式计算该轴的刚度。对于利用两顶尖支承的轴来讲，可以视为简支梁计算该轴的刚度。但是如何确定机床的刚度？难道也期望利用现成的理论公式去计算头架、尾座和刀架的刚度吗？

2. 工艺系统的弹性变形对加工精度的影响

（1）切削力作用点位置变化引起的加工误差（假定切削力大小不变）

1）首先假定工件刚度无穷大，顶尖呈简支状态，不会对工件施加转矩。此时，工艺系统的变形只考虑机床的变形。当车刀纵向移动到图 5-23 所示位置时，在 F_A、F_B 和 F_p 三个力的作用下，头架、尾座和刀架的变形分别为 $y_{tj} = \overline{AA'}$、$y_{wz} = \overline{BB'}$ 和 $y_{dj} = \overline{CC'}$。轴的中心线 AB 移至 $A'B'$，车刀切削点 C 的位移 y_x 可以表示成

$$y_x = y_{tj} + \Delta x = y_{tj} + (y_{wz} - y_{tj})\frac{x}{L} \tag{5-15}$$

那么，机床的总变形为

$$y_{jc} = y_x + y_{dj} \tag{5-16}$$

根据刚度的定义，有

$$y_{tj} = \frac{F_A}{K_{tj}} = \frac{F_p}{K_{tj}}\left(\frac{L-x}{L}\right),\quad y_{wz} = \frac{F_B}{K_{wz}} = \frac{F_p}{K_{wz}}\frac{x}{L},\quad y_{dj} = \frac{F_p}{K_{dj}}$$

其中，K_{tj}、K_{wz}、K_{dj} 分别代表头架的刚度、尾座的刚度和刀架的刚度。

headstock, tailstock and toolpost are $y_{tj} = \overline{AA'}$, $y_{wz} = \overline{BB'}$ and $y_{dj} = \overline{CC'}$. At this time, the central line AB of shaft moves to $A'B'$, the displacement y_x at the current cutting point C can be expressed as

$$y_x = y_{tj} + \Delta x = y_{tj} + (y_{wz} - y_{tj})\frac{x}{L} \quad (5\text{-}15)$$

Then the general deformation y_{jc} of machine tool can be expressed as

$$y_{jc} = y_x + y_{dj} \quad (5\text{-}16)$$

According to the definition of rigidity, there is

$$y_{tj} = \frac{F_A}{K_{tj}} = \frac{F_p}{K_{tj}}\left(\frac{L-x}{L}\right), \quad y_{wz} = \frac{F_B}{K_{wz}} = \frac{F_p}{K_{wz}}\frac{x}{L}, \quad y_{dj} = \frac{F_p}{K_{dj}}$$

where, K_{dj}, K_{wz}, K_{dj} stand for the rigidity of headstock, tailstock and toolpost respectively.

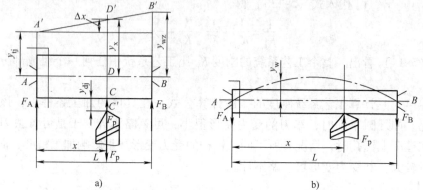

Fig. 5-23 Deformation of process system 工艺系统变形
a) Turning a short and thick shaft 车削短粗轴 b) Turning a slender shaft 车削细长轴

Substituting them into Eq. (5-16), the general deformation y_{jc} can be obtained.

$$y_{jc} = F_p\left[\frac{1}{K_{tj}}\left(\frac{L-x}{L}\right)^2 + \frac{1}{K_{wz}}\left(\frac{x}{L}\right)^2 + \frac{1}{K_{dj}}\right] = y_{jc}(x) \quad (5\text{-}17)$$

Eq. (5-17) illustrates that the deformation of machine tool is the function of the acting point x of cutter. The deformation of machine tool must cause the variation of shaft diameter, thus leading to the axial form error, as shown in Fig. 5-24.

【Example 5-1】 A lathe with $K_{tj} = 6 \times 10^4 \text{N/mm}$, $K_{wz} = 5 \times 10^4 \text{N/mm}$, $K_{dj} = 4 \times 10^4 \text{N/mm}$ is used to turn a shaft ($L = 600\text{mm}$). Set $F_p = 300\text{N}$, the displacements of process system along the shaft are listed in Tab. 5-2.

It is clear that the shaft machined by turning is in the shape of saddle. The cylindricity of the shaft is $(0.0135 - 0.0102)\text{mm} = 0.0033\text{mm}$. As yet, the deformation of the workpiece has been neglected.

2) Suppose the workpiece is a slender shaft, whereas machine tool rigidity K_m is infinite, then the deformation of machine tool can be neglected. In this case, the deformation of workpiece is thought to be the deformation of process system. For the case of the workpiece held between centers, based on the displacement calculation formula in material mechanics, there is:

$$y_w = \frac{F_p}{3EI}\frac{(L-x)^2 x^2}{L} \quad (5\text{-}18)$$

【Example 5-2】 Set $F_p = 300\text{N}$, the shaft is $\phi 30\text{mm}$ in diameter and 600mm long, $E = 2 \times$

把它们代入式（5-16），就可以得到机床的总变形，即

$$y_{jc} = F_p \left[\frac{1}{K_{tj}} \left(\frac{L-x}{L} \right)^2 + \frac{1}{K_{wz}} \left(\frac{x}{L} \right)^2 + \frac{1}{K_{dj}} \right] = y_{jc}(x) \tag{5-17}$$

式（5-17）说明机床的变形是切削力作用点 x 的函数。机床的变形必然引起轴直径的变化，从而产生轴向形状误差，如图5-24所示。

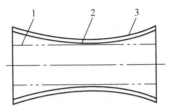

Fig. 5-24　Axial form error of shaft machined between centers 工件在顶尖上车削后的形状
　　　　1——Ideal shape without y_{jc} 没有机床变形的理想形状
　　　　2——Axial shape with y_{tj} and y_{wz} 考虑头架、尾座变形后的形状
　　　　3——Axial shape with y_{tj}, y_{wz} and y_{dj} 考虑头架、尾座和刀架变形后的形状

【例5-1】　在一台头架、尾座和刀架刚度分别为 $K_{tj} = 6 \times 10^4 \text{N/mm}$，$K_{wz} = 5 \times 10^4 \text{N/mm}$，$K_{dj} = 4 \times 10^4 \text{N/mm}$ 的车床上车削长度 $L = 600\text{mm}$ 的轴，设 $F_p = 300\text{N}$，则沿工件长度上工艺系统的位移见表5-2。

Table. 5-2　Deformations in different places along the shaft 沿轴向不同位置时的变形量

（单位：mm）

x	0（at headstock 头架处）	$L/6$	$L/3$	$L/2$	$2L/3$	$5L/6$	L（at tailstock 尾座处）
y_x	0.0125	0.0111	0.0104	0.0102	0.0107	0.0118	0.0135

显然，车削后的轴呈现鞍形。轴的圆柱度误差为 $(0.0135 - 0.0102)\text{mm} = 0.0033\text{mm}$。然而，这里并没有考虑工件的变形。

2）假设工件是一根细长轴，而机床的刚度 K_m 为无穷大，那么机床的变形就可以忽略。在此情况下，工艺系统的变形完全取决于工件的变形。根据材料力学关于简支梁变形的计算公式，切削点的变形量为

$$y_w = \frac{F_p (L-x)^2 x^2}{3EI L} \tag{5-18}$$

式中　E——工件材料的弹性模量；
　　　I——工件截面的惯性矩。

【例5-2】　设 $F_p = 300\text{N}$，工件尺寸为 $\phi 30\text{mm} \times 600\text{mm}$，$E = 2 \times 10^5 \text{N/mm}^2$，则沿工件长度上的变形见表5-3。

105N/mm². The deformations of the shaft in longitudinal direction are listed in Tab. 5-3.

It is clear that the shaft machined by turning is in the shape of drum. The cylindricity error of the shaft is (0.17-0)mm = 0.17mm.

3) General rigidity of process system. Taking all the factors into consideration, the general rigidity K_s of process system can be obtained by Eq. (5-19).

$$y_s = y_m + y_w$$

$$y_s = F_p \left[\frac{1}{K_{tj}} \left(\frac{L-x}{L} \right)^2 + \frac{1}{K_{wz}} \left(\frac{x}{L} \right)^2 + \frac{1}{K_{dj}} + \frac{1}{K_w} \right] = \frac{F_p}{K_s}$$

$$\frac{1}{K_s} = \frac{1}{K_{tj}} \left(\frac{L-x}{L} \right)^2 + \frac{1}{K_{wz}} \left(\frac{x}{L} \right)^2 + \frac{1}{K_{dj}} + \frac{(L-x)^2 x^2}{3EIL} \tag{5-19}$$

(2) Machining error caused by the variation of the cutting force

In fact, the blank of a workpiece does not have an exact geometric form. For example, the sections of a cylinder would not be exactly round; longitudinally, the blank may be barrel shape or conical. Similarly, a surface to be face-milled will not initially be perfectly flat. These mean that the machining allowance of the blank is uneven. The uneven machining allowance or hardness of the blank would cause the variation of cutting force. Consequently, the forced deformation between the tool and the workpiece also varies, and initial form errors will be "duplicated" onto the machined surface.

It is known that the cutting force, F_p, is the function of the depth of cut a_p and the feed f. Generally this relationship is expressed as

$$F_p = C_{F_p} a_p^{x_{F_p}} f^{y_{F_p}} K_{F_p} \tag{5-20}$$

where, C_{F_p}, K_{F_p} —Coefficient related to cutting conditions;

x_{F_p}, y_{F_p} —Exponents.

Suppose the feed f and other cutting conditions are kept in constant in one cutting stroke, then

$$C_{F_p} f^{y_{F_p}} K_{F_p} = C$$

As $x_{F_p} \approx 1$ in turning operation, the Eq. (5-20) can be expressed as

$$F_p = C a_p \tag{5-21}$$

Let us now consider, as an example, a radial cut as shown in Fig. 5-25. The nominal depth of cut varies between a_{p1} and a_{p2}.

The difference between these two values can be denoted as the maximum initial form error of the blank Δ_b:

$$\Delta_b = a_{p1} - a_{p2}$$

Therefore, $F_{p1} = C(a_{p1} - y_1)$, $F_{p2} = C(a_{p2} - y_2)$

As $y_1 \ll a_{p1}$, $y_2 \ll a_{p2}$, then,

$$F_{p1} = Ca_{p1}, \quad F_{p2} = Ca_{p2}$$

The roundness error of the workpiece is

$$\Delta_w = y_1 - y_2 = \frac{F_{p1}}{K_s} - \frac{F_{p2}}{K_s} = \frac{C}{K_s}(a_{p1} - a_{p2}) = \frac{C}{K_s}\Delta_b \tag{5-22}$$

Table. 5-3 Deformations int different place along the shaft 沿轴向不同位置时的变形量

(单位: mm)

x	0 (at headstock 头架处)	$L/6$	$L/3$	$L/2$	$2L/3$	$5L/6$	L (at tailstock 尾座处)
y_x	0	0.052	0.132	0.17	0.132	0.052	0

显然,车削后的轴呈现鼓形。轴的圆柱度误差为 $(0.17-0)$ mm $=0.17$ mm。

3) 工艺系统的总刚度。考虑所有的因素,则工艺系统的总刚度可由式(5-19)表示。

$$y_s = y_m + y_w$$

$$y_s = F_p \left[\frac{1}{K_{tj}} \left(\frac{L-x}{L} \right)^2 + \frac{1}{K_{wz}} \left(\frac{x}{L} \right)^2 + \frac{1}{K_{dj}} + \frac{1}{K_w} \right] = \frac{F_p}{K_s}$$

$$\frac{1}{K_s} = \frac{1}{K_{tj}} \left(\frac{L-x}{L} \right)^2 + \frac{1}{K_{wz}} \left(\frac{x}{L} \right)^2 + \frac{1}{K_{dj}} + \frac{(L-x)^2 x^2}{3EIL} \tag{5-19}$$

(2) 切削力大小变化引起的加工误差 事实上,毛坯不可能有完全准确的形状。例如,圆柱体的横截面不会是准确的圆截面,轴向上也可能会是鼓形或锥形。同样,要铣削加工的平面起初也不会很平。这些例子都意味着毛坯的加工余量是不均匀的。毛坯的加工余量不均匀或硬度不均匀都会引起切削力变化。因此,刀具与工件之间的受力变形也就随之变化,毛坯的原始误差将会"复映"在加工后的表面上。

切削分力 F_p 是切削深度 a_p 和进给量 f 的函数。通常它们之间的关系可以表示为

$$F_p = C_{F_p} a_p^{x_{F_p}} f^{y_{F_p}} K_{F_p} \tag{5-20}$$

式中 C_{F_p}、K_{F_p}——与切削条件有关的系数;

x_{F_p}、y_{F_p}——指数。

假定进给量 f 和其他切削条件在一次走刀中保持不变,那么

$$C_{F_p} f^{y_{F_p}} K_{F_p} = C$$

由于在车削时 $x_{F_p} \approx 1$,因此式(5-20)可以写成

$$F_p = C a_p \tag{5-21}$$

现以径向车削为例来说明,如图5-25所示。其名义切削深度在 a_{p1} 和 a_{p2} 之间变化。

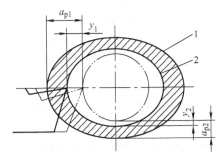

Fig. 5-25 Duplication of the blank error 毛坯误差复映

这两个值之差可以用毛坯的最大原始误差 Δ_b 来表示

$$\Delta_b = a_{p1} - a_{p2}$$

因此,$F_{p1} = C(a_{p1} - y_1)$,$F_{p2} = C(a_{p2} - y_2)$。因为 $y_1 \ll a_{p1}$,$y_2 \ll a_{p2}$,那么

$$F_{p1} = C a_{p1},\ F_{p2} = C a_{p2}$$

It can be seen that it is the blank roundness error $\Delta_b = a_{p1} - a_{p2}$ that cause the workpiece to produce roundness error $\Delta_w = y_1 - y_2$. Moreover, the larger the blank error Δ_b is, the larger the error Δ_w of the workpiece. This phenomenon is called "blank error duplication" in machining process. In order to measure the degree of error duplication, an error duplication coefficient ε is introduced here.

$$\varepsilon = \Delta_w / \Delta_b = C / K_s \tag{5-23}$$

Generally, $\Delta_w < \Delta_b$, so $\varepsilon < 1$. ε reflects quantificationally the reduction extent of blank error. It can be seen from Eq. (5-23) that the higher the K_s, the smaller the ε, so the smaller the machining error. In order to reduce error duplication, one way is to improve the rigidity K_s of the process system, the other way is to add feed times. Suppose the error duplication coefficient of every feed is $\varepsilon_1, \varepsilon_2, \cdots, \varepsilon_n$ then the overall coefficient is

$$\varepsilon_\Pi = \prod \varepsilon_i (i = 1, 2, \cdots, n)$$

As $\varepsilon_i < 1$, $\varepsilon_\Pi \ll 1$.

Obviously, to add feed times can reduce error duplication coefficient and improve machining accuracy, but the productivity is decreased. Therefore, to improve the rigidity of the process system is an intellectual choice, and has a significance for the reducing ε.

As analyzed above, when the blanks have some form errors, such as roundness, cylindricity, straightness, or mutual position errors, such as coaxiality, radial runout, these errors may still appear on the machined workpieces, only they are smaller than the errors before machining.

3. Influence of deformation caused by other forces on machining accuracy

In addition to cutting force, there are many other forces, such as inertial force, clamping force, weight, and so on, which would also make the process system produce deformation, causing machining errors.

(1) Machining errors caused by inertial force In high speed machining process, if the imbalance component in process system rotates at high speed, the centrifugal force would be produced. The magnitude of centrifugal force in the error sensitive direction varies periodically with the rotating angle of the component, and the deformation caused by the force would vary, thus causing the radial form error. In order to decrease the influence of the inertial force, a balance block is set in the symmetrical orientation to the imbalance mass of workpiece and fixture so as to cancel out the centrifugal forces of both sides. If necessary, the speed of rotation can be reduced properly to decrease the influence of the centrifugal force.

(2) Machining errors caused by clamping force In a clamping workpiece, if the workpiece to be machined has a low rigidity, or the clamping force is exerted on the improper part of the workpiece, it would make the workpiece produce deformation, thus causing a machining error. That is true of the thin-wall sleeve and sheet metal in particular. The illustration of the machining error caused by clamping force is shown in Fig. 5-26.

(3) Machining errors caused by the weight of the component itself In process system, the dead weight of the component itself would cause the deformation of itself. For example, the weight of the headstock or tool rest mounted on the cross rail in a large-size vertical lathe, planer, planer type milling machine would make the cross rail produce deformation (bending). Similarly, the radial

于是，工件的圆度误差为

$$\Delta_w = y_1 - y_2 = \frac{F_{p1}}{K_s} - \frac{F_{p2}}{K_s} = \frac{C}{K_s}(a_{p1} - a_{p2}) = \frac{C}{K_s}\Delta_b \tag{5-22}$$

由此可知，正是毛坯的圆度误差 Δ_b 使工件产生相应的圆度误差 Δ_w，且毛坯误差 Δ_b 越大，造成的工件误差也越大。这种现象称为加工过程中的误差复映现象。为了衡量误差复映程度，这里引入误差复映系数 ε，即

$$\varepsilon = \Delta_w / \Delta_b = C / K_s \tag{5-23}$$

通常，$\Delta_w < \Delta_b$，因此，$\varepsilon < 1$。ε 定量地反映了毛坯误差经加工之后的减小程度。由式（5-23）看出，工艺系统的刚度 K_s 越大，复映系数 ε 越小，加工误差就越小。为减少误差复映，一种方法是提高工艺系统刚度，另一种方法是增加走刀次数。假设每次走刀的误差复映系数是 $\varepsilon_1, \varepsilon_2, \cdots, \varepsilon_n$，那么总的误差复映系数为

$$\varepsilon_\Pi = \prod \varepsilon_i \quad (i = 1, 2, \cdots, n)$$

由于 $\varepsilon_i < 1$，所以 $\varepsilon_\Pi \ll 1$。

显然，增加走刀次数可减小误差复映，提高加工精度，但生产率降低了。因此，提高工艺系统刚度是一个明智的选择，对减小误差复映系数具有重要意义。

由以上分析可知，当工件毛坯有一些形状误差，如圆度误差、圆柱度误差、直线度误差等，或相互位置误差，如同轴度误差、径向圆跳动误差等时，加工后仍然会有同类型的加工误差出现，只是这些误差比加工前小了。

3. 其他力引起的变形对加工精度的影响

除了切削力之外，还有许多其他力，如夹紧力、惯性力、重力等，它们都会引起工艺系统变形，从而造成加工误差。

（1）惯性力引起的加工误差　在高速切削过程中，工艺系统中如果存在高速旋转的不平衡构件，就会产生离心力，它在误差敏感方向的分力大小随构件的转角变化呈周期性的变化，由它所引起的变形也相应地变化，从而造成工件的径向形状误差。为减小惯性力的影响，可在工件与夹具不平衡质量对称的方位配置一平衡块，使两者的离心力互相抵消。必要时还可适当降低转速，以减小离心力的影响。

（2）夹紧力引起的加工误差　被加工工件在装夹过程中，由于刚度较低或夹紧力着力点位置不当，都会引起工件的变形，造成加工误差。薄壁套、薄板等低刚度零件尤其如此。图5-26所示为夹紧力引起的误差。

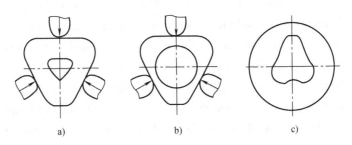

Fig5-26　Error caused by clamping force　夹紧力引起的误差

（3）零部件本身重力引起的加工误差　在工艺系统中，由于零部件的自重也会使其自身

arm of the radial drill press under the action of the headstock would produce deformation, causing the axis of the spindle to be not vertical to the worktable; the boring bar of a horizontal boring machine would produce drooping with the elongation of the boring bar. All these deformations caused by the weight of the component itself would cause the workpiece to produce errors.

For the machining of large-size workpieces, the deformation caused by the weight of the workpiece to be machined may be sometimes the primary reason for the shape error of the workpiece. Therefore, when setting the large-size workpieces, the locators should be arranged properly so as to reduce the deformation caused by the weight of the workpiece, thus reducing machining error.

5.2.4 Errors caused by thermal deformation of process systems

1. Introduction

In machining process, influenced by different thermal sources, process system would produce various complex deformations. The deformation would disrupt the accurate position and movement of a workpiece relative to a cutter, thus causing machining errors. In high speed machining, super-precision machining and automatic machining, the issues involving the thermal deformation of process system became more severe and have been the important research projects in machining industry.

The main thermal sources in process system include frictional heat, cutting heat and ambient temperature etc. Under the effect of different thermal sources, the temperature of each component in process system rises higher and higher. At the same time, the heat produced disperses all around through different heat transferring ways. When the amount of heat equals to the dispersed heat, process system has reached a heat balance state. At this time, the thermal deformation of process system becomes relative stable. Generally, after a machine tool runs in a certain period of time, the temperature gradually tends to be stable, and its working accuracy also becomes relative stable. Therefore, precision machining should be performed in a heat balance state.

2. Errors caused by thermal deformation of a machine tool

Because of the nonuniformity of thermal sources in machine tool and the complexity of machine tool structure, the deformation degree of each component is different, thus damaging the intrinsic geometric precision of the machine tool, decreasing the machining accuracy of the machine tool. As different machine tools have a great disparity in structure and working condition, and their thermal sources are also different, the machining errors caused by thermal deformation are different.

For a lathe, milling machine, drilling machine and boring machine, the friction heat of bearings and oil tank in headstock are the primary thermal sources, which would cause the temperature of both the headstock and the bed or column to go up and have the spindle lifted and inclined upward. The thermal deformation trend of a lathe is shown in Fig. 5-27a. Its thermal deformation would finally lead to the parallelism error between the axis of the spindle and guideways. After turning a shaft on this kind of lathe, you would found the machined shaft has the error in cylindricity. The primary transmission system of a horizontal milling machine is a primary thermal source of this machine. As the left wall of a gearbox has higher temperature, it lead to the inclination of the axis of the spindle, as shown in Fig. 5-27b.

产生变形。如大型立车、龙门铣床、龙门刨床的刀架横梁等,由于主轴箱或刀架的重力而产生变形。摇臂钻床的摇臂在主轴箱自重的影响下产生变形,造成主轴轴线与工作台不垂直;铣镗床镗杆伸长而下垂变形等,它们都会造成加工误差。

对于大型工件的加工,工件自重引起的变形有时成为产生加工形状误差的主要原因。因此,在装夹大型工件时,恰当地布置支承可减小工件自重引起的变形,从而减小加工误差。

5.2.4 工艺系统热变形引起的加工误差

1. 概述

在机械加工过程中,工艺系统在各种热源的影响下,产生复杂的变形,从而破坏工件与刀具之间的相互位置和相对运动的准确性,引起加工误差。在高速加工、超精密加工以及自动化加工中,工艺系统热变形问题变得尤为突出,已成为机械加工领域的重要研究课题。

工艺系统的主要热源是切削热、摩擦热、环境温度等。在各种热源的作用下,工艺系统各组成部分的温度逐渐升高。同时,它们也通过各种传热方式向周围散发热量。当单位时间内传入和散发的热量相等时,工艺系统达到了热平衡状态,而工艺系统的热变形也就达到某种程度的稳定。系统一般在工作一定时间后,温度才逐渐趋于稳定,其精度也比较稳定。因此,精密加工应在热平衡状态下进行。

2. 机床热变形引起的加工误差

由于机床热源的不均匀性及其结构的复杂性,使机床各部分的变形程度不等,从而破坏了机床原有的几何精度,降低了机床的加工精度。由于各类机床的结构和工作条件相差很大,其主要热源各不相同,热变形引起的加工误差也不相同。

车、铣、钻、镗类机床,主要热源是主轴箱轴承的摩擦热和主轴箱中油池的发热,使主轴箱及与它相连接部分(如床身或立柱)的温度升高,从而引起主轴的抬高和倾斜。图5-27a所示为车床热变形趋势。其热变形最终导致主轴回转轴线与导轨的平行度误差,使加工后的零件产生圆柱度误差。图5-27b所示为万能铣床的热源也是主传动系统,由于左箱壁温度高也导致主轴线升高并倾斜。

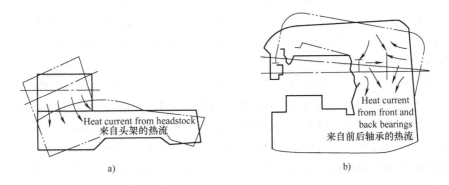

Fig. 5-27 Thermal deformation trend of machine tools 机床的热变形趋势
a) Lathe 车床　b) Horizontal milling machine 卧式铣床

Such machine tools as grinding machines generally have hydraulic drive system and high-speed grinding head. The chief sources of heat are the spindle bearings of the grinding wheel and hydraulic system. The deformations caused by heat sources are the displacement of the wheel frame, the displacement of headstock, and the deformation of guideways. The displacement of the wheel frame has a direct influence on the dimension accuracy of the workpiece ground. For such large-size machine tools with a long bed as guideway grinder, external cylindrical grinding machine, planer-type milling machine, vertical lathe, etc., the thermal deformation of the machine bed would be the primary influencing factor on machining accuracy.

3. Errors caused by thermal deformation of a workpiece

The thermal deformation of a workpiece is caused chiefly by cutting heat. But the ambient temperature can't be ignored for large-size or precise workpieces. Because of the differences in structural dimension, workpieces are heated in two ways.

(1) The workpiece is heated evenly. For some workpieces with simple shape and symmetrical structure, such as shafts, sleeves, and so on, the cutting heat can be transferred into the workpiece evenly in machining (turning, grinding). The thermal deformation of a workpiece can be estimated as follows:

$$\Delta L = \alpha L \Delta t \quad (5\text{-}24)$$

where, α—Thermal expansion coefficient of material, $1/℃$;

L—Dimension (length or diameter) of the workpiece in the direction of thermal deformation, mm;

Δt—Temperature rise of the workpiece, ℃.

When turning the shorter shaft, its axial thermal deformation is very small and can be ignored. When turning the long shaft, there is no temperature rise at the beginning. With the cutting going on, the temperature of the shaft will rise, and the shaft will expand, i.e. the diameter of the shaft will increase gradually. Therefore, the depth of cut will be gradually larger and larger. Consequently, the shaft machined has cylindricity error after it cools. When machining lead screws, the thermal elongation of the workpiece becomes the primary factor which influences on the pitch error of lead screws. For a lead screw with 400mm long, if the temperature rises 1℃ for every grinding, the ground lead screw will elongate 4.7μm. It is required that the pitch error of a lead screw with grade 5 in precision should not be larger than 5μm. It follows that thermal deformation has a quite influence on the machining accuracy of the workpiece.

(2) The workpiece is heated unevenly. In milling or grinding the plane on the plate, as the single side of the plate is heated, the temperature difference between upper and lower sides causes convexity in the middle of the plate. The convex part being cut, the machined surface becomes a concavity after cooling, thus leading to flatness error. Generally, if the temperature difference between upper and lower sides is 1℃, it will cause the flatness error with 0.01mm. When the workpiece is heated unevenly, its convex quantity increases sharply with the increase of its length. The smaller the thickness of the plate, the larger the convex quantity. The difference Δt of temperature must be controlled in order to reduce the error caused by thermal deformation.

磨床类机床通常都有液压传动系统并配有高速磨头。它的主要热源为砂轮主轴轴承的发热和液压系统的发热。引起的变形主要表现为砂轮架的位移、工件头架的位移和导轨的变形。砂轮架的位移，直接影响被磨工件的尺寸精度。对于大型机床如导轨磨床、外圆磨床、立式车床、龙门铣床等的长床身部件，机床床身的热变形则是影响加工精度的主要因素。

3. 工件热变形引起的加工误差

工件的热变形主要是由切削热引起的。对于大型或精密零件，环境温度的影响也不可忽视。由于工件结构尺寸的差异，工件受热有两种情况。

（1）工件均匀受热　对于一些形状简单、对称的零件，如轴、套筒等，加工时（如车削、磨削）切削热能较均匀地传入工件。工件热变形量可按下式估算

$$\Delta L = \alpha L \Delta t \tag{5-24}$$

式中　α——工件材料的热膨胀系数（1/℃）；
　　　L——工件在热变形方向的尺寸（长度或直径）（mm）；
　　　Δt——工件温升（℃）。

当车削较短的轴类零件时，其轴向热变形很小，可忽略不计。当车削较长的轴类零件，开始切削时工件温升为零。随着切削的进行，工件温度逐渐升高而使直径逐渐增大，切削深度也随着进给逐渐增大。结果，加工完的工件冷却后会产生圆柱度误差。加工丝杠时，工件的热伸长成为影响螺距误差的主要因素。对于一根长度为400mm的丝杠，如果每磨削一次温度升高1℃，则被磨削丝杠将伸长4.7μm。而5级丝杠的螺距累积误差在400mm长度上不允许超过5μm。可见热变形对工件加工精度影响之大。

（2）工件不均匀受热　在铣削、磨削平板零件时，平板单面受热，上下平面间产生温差导致平板呈现中凸。凸起部分被切去，冷却后加工表面下凹，使工件产生平面度误差。一般地，工件上下表面温差1℃，就会产生0.01mm的平面度误差。工件不均匀受热时，工件凸起量随工件长度的增加而急剧增加。工件厚度越薄，工件凸起量就越大。要减小变形误差，必须控制温差Δt。

4. 刀具热变形引起的加工误差

刀具热变形主要是由切削热引起的。传给刀具的热量虽不多，但由于刀具切削部分体积小，切削部分仍产生很高的温升。如高速钢刀具车削时刃部的温度可高达700~800℃，而硬质合金刀刃部可达1000℃以上。这样，不但刀具热伸长影响加工精度，而且刀具的硬度也会下降。

图5-28所示为车刀热伸长与切削时间的关系。ΔL表示车刀的热伸长量，T_1指的是切削时间，T_2是停止切削的间隔时间。曲线A代表了连续车削时车刀的热变形情况。显然，在刀具达到热平衡前加工的工件必然存在形状误差。曲线B表示车刀停止车削后刀具的冷却变形过程。当加工一批短小轴类工件时，刀具间断切削（如装卸工件），刀具受热和冷却是交替进行的，热变形情况如曲线C所示。对每一个工件来说，产生的形状误差较小；而对一批工件来说，在刀具达到热平衡前加工出的一批工件都有尺寸误差。

4. Errors caused by thermal deformation of cutters

The thermal deformation of cutters arises from cutting heat. Although the quantity of heat transferred to cutter is not too much, as the volume of a edge part is small, the temperature rise of the cutting edge is still very large. For instance, the cutting temperature of the lathe cutter made of HSS can rise up to 700 – 800 ℃, and the temperature of the edge part made of carbide alloy can be above 1000 ℃. In this way, high temperature would not only lead to the decrease of cutter hardness, but the thermal elongation of the cutter, which will influence the machining accuracy.

Fig. 5-28 shows the relation of the thermal elongation of the lathe cutter to cutting time. ΔL stands for the thermal elongation of the cutter. T_1 means the cutting time, and T_2 means the interval time for stop cutting. The curve A stands for the thermal deformation trend of the cutter in continuous cutting. Clearly, the workpieces machined before the thermal balance of the cutter must have a certain form errors. Curve B illustrates the cooling deformation of the cutter after it stops cutting. When machining a batch of short shafts, as the cutting action is intermittent (because of loading/unloading workpiece), the heating and cooling processes of the cutter carry on alternately. This thermal deformation of the cutter is expressed in the curve C. The form error of every machined workpiece is very small; but for a batch of workpieces machined before the thermal balance of the cutter, they must have the dimensional error.

5.2.5 Errors caused by inner stress of the workpiece

Inner stress (or residual stress) of a workpiece means the stress remained in the workpiece after the external load is removed. The internal structure of the workpiece with residual stress is often in an unstable state, and easy to lose original stress state under the effect of normal temperature, especially some external factors. With the re-distribution of inner stress, the workpiece would produce corresponding deformation, thus destroying original accuracy.

1. Reasons for inner stress of the workpiece

(1) Inner stress produced in the manufacturing of the blank and heat treatment In the process of casting, forging, welding, and heat treatment, if each part of the blank is heated unevenly, or cooled at different speed, or there is a volume variation arising from the transformation of metallographic structure, quite a large inner stress within blank will be produced. The more complex the blank structure, or the more uneven the wall thickness of the blank, or the larger the difference in cooling conditions, the larger the residual stress in blank. Initial stress is in the state of relative balance temporarily. When a layer of metal is cut from a surface of the workpiece, as the balance mentioned above is broken, the inner stress within the workpiece will be redistributed, and then, the workpiece will produce a visible deformation.

A casting with inner stress temporarily being in the state of relative balance is shown in Fig. 5-29a. Once an opening is cut on the wall C, as shown in Fig. 5-29b, original balance state is broken, and the casting produces flexural deformation.

(2) Inner stress caused by cold alignment A force F is exerted on a flexual shaft, making the shaft produce plastic deformation (Fig. 5-30a) so as to align the flexual shaft. The distribution of in-

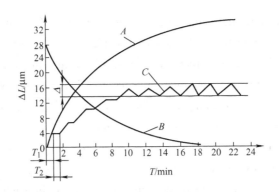

Fig. 5-28 Relation between the thermal elongation of the lathe cutter and the cutting time
车刀热伸长与切削时间的关系

5.2.5　工件内应力引起的误差

零件的内应力（又称残余应力）是指当外部载荷去除后仍保留在零件内部的应力。具有内应力的零件，其内部组织往往处于一种很不稳定的状态，在常温下特别是在外界某种因素的影响下很容易失去原有状态，使内应力重新分布，零件产生相应的变形，从而破坏了原有的精度。

1. 工件内应力产生的原因

（1）毛坯制造和热处理过程中产生的内应力　在铸、锻、焊及热处理过程中，由于零件各部分受热不均匀、或冷却速度不同以及金相组织转变的体积变化，使毛坯内部产生相当大的内应力。毛坯的结构越复杂、壁厚越不均匀、散热条件差别越大，毛坯内部产生的内应力也越大。初期内应力暂时处于相对平衡状态，但当从某表面切去一层金属后，就打破了这种平衡，使得内应力重新分布，于是零件就出现了明显的变形。

图 5-29a 所示为一个内应力暂时处于平衡状态的铸件。当在壁 C 上切一个口子后，如图 5-29b 所示，使得原来的应力平衡状态被打破，铸件发生弯曲变形。

（2）冷校直引起的内应力　在一弯曲的细长轴上施加作用力 F，使其弯曲并产生塑性变形，如图 5-30a 所示，以达到校直的目的。在力 F 的作用下轴内部内应力的分布如图 5-30b 所示。当力 F 去掉后内应力重新分布，如图 5-30c 所示。显然，冷校直虽能减小弯曲，但工件却处于不稳定状态，如果再次加工，又将产生新的变形。因此，高精度丝杠的加工，不允许用冷校直的方法来减小弯曲变形，而是用多次人工时效来消除残余内应力。

（3）切削加工产生的内应力　切削（含磨削）过程中产生的力和热，也会使被加工工件的表面层产生内应力。这种内应力的分布情况由加工时的工艺因素决定。（详见本章第 5 节）

2. 减少内应力的措施

（1）合理设计零件结构　在零件结构设计中，尽量做到结构对称，壁厚均匀，以消除内应力产生的隐患。

ner stress within the shaft under the action of F is shown in Fig. 5-30b. After F is removed, the inner stress within the shaft is redistributed, as shown in Fig. 5-30c. Clearly, cold alignment can reduce the bending degree, but the workpiece after cold alignment still in an unstable state. If it is machined again, it will produce new deformation. Therefore, cold alignment is not permitted to reduce the flexual deformation in machining precision lead screw, but artificial aging should be used repeatedly to eliminate the residual stress in the lead screw.

(3) Inner stress produced in machining process The cutting force and cutting heat produced in cutting (or grinding) would also cause the inner stress in the surface layer of the workpiece. The distribution of this kind inner stress depends on specific process factors in machining (see section 5.4 of this chapter for details).

2. Measures to reduce inner stress

(1) To design the structure of the workpiece reasonably In the design of the workpiece, the structure should be symmetrical and the wall thickness should be even as far as possible in order to eliminate the hidden peril caused by inner stress.

(2) To arrange manufacturing process of the workpiece reasonably The prior heat treatments, such as annealing, normalizing, aging, etc., should be arranged before machining in the workshop. For the key workpieces, the aging treatment should be arranged aptly after rough machining; and for some precise workpieces, the aging treatment should be arranged several times between operations. The rough machining and finish machining of key surfaces should be arranged separately so that there is enough time after rough machining to allow inner stress to be redistributed so as to reduce the influence of it on finish machining.

5.3 Statistic analysis of machining errors

5.3.1 Categories of machining errors

According to the law of errors arisen in machining a batch of workpieces, machining errors are classified as systematic errors and random errors. The reason for classifying machining errors is that the machining errors with different natures have different ways to treat them.

1. Systematic errors Δ_s

(1) Constant systematic errors Δ_{cs} When machining a lot of workpieces in sequence, the magnitude and direction of machining errors keep a constant on the whole. This kind of error is called constant systematic error. Such errors as the machining principle error, manufacturing errors of machine tool, cutting tool, fixture and measuring tool, and the errors caused by forced deformation of process system belong to constant systematic errors.

(2) Variable systematic errors Δ_{vs} When machining a lot of workpieces in sequence, the magnitude and direction of machining errors vary in the light of a certain law. This kind of error is called variable systematic error. The errors caused by thermal deformation of process system before thermal balance and by tool wear belong to variable systematic errors.

（2）合理安排零件制造工艺　毛坯在进入机械加工之前，应安排预备热处理，如退火、正火、时效等。重要零件在粗加工后须适当安排时效处理；对于一些精密零件，应在工序间多安排时效处理。对工件的重要表面应注意粗、精加工分开，使粗加工后有一定的间隔时间让内应力重新分布，以减少对精加工的影响。

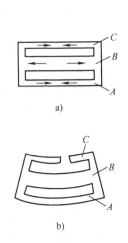

Fig. 5-29　Deformation caused by inner stress of casting
铸件内应力引起的变形

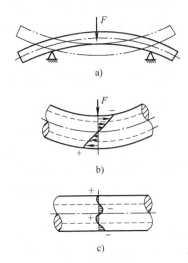

Fig. 5-30　Inner stress caused by cold alignment
冷校直引起的内应力

5.3　加工误差的统计分析

5.3.1　加工误差的分类

根据加工一批工件时误差出现的规律，加工误差可分为系统误差和随机误差。之所以要对加工误差重新分类，是因为不同性质的加工误差，其解决的途径也不同。

1. 系统误差 Δ_s

（1）常值系统误差 Δ_{cs}　在顺序加工一批工件时，误差的大小和方向保持不变，这种误差称为常值系统误差。加工原理误差，机床、刀具、夹具、量具的制造误差，工艺系统静力变形引起的加工误差等都属于常值系统误差。

（2）变值系统误差 Δ_{vs}　在顺序加工一批工件时，误差的大小和方向按一定规律变化，这种误差称为变值系统误差。工艺系统（特别是机床、刀具）在热平衡前的热变形、刀具磨损均属于变值系统误差。

2. 随机误差 Δ_r

在顺序加工一批工件时，若误差的大小和方向呈无规律的变化，称为随机误差。如毛坯误差的复映、定位误差、夹紧误差、内应力引起的误差、多次调整的误差等都属于随机误差。随机误差从表面上看似乎没有什么规律，但应用数理统计方法，可以找出一批工件加工误差的总体规律。

2. Random errors Δ_r

When machining a lot of workpieces in sequence, the magnitude and direction of machining errors vary irregularly. This kind of error is called random error. Blank error duplication, location error, clamping error, the errors caused by inner stress, and errors produced in several adjustments belong to random errors. Apparently, it seems that random errors have not any law, but the general law of machining errors produced in machining a lot of workpieces can be found by means of mathematical statistics.

5.3.2 Distribution curve method

The distribution curve method means the method to analyze and estimate machining errors by means of dimension (or error) distribution curve, which is drawn based on the practical dimensions or errors acquired by measuring a batch of machined workpieces.

1. Practical distribution curve

The procedures to draw a practical distribution curve can be illustrated through the following example.

【**Example 5-3**】 Investigate the diameter distribution situation of pin holes in a batch of pistons machined with finish boring. Given that the diameter of the pin hole specified on the part drawing is $\phi 28_{-0.015}^{0}$ mm. The pin holes have been machined in finish boring on a horizontal boring machine. Take $n = 100$ workpieces at random as sample. It is found by measuring that the holes' diameters are different. This phenomenon is called "dimension dispersal". Record all measured data and group them according to the hole size, the grouping numbers (k) should be proper (refer to Tab. 5-4). The dimensions in every group are in the range of interval. The numbers of workpieces in the same group is called frequence m. The percentage of the frequence m to the total numbers n of workpieces is called frequency, that is:

$$\text{Frequency} = m/n \times 100\%$$

Table 5-4 Selection of sample and group number 样本与组数的选择

Numbers of sample 样本的数量	Grouping numbers 分组数
50 ~ 100	6 ~ 10
100 ~ 250	7 ~ 12
250 以上	10 ~ 20

Take $k = 6$. The dimension interval between two groups is: $h = (X_{max} - X_{min})/(k - 1) = 0.002$ mm. The frequence and frequency are shown in Tab. 5-5.

Take the dimension or error of the workpiece as "abscissa"—axis of abscissa, and the frequency of each group as "ordinate"—vertical axis. The practical distribution curve (polygon diagram) then can be drawn out, as shown in Fig. 5-31.

$$\text{Dispersal range} = X_{max} - X_{min} = (28.004 - 27.992)\text{mm} = 0.012\text{mm}$$

In order to analyze further the machining accuracy of this operation, the tolerance band of this operation is also marked on the practical distribution curve. In Fig. 5-31, the mean value $\overline{X}$ of the

5.3.2 分布曲线分析法

分布曲线分析法指的是一种用尺寸或误差的分布图去分析和判断加工误差的方法。而尺寸或误差分布图是根据一批工件加工后其实际尺寸或误差的测量结果绘制出来的。

1. 实际分布曲线

通过实例来说明实际分布曲线的绘制步骤。

【例 5-3】 考查一批精镗活塞销孔直径的分布情况。已知零件图上规定的活塞销孔直径为 $\phi 28_{-0.015}^{0}$ mm。活塞销孔已经在卧式镗床上精镗过。随机抽取 $n=100$ 个工件（称为样本）进行测量，发现这些孔的尺寸各不相同，这种现象称为"尺寸分散"。把测得的数据记录下来，并按尺寸大小将整批工件进行分组，分组数应适当（参照表5-4）。每一组中的零件尺寸处在一定的间隔范围内。同一尺寸间隔内的零件数量称为频数 m，频数与该批零件总数 n 之百分比称为频率，即

$$频率 = (m/n) \times 100\%$$

分组数 $k=6$。分组间隔取 $h=(X_{\max}-X_{\min})/(k-1)=0.002$ mm。频数和频率见表5-5。

Table 5-5 Measured results of pin hole of piston 活塞销孔测量结果

Group No. 组别	Dimension range 尺寸范围	Mid-value of group 组中值 X_i	Frequence 频数 m_i	Frequency 频率 m_i/n
1	27.992~27.994	27.993	4	4/100
2	27.994~27.996	27.995	16	16/100
3	27.996~27.998	27.997	32	32/100
4	27.998~28.000	27.999	30	30/100
5	28.000~28.002	28.001	16	16/100
6	28.002~28.004	28.003	2	2/100

以工件尺寸或误差为横坐标，以频数或频率为纵坐标，即可作出该工序工件加工尺寸的实际分布图（折线图），如图5-31所示。

分散范围 = 最大孔径 − 最小孔径 = (28.04 − 27.992) mm = 0.012 mm

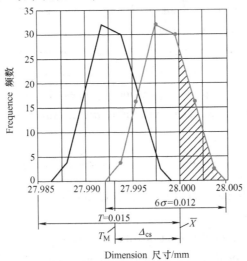

Fig. 5-31 Practical distribution curve of hole diameters 销孔直径实际分布图

sample denotes the center of the dispersal range, and depends on the magnitude of the adjusting dimension and constant systematic error. It can be seen from Fig. 5-31 that the center of the tolerance range $T_M = (28 - 0.015/2)$ mm $= 27.9925$ mm, and the constant systematic error $\Delta_{cs} = (27.9979 - 27.9925)$ mm $= 0.0054$ mm, then $\overline{X}$ can be calculated as follows:

$$\overline{X} = \frac{1}{n}\sum_{i=1}^{n} X_i m_i = 27.9979 \text{mm} \tag{5-25}$$

The sample standard deviation is the evaluation value of total standard deviation σ (also called root-mean-square deviation of total). σ reflects the dimension dispersal extent of the batch of workpieces, and depends on the sizes of variable systematic error and random error.

$$\sigma = \sqrt{\frac{1}{n}\sum_{i=1}^{n}(X_i - \overline{X})^2} = 0.002244 \text{mm} \tag{5-26}$$

It can be seen from Fig. 5-31 that the dimensions of part of workpieces have gone beyond the tolerance range (28.000 – 28.004mm, with 18%), that is, rejects have been produced (the part with shade lines). On the other hand, it is also found from Fig. 5-31 that $T > 6\sigma = 0.012$ mm. This means the machining accuracy of this system can meet the tolerance requirements. The reason why rejects appear is that the center of the distribution range deviates too far from the center of the tolerance range. The deviation is the constant systematic error $\Delta_{cs} = (27.9979 - 27.9925)$ mm $= 0.0054$ mm. If we set the center of the distribution range to the center of the tolerance range, the whole batch of workpieces dimensions would fall into the tolerance range.

In order to study further the machining accuracy of this operation, some theoretical distribution curves in mathematic statistics are often used to substitute practical distribution curves so as to find the relation between the frequency density and the machining dimension (or error).

2. Normal distribution curve

When drawing the dimension distribution diagram of a batch of workpieces, with the increase of the workpiece numbers and the smaller of the interval between two groups, the polygon diagram is approaching the smooth curve. In studying machining errors, we often use the normal distribution curve to replace the practical distribution curve.

(1) Normal distribution curve equation   Abundant experiments show that when machining a batch of workpieces by means of adjusting method, if there is no apparent variable systematic error, the dimensions of machined workpieces approximately obey the normal distribution curve, i. e. Gauss curve, as shown in Fig. 5-32. The probability density function of normal distribution is expressed as follows.

$$y = \frac{1}{\sigma\sqrt{2\pi}} e^{-\frac{1}{2}\left(\frac{x-\mu}{\sigma}\right)^2} \quad (-\infty < x < +\infty, \sigma > 0) \tag{5-27}$$

where, y—probability density function of normal distribution;

x—random variable;

μ—arithmetic mean value of normal distribution total;

σ—root-mean-square deviation of normal distribution total.

The probability density function of normal distribution has two characteristic parameters: μ and

为了进一步分析该工序的加工精度情况,可在实际分布图上标出该工序的加工公差带位置。在图 5-31 中,样本平均值 $\overline{X}$ 表示该样本的分散范围中心,它主要取决于调整尺寸的大小和常值系统误差。由图 5-31 可见,公差范围中心 $T_M = (28 - 0.015/2)\text{mm} = 27.9925\text{mm}$,常值系统误差 $\Delta_{cs} = (27.9979 - 27.9925)\text{mm} = 0.0054\text{mm}$,于是,$\overline{X}$ 可以按下式计算

$$\overline{X} = \frac{1}{n}\sum_{i=1}^{n} X_i m_i = 27.9979\text{mm} \tag{5-25}$$

样本的标准偏差是总体标准偏差 σ(又称均方根偏差)的估计。σ 反映了该批工件的尺寸分散程度,它是由变值系统误差和随机误差决定的。

$$\sigma = \sqrt{\frac{1}{n}\sum_{i=1}^{n}(X_i - \overline{X})^2} = 0.002244\text{mm} \tag{5-26}$$

由图 5-31 看出,部分工件的尺寸超出了公差范围(28.000~28.004mm,约占 18%),即出现了废品(曲线图中阴影部分)。但另一方面也发现,$T > 6\sigma = 0.012\text{mm}$,说明该系统的加工精度可以满足公差要求。之所以出现了废品,是由于分散范围中心远离了公差范围中心,即有常值系统误差 $\Delta_{cs} = (27.9979 - 27.9925)\text{mm} = 0.0054\text{mm}$ 存在。如果能设法将分散中心调整到公差范围中心,工件就完全合格。

为了进一步研究工序的加工精度问题,常应用数理统计学中的一些理论分布曲线代替实际分布曲线,以便找出频率密度与加工尺寸(或误差)之间的关系。

2. 正态分布曲线

当绘制一批工件尺寸的分布曲线时,随着工件数量的增加和组间距的减小,折线图会渐渐地向曲线逼近。在研究加工误差时,经常使用正态分布曲线代替实际分布曲线。

(1) 正态分布曲线方程　大量实验表明,用调整法加工一批工件,当加工中不存在明显的变值系统误差时,则加工后零件的尺寸近似服从正态分布曲线(即高斯曲线),如图 5-32 所示。其概率密度的函数表达式是

$$y = \frac{1}{\sigma\sqrt{2\pi}} e^{-\frac{1}{2}\left(\frac{x-\mu}{\sigma}\right)^2} \quad (-\infty < x < +\infty, \sigma > 0) \tag{5-27}$$

式中　y——分布概率密度;
$\quad\quad x$——随机变量;
$\quad\quad \mu$——正态分布总体的算术平均值;
$\quad\quad \sigma$——正态分布的均方差。

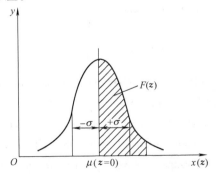

Fig. 5-32　Normal distribution curve 正态分布曲线

σ. It can be seen from Eq. (5-27) and Fig. 5-32 that when $x = \mu$, $y_{max} = 1/\sigma\sqrt{2\pi}$. This is a maximum of the curve, and also the dispersal center of the curve. Generally, μ and σ of normal distribution total are not known, but they can be found by using the estimated value of the mean value $\bar{x}$ and standard deviation σ of the sample. That is, μ of total is replaced by $\bar{x}$ of the sample, and σ of total is replaced by σ of the sample.

(2) Shape of the curve Fig. 5-33a shows that when σ remains unchanged, the shape of the curve keeps unchanged. And the curve would move along x axis with the change of $\bar{x}$. Fig. 5-33b shows that if $\bar{x}$ remains unchanged, the position of the curve keeps unchanged. The shape of the curve would change with the variation of σ. Therefore, σ symbolizes the shape of the distribution curve, i.e. the dimension dispersal feature (machining accuracy), and $\bar{x}$ symbolizes the position of the curve.

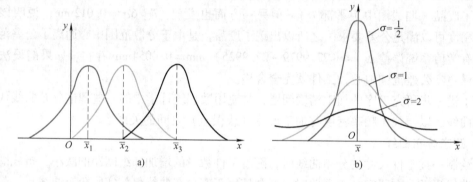

Fig. 5-33 Effect of $\bar{x}$ and σ on normal distribution curve $\bar{x}$、σ 对正态分布曲线的影响

(3) Practical dispersal range of the curve The normal distribution with $\mu = 0$, $\sigma = 1$ is named as standard normal distribution, denoted by $x(z) - N(0, 1)$. According to the definition of normal distribution, normal distribution function is the integral of probability density function of normal distribution. That is,

$$F(x) = \frac{1}{\sigma\sqrt{2\pi}} \int_{-\infty}^{x} e^{-\frac{1}{2}(\frac{x-\mu}{\sigma})^2} dx \qquad (5-28)$$

The area covered by normal distribution curve is $F(x) = \int_{-\infty}^{+\infty} y dx = 1$, which stands for the total numbers of the workpiece, i.e. 100%. Set $z = \frac{x-\mu}{\sigma}$, then

$$F(z) = \frac{1}{\sqrt{2\pi}} \int_{0}^{z} e^{-\frac{z^2}{2}} dz \qquad (5-29)$$

Non-standard normal distribution can be changed into standard normal distribution by means of Eq. (5-29) to make calculations (see Fig. 5-32). $F(z)$ stands for the area of hatched section in Fig. 5-32. The values of $F(z)$ with different z can be looked up from Tab. 5-6.

When $x-\bar{x} = 3\sigma$, it can be looked up from Tab. 5-6 that $F(3) = 49.865\%$, $2F(3) = 99.73\%$ (areas covered by the curve in the range of $\pm 3\sigma$), that is to say, the probability presented outside $\pm 3\sigma$ is only 0.27%, which can be ignored. So it is thought that the practical distribution range of

正态分布的概率密度函数有两个特征参数。由式（5-27）与图 5-32 可以看出，当 $x = \mu$ 时，$y_{\max} = 1/\sigma\sqrt{2\pi}$，这是曲线的最大值，也是曲线的分布中心。正态分布总体的 μ 和 σ 通常是不知道的，但可以通过它的样本平均值 $\bar{x}$ 和样本标准偏差 σ 来估计。即用样本的 $\bar{x}$ 代替总体的 μ，用样本的 σ 代替总体的 σ。

（2）正态分布曲线的形状 当 σ 不变时，改变 $\bar{x}$ 的值，分布曲线沿横坐标移动，但形状不变，如图 5-33a 所示。σ 是表征分布曲线形状的参数。当 $\bar{x}$ 不变时改变 σ，曲线形状发生变化，如图 5-33b 所示。可见，σ 表征分布曲线的形状，即尺寸分散特性（加工精度）。$\bar{x}$ 表征分布曲线的位置。

（3）分布曲线的实际分散范围 平均值 $\mu = 0$，标准差 $\sigma = 1$ 的正态分布称为标准正态分布，记为：$x(z) \sim N(0, 1)$。由分布函数的定义可知，正态分布函数是正态分布概率密度函数的积分，即

$$F(x) = \frac{1}{\sigma\sqrt{2\pi}} \int_{-\infty}^{x} e^{-\frac{1}{2}(\frac{x-\mu}{\sigma})^2} dx \tag{5-28}$$

正态分布曲线下的面积 $F(x) = \int_{-\infty}^{+\infty} y dx = 1$ 代表了全部零件数（即100%）。令 $z = \frac{x-\mu}{\sigma}$，则

$$F(z) = \frac{1}{\sqrt{2\pi}} \int_{0}^{z} e^{-\frac{z^2}{2}} dz \tag{5-29}$$

利用式（5-29），可将非标准正态分布转换成标准正态分布进行计算（图5-32）。$F(z)$ 为图 5-32 中有阴影部分的面积。不同 z 值的 $F(z)$ 可由表 5-6 查出。

当 $x - \bar{x} = 3\sigma$ 时，由表 5-6 查得 $F(3) = 49.865\%$，$2F(3) = 99.73\%$（曲线在 $\pm 3\sigma$ 范围内覆盖的面积），也就是说，出现在 $\pm 3\sigma$ 范围以外的概率仅有 0.27%，可忽略不计。因此，可以认为正态分布曲线的实际分散范围为 6σ。

Table 5-6 $F(z) = \frac{1}{\sqrt{2\pi}} \int_{0}^{z} e^{-\frac{z^2}{2}} dz$

z	F(z)	z	F(z)	z	F(z)	z	F(z)	z	F(z)
0.02	0.0080	0.36	0.1406	0.66	0.2454	1.10	0.3643	2.10	0.4821
0.04	0.0160	0.38	0.1480	0.68	0.2517	1.15	0.3749	2.20	0.4861
0.06	0.0239	0.40	0.1554	0.70	0.2580	1.20	0.3849	2.30	0.4893
0.08	0.0319	0.42	0.1628	0.72	0.2642	1.25	0.3944	2.40	0.4918
0.09	0.0359	0.43	0.1664	0.74	0.2703	1.30	0.4032	2.50	0.4938
0.10	0.0398	0.44	0.1700	0.76	0.2764	1.35	0.4115	2.60	0.4953
0.12	0.0478	0.45	0.1736	0.78	0.2823	1.40	0.4192	2.70	0.4965
0.14	0.0557	0.46	0.1772	0.80	0.2881	1.45	0.4265	2.80	0.4974
0.16	0.0636	0.47	0.1808	0.82	0.2939	1.50	0.4332	2.90	0.4981
0.18	0.0714	0.48	0.1844	0.84	0.2995	1.55	0.4394	3.00	0.49865
0.19	0.0753	0.49	0.1879	0.86	0.3051	1.60	0.4452	3.20	0.49931
0.20	0.0793	0.50	0.1915	0.88	0.3106	1.65	0.4505	3.40	0.49966
0.22	0.0871	0.52	0.1985	0.90	0.3159	1.70	0.4554	3.60	0.499841
0.24	0.0948	0.54	0.2054	0.92	0.3212	1.75	0.4599	3.80	0.499928
0.26	0.1026	0.56	0.2123	0.94	0.3264	1.80	0.4641	4.00	0.499968
0.28	0.1103	0.58	0.2190	0.96	0.3315	1.85	0.4678	4.50	0.499997
0.30	0.1179	0.60	0.2257	0.98	0.3365	1.90	0.4713	5.00	0.49999997
0.32	0.1255	0.62	0.2324	1.00	0.3413	1.95	0.4744	—	—
0.34	0.1331	0.64	0.2389	1.05	0.3531	2.00	0.4772	—	—

normal distribution curve is 6σ.

Generally speaking, the size of 6σ represents the machining accuracy of a machining method under specified condition. Under ordinary conditions, the relationship between T and 6σ should be kept as follows:

$$6\sigma \leqslant T \quad (5\text{-}30)$$

If there is constant systematical error Δ_{cs}, the condition without rejects should be: $T \geqslant 6\sigma + 2\Delta_{cs}$, as shown in Fig. 5-34.

3. Applications of distribution curves

(1) To determine whether the machining method can ensure the machining accuracy (see Fig. 5-34) As mentioned above, if the machining dimension obeys normal distribution, 6σ stands for the process capability. In practice, the process capability coefficient C_p is used to measure the process capability of a machining method. C_p is calculated as follows.

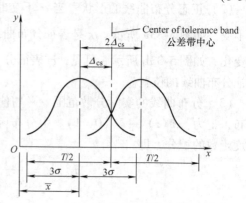

Fig. 5-34 Situation of $T > 6\sigma$ $T > 6\sigma$ 的情形

$$C_p = T/6\sigma \quad (5\text{-}31)$$

According to the value of C_p, process capability is classified as five grades, as shown in Tab. 5-7. Generally, process capability can't be lower than grade 2.

(2) To distinguish Δ_r from Δ_s according to the main factors affecting machining accuracy If $T/6\sigma < 1$, rejects must appear and are caused mainly by random errors Δ_r. If $T/6\sigma \geqslant 1$ and rejects have been produced, the rejects are caused mainly by systematic errors; if the dimensions distribution obey normal distribution at this time, that means there is no apparent variable systematic errors. That the center $\bar{x}$ of the dispersal range is not consistent with the center of the tolerance band means there is a constant systematical error.

(3) To determine the rejection rate

【Example 5-3】(continue):

$$C_p = T/6\sigma = 1.11$$

$$\frac{x - \bar{x}}{\sigma} = \frac{28 - 27.9979}{0.002244} = 0.9358$$

By looking up related table similar to the Tab. 5-6, $F(0.9358) = 0.3253 = A$, then the rejection rate P_{rr} and percent of pass P_{pp} can be found.

$$P_{rr} = 0.5 - A = 17.47\%$$
$$P_{pp} = 0.5 + A = 82.53\%$$

【Discussion】 Now that $C_p = T/6\sigma = 1.11 > 1$, why do the rejects still appear? What are the primary reasons? Can you find a good way to reduce or eliminate the rejects?

(4) Short comings of distribution curve

一般来说，6σ 的大小代表某加工方法在一定条件下所能达到的加工精度。通常情况下，应该使所选择加工方法的标准差 σ 与公差带宽度 T 之间具有下列关系

$$6\sigma \leq T \tag{5-30}$$

如果存在常值系统误差 Δ_{cs}，则不出废品的条件应该是：$6\sigma + 2\Delta_{cs} \leq T$，如图 5-34 所示。

3. 分布曲线的应用

（1）确定加工方法是否能够满足加工精度要求（图 5-34） 如前所述，当加工尺寸服从正态分布时，工艺能力为 6σ。在生产实际中，使用工序能力系数 C_p 来衡量工艺能力的大小。C_p 按下式计算

$$C_p = T / 6\sigma \tag{5-31}$$

根据 C_p 的大小，把工艺能力分为五级，见表 5-7。一般情况下，工艺能力不应低于二级。

Table5-7 Process capability grade 工艺能力等级

C_p	$C_p > 1.67$	$1.67 \geq C_p > 1.33$	$1.33 \geq C_p > 1.0$	$1.0 \geq C_p > 0.67$	$0.67 \geq C_p$
Grade of capability 工艺能力等级	Special grade 特级	Grade 1 一级	Grade 2 二级	Grade 3 三级	Grade 4 四级
Capability estimation 能力判断	Over high 过高	Enough 足够	After a fashion 勉强	Obvious deficiency 明显不足	Very deficient 非常不足

（2）区别影响加工精度的主要因素是随机误差 Δ_r 还是系统误差 Δ_s 如果 $T/6\sigma < 1$，则必然出废品，且主要是由随机误差造成的。如果 $T/6\sigma \geq 1$，且有废品出现，则主要是由系统误差引起的；如果此时尺寸分布服从正态分布，则说明加工过程中没有明显的变值系统误差。尺寸分散中心 $\overline{X}$ 与公差带中心不重合就说明存在常值系统误差。

（3）计算废品率

【例 5-3】 （接上）

$$C_p = T / 6\sigma = 1.11$$

$$\frac{x - \overline{x}}{\sigma} = \frac{28 - 27.9979}{0.002244} = 0.9358$$

通过查阅类似于表 5-6 的详细表格可得 $F(0.9358) = 0.3253 = A$，由此可以求出这批零件的废品率 P_{rr} 和合格率 P_{pp}

$$P_{rr} = 0.5 - A = 0.5 - 0.3253 = 0.1747 = 17.47\%$$

$$P_{pp} = 0.5 + A = 0.5 + 0.3253 = 0.8253 = 82.53\%$$

【讨论】既然 $C_p = T / 6\sigma = 1.11 > 1$，为什么会出废品？主要是由什么原因造成的？有什么办法可以减少甚至消除废品？

（4）分布曲线分析法的缺点

1）分布曲线分析法不能把随机误差与变值系统误差区分开，主要是由于分析时没有考虑到工件加工的先后顺序。

2）只能在一批工件加工完毕后才知道尺寸分布情况，因此不能揭示加工过程中的误差变化规律。

1) It can't distinguish the variable systematical error from the random error because it didn't consider the machining sequence.

2) Only after a batch of workpieces are completely finished can draw the distribution curve. It can't reveal the error variation law in the process.

5.3.3 Point diagram method

A point diagram is drawn out in the light of machining sequence of a workpiece to reflect the variation of the dimension and to reveal the total feature of the error variation in whole machining process. Taking the machining sequence number of the workpiece as abscissa and the dimension of the workpiece as ordinate, the point diagram reflects completely the relation of machined dimension variation over time, as shown in Fig. 5-35.

Fig. 5-35 is drawn based on the measured diameters of workpieces which are machined by an automatic lathe. Curves AA and BB represent the upper and lower limits of points respectively. Mean curve OO reflects the variation of variable systematical errors. The width between curve AA and curve BB represents the "dimension dispersal" which is caused by random errors in machining process. When workpieces are inspected one by one till No. 50, it is found that the dimension of the No. 50 exceeds the upper limit of tolerance.

After setting the cutting tool, a constant systematical error Δ_{cs} appears (Δ_{cs} equals to the distance in vertical direction between the point O and the point O'). Δ_{cs} only affects the position of the curve (up or down), and does not affect the shape of the curve or dimension dispersal range. Therefore, the point diagram can be used in machining process to estimate the variable trend of the dimension, and to determine when the machine tool is required to reset. However, if the point diagram is directly used to control the machining process, the workpieces have to be measured one by one, leading to lots of consuming in manpower and material resources. Now, many factories in mass production adopt another kind of point diagram, $\overline{X} - R$ Chart (mean value—dispersal range chart) to control the operation quality.

$\overline{X} - R$ chart is a joint control chart. $\overline{X}$ chart reflects the variable trend of systematical errors, and R chart reflects the variable trend of random errors. $\overline{X} - R$ chart is drawn based on random sampling in light of small-sample sequence. Its abscissa is the grouping sequence number of small sample collected in chronological order, the ordinate is $\overline{X}$ and R of the sample separately, where R is the difference between the maximum and minimum in the sample, named as extreme difference.

In machining process, it is required to take out a small sample including $m = 2 - 10$ workpieces at random every a certain time, and then to find the mean $\overline{X}_i$ and R_i of the sample. After a while, several groups of small samples (for instance, k groups, $k = 25$ generally) can be acquired. In this way, the $\overline{X}$ chart and R chart can be plotted separately, as shown in Fig. 5-36.

There are three lines, i.e. center line (CL), upper control line (UCL) and lower control line (LCL), as shown in Fig. 5-37. Fig. 5-37 is just a $\overline{X} - R$ joint control chart.

5.3.3 点图分析法

点图是按零件加工的先后顺序绘制出来的，用以反映尺寸的变化，揭示整个加工过程中误差变化的全貌。点图以加工零件序号为横坐标，以加工后测量所得尺寸为纵坐标，充分反映了加工尺寸的变化与时间的关系，如图 5-35 所示。

图 5-35 是根据在自动车床上所加工工件直径的测量结果而绘制的。曲线 AA、BB 分别表示所有点子的上限和下限。平均值线 OO 反映了变值系统误差的变化。曲线 AA 和 BB 之间的宽度代表了随机误差作用下加工过程的尺寸分散。当工件一个一个检查到第 50 号时，发现尺寸超出了公差上限。

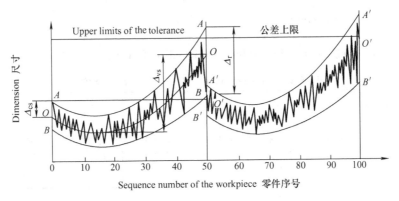

Fig. 5-35 Illustration of a point diagram 点图示例

在进行了一次调刀后，出现了常值系统误差 Δ_{cs}。（Δ_{cs} 就是 O 点到 O' 点在垂直方向的距离）。Δ_{cs} 只影响曲线的位置（上或下），并不影响曲线的形状或分散范围的大小。所以，点图可以在加工过程中用来估计工件尺寸的变化趋势，并决定机床重新调整的时间。但是如果直接用点图控制加工过程，就必须测量每一个工件，这将造成很大的人力和物力浪费。因此，许多工厂在大量生产中采用的是另一种点图，即 $\bar{X} - R$ 图（平均值和分散范围图），进行工序的质量控制。

$\bar{X} - R$ 图是一种联合控制图。$\bar{X}$ 点图反映的是系统误差的变化趋势，而 R 点图反映的是随机误差的变化趋势。绘制 $\bar{X} - R$ 图是以小样本顺序随机抽样为基础的。$\bar{X} - R$ 图的横坐标是按时间先后采集的小样本的组序号，纵坐标为样本的平均值 $\bar{X}$ 和极差 R。

在加工过程中，每隔一定时间抽取容量 $m = 2 \sim 10$ 件的一个小样本，求出小样本的平均值 $\bar{X}_i$ 和极差 R_i。经过若干时间后，就可取得若干组（如 k 组，通常取 $k = 25$）小样本。这样，就可分别作 $\bar{X}$ 点图和 R 点图，如图 5-36 所示。

在 $\bar{X} - R$ 图上有三根线，即中心线和上、下控制线，如图 5-37 所示。图 5-37 就是 $\bar{X} - R$ 联合控制图。

$\bar{X}$ 图的中心线
$$CL = \frac{1}{k} \sum_{i=1}^{k} \bar{X}_i \tag{5-32}$$

Center line of $\overline{X}$ chart
$$CL = \frac{1}{k}\sum_{i=1}^{k}\overline{X}_i \tag{5-32}$$

Upper control line of $\overline{X}$ chart
$$UCL = \overline{X} + AR \tag{5-33}$$

Lower control line of $\overline{X}$ chart
$$LCL = \overline{X} - AR \tag{5-34}$$

Center line of R chart
$$CL = \frac{1}{k}\sum_{i=1}^{k}R_i \tag{5-35}$$

Upper control line of R chart
$$UCL = D_1\overline{R} \tag{5-36}$$

Lower control line of R chart
$$LCL = D_2\overline{R} \tag{5-37}$$

where, A, D_1, D_2—coefficients, their values are listed in Tab. 5-8.

Table5-8 Values of A, D_1, D_2 (A、D_1、D_2 的数值)

Piece/group 件数/组	3	4	5	6	7	8	9
A	1.0231	0.7285	0.5768	0.4833	0.4193	0.3726	0.3367
D_1	2.5742	2.2819	2.1145	2.0039	1.9242	1.8641	1.8162
D_2	0	0	0	0	0.0758	0.1359	0.1838

5.4 Factors influencing surface quality

5.4.1 Factors influencing surface roughness

1. Factors of influencing surface roughness in cutting process

(1) Influence of geometric parameters of a cutter The geometric factor resulting in roughness is due to a part of the cutting layer remaining on the machined surface generated by the feed movement of the cutting tool relative to the workpiece, as shown in Fig. 5-38. The primary factors of affecting surface roughness are cutting edge angle κ_r, minor cutting edge angle κ_r', nose radius r_ε and feed f.

For the cutter with a straight cutting edge (Fig. 5-38a), the height of residual area H is
$$H = f/(\cot\kappa_r + \cot\kappa_r') \tag{5-38}$$

For the cutter with a nose radius (Fig. 5-38b), the height of residual area H is
$$H = f^2/8r_\varepsilon \tag{5-39}$$

In fact, the rake angle γ_o and clearance angle have an influence on surface roughness to some extent. The increase of γ_o can reduce the plastic deformation of material, which is helpful to reduc-

$\overline{X}$ 图的上控制线	$UCL = \overline{\overline{X}} + A\overline{R}$	(5-33)
$\overline{X}$ 图的下控制线	$LCL = \overline{\overline{X}} - A\overline{R}$	(5-34)
R 的中心线	$CL = \dfrac{1}{k}\sum\limits_{i=1}^{k} R_i$	(5-35)
R 的上控制线	$UCL = D_1 \overline{R}$	(5-36)
R 的下控制线	$LCL = D_2 \overline{R}$	(5-37)

式中 A、D_1、D_2——系数，其数值列于表 5-8 中。

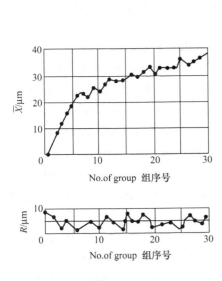

Fig. 5-36 $\overline{X}$ chart and R chart
$\overline{X}$ 点图和 R 点图

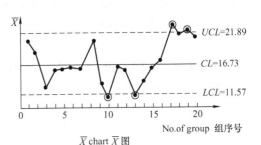
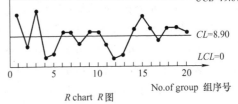

Fig. 5-37 $\overline{X} - R$ chart
$\overline{X} - R$ 图

5.4 影响加工表面质量的因素

5.4.1 影响表面粗糙度的因素

1. 切削加工中影响表面粗糙度的因素

（1）刀具几何形状的影响 切削加工后的表面粗糙度是刀具相对工件作进给运动时，在加工表面上遗留下来的切削层残留面积，如图 5-38 所示。影响表面粗糙度的主要因素有刀尖圆弧半径 r_ε、主偏角 κ_r、副偏角 κ_r' 及进给量 f 等。

当用尖刀刃切削时（图 5-38a），切削层残留面积高度为

$$H = f/(\cot\kappa_r + \cot\kappa_r') \tag{5-38}$$

当用圆弧刀刃切削时（图 5-38b），切削层残留面积高度为

ing surface roughness. To increase α_o properly can reduce the friction between the flank and the workpiece, which is also helpful to reducing surface roughness.

(2) Influence of cutting variables Of the three cutting variables, f has the largest influence on surface roughness. But the influence of the cutting speed v on surface roughness can't be ignored. Why are the higher cutting speeds usually used for finishing plastic metals? We learned before that build-up edge (BUE) can be avoided at higher cutting speeds.

(3) Influence of other factors The plasticity of material has larger influence on surface roughness. The better the plasticity of material, the easier the BUE and scale form, and the rougher the machined surface. Quenching and tempering treatment of material is beneficial for improving the surface roughness. Vibration produced in cutting process can increase the surface roughness of the workpiece. To choose the proper cutting fluid is also beneficial for reducing the surface roughness of the workpiece.

2. Factors of influencing on surface roughness in grinding process

Grinding is different from cutting in many aspects. The surface roughness is caused by the scratches which are generated by the relative scrape between the grinding surface and the abrasive grits distributing on the surface of the grinding wheel. Grinding velocity is much higher than ordinary cutting speed. Most of abrasive grains have the negative rake angle. As the grinding temperature is very high, whereas the grinding layer is very thin, the phase transition and burn appear easily on the surface layer of the workpiece. Therefore, the plastic deformation in grinding process is much larger than that in ordinary cutting process. Factors of influencing on surface roughness in grinding process can be considered from the following three aspects.

(1) Grinding wheel The finer grain size is helpful to reducing roughness of the ground surface. But over small grain size would cause the chip to block the spaces of among the grains. Instead, this is harmful to reducing surface roughness, and easy to cause the burn of the workpiece. The grade of the grinding wheel should be proper. Over high or over low grade is unfavorable to reducing surface roughness. Medium-soft wheel is often used in production. The dressing quality of the grinding wheel is also an important factor to improve surface roughness.

(2) Grinding variables Grinding variables include usually the grinding speed v, the workpiece speed v_w, the depth of grinding a_p and the longitudinal feed f. Generally, surface roughness decreases with the increase of the grinding speed, but increases with the increase of the workpiece speed, depth of grinding and longitudinal feed. The depth of grinding has quite a large effect on surface roughness. The relation between grinding variables and surface roughness is shown in Fig. 5-39. To increase the times of finish grinding without radial feed can decrease surface roughness of a ground workpiece.

(3) Other factors When grinding the materials with low hardness, good plasticity, poor thermal conductivity and poor grindability, roughness of ground surface is large. To select proper grinding fluid and use effective cooling method are helpful to reducing surface roughness of a ground workpiece and preventing the grinding surface from burning.

$$H = f^2/8r_\varepsilon \tag{5-39}$$

事实上，刀具前角 γ_o 和后角 α_o 对表面粗糙度也有一定的影响。如增大前角 γ_o 可以减小材料的塑性变形，有利于降低粗糙度。适当增大后角 α_o 可以减小后面与工件的摩擦，也有利于降低表面粗糙度值。

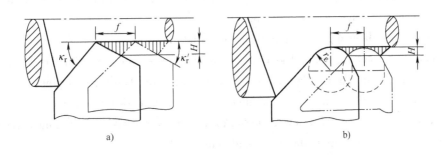

Fig. 5-38　Residual area of a cutting layer
切削层残留面积

(2) 切削用量的影响　切削用量三要素中，进给量 f 对表面粗糙度影响最大。但切削速度 v 的影响也不能忽视。为什么在精加工塑性金属材料时通常采用高速切削呢？是因为高速切削可以避免形成积屑瘤。

(3) 其他因素的影响　材料的塑性对表面粗糙度的影响也很大。材料的塑性越好，越易形成积屑瘤和鳞刺，加工后的表面就越粗糙。对材料进行调质处理有助于改善加工后的表面粗糙度。切削加工中的振动使工件的表面粗糙度增大。合理选择切削液也有利于减小加工表面的粗糙度。

2. 磨削加工中影响表面粗糙度的因素

磨削加工与切削加工有许多不同之处。磨削时由于分布在砂轮表面上的磨粒与被磨削表面间作相对滑擦形成的划痕，构成了表面粗糙度。磨削速度比一般切削加工速度高得多，磨粒大多为副前角，磨削区温度很高，磨削层很薄，工件表层金属极易产生相变和烧伤。所以，磨削过程的塑性变形要比一般切削过程大得多。磨削中影响表面粗糙度的因素可从以下三个方面考虑。

(1) 砂轮方面　砂轮的粒度越小，越有利于降低表面粗糙度。但粒度过小，砂轮容易堵塞，反而使表面粗糙度增大，还易引起烧伤。砂轮硬度应大小合适，砂轮过硬或过软，都不利于降低表面粗糙度。在生产中常采用中软砂轮。砂轮的修整质量是改善表面粗糙度的重要因素。

(2) 磨削用量　磨削用量包括砂轮速度 v、工件速度 v_w、磨削深度 a_p 和纵向进给量 f。提高砂轮速度有利于降低表面粗糙度。工件速度、磨削深度和纵向进给量增大，均会使表面粗糙度增大。其中磨削深度对表面粗糙度的影响相当大。磨削用量与表面粗糙度的关系如图 5-39 所示。增加"无进给光磨"次数可以降低磨削工件的表面粗糙度。

5.4.2 Factors influencing physical and mechanical properties of surface layer

1. Main factors of influencing work-hardening

The plastic deformation produced in cutting or grinding process would make the crystal lattices in the surface layer of metal appear aberrant, the grains in the surface layer suffer shearing slip, elongating, and even breaking, thus increasing the hardness and strength of the surface layer of metal. The phenomenon is called work-hardening.

Main factors of influencing work-hardening are as follows.

(1) Geometry of cutting tool   The work-hardening degree increases with the decrease of the rake angle. The rounded cutting edge radius increases, the work-hardening degree increases. The wear amount VB of tool flank has a great influence on the work-hardening degree.

(2) Workpiece material   The lower the hardness of work material, or the larger the plasticity of work material, the more serious the work-hardening will be.

(3) Machining variables   In cutting, the increase of the feed f will lead to the increase of the work-hardening degree, but the effect of the back engagement a_p on work-hardening is not obvious. In grinding, the work-hardening degree increases with the increase of the feed and back engagement. As the cutting temperature increases with the increase of the cutting (or grinding) speed, and the increase of the temperature is helpful for the recovery of the work-hardening layer, work-hardening degree will decrease.

2. Factors of influencing the phase variation of the machined surface

As the cutting temperature is not very high in ordinary cutting, the metallurgical structure of the machined surface would not change. The most of energy consumed in grinding is transformed into grinding heat, and about 70% of total quantity of heat is transferred to the workpiece, thus leading to the phase variation of the machined surface. The variation of metallurgical structure will cause the decrease of the strength and hardness of metal, the occurrence of residual stress, and even cracks. Above phenomena are called grinding burn.

The basic reason for grinding burn is over-high grinding temperature. Therefore, the basic approach to avoiding and lightening grinding burn is to reduce the quantity of heat produced and quicken the radiating speed as far as possible. The specific measures are: ①to choose proper grinding wheel; ②to reduce the grinding speed and the depth of cut properly, and to increase the workpiece rotation speed and the axial feed properly; ③to adopt efficient cooling methods so as to reduce the temperature in grinding region effectively.

3. Residual stress in machined surface layer

As the shape, or volume, or metallurgical structure of the metal in the workpiece surface layer would change in the machining process, some stresses will remain in the surface layer. The causes for residual stress are as follows.

(1) Cold plastic deformation   In machining, the surface layer of the workpiece experiences intensively plastic deformation. The metal in surface layer is stretched in the direction of the cutting speed, whereas the subsurface metal would hinder the deformation of the surface metal. Therefore,

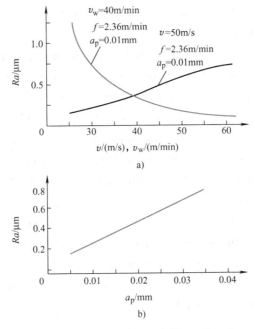

Fig. 5-39 Relation between grinding variables and surface roughness
磨削用量与表面粗糙度的关系

（3）其他方面　工件材料硬度越小、塑性大、导热性差，磨削性差，磨削后的表面粗糙度大。采用合适的切削液和有效的冷却方式有利于降低表面粗糙度，防止表面烧伤。

5.4.2　影响表面层物理力学性能的因素

1. 影响加工硬化的主要因素

切削（磨削）过程中产生的塑性变形，会使表层金属的晶格发生畸变，晶粒间产生剪切滑移，晶粒被拉长，甚至破碎，从而使表层金属的硬度和强度提高，这种现象称为加工硬化或冷作硬化。

影响表层金属加工硬化的主要因素有：

（1）刀具的几何形状　前角减小，加工硬化程度增大。刀具刃口钝圆半径增大，加工硬化程度增大。刀具后面的磨损量对加工硬化程度影响也很大。

（2）工件材料　工件材料硬度越小，或塑性越大，加工硬化倾向越大，硬化程度越严重。

（3）切削用量　切削加工时，进给量 f 增大会使加工硬化程度增大，而切削深度 a_p 影响不明显。但磨削时，工件转速增大，加工硬化程度增大；磨削深度 a_p 增大，加工硬化程度增大。切（磨）削速度 v 增大，会使切削温度升高，加工硬化程度将会减小。

2. 影响表面层金属相变的因素

由于一般切削加工时的切削温度不是很高，表层金属的金相组织没有质的变化。而磨削加工时所消耗的能量绝大部分要转化为热，且有约70%以上的热量传给工件，使加工表面层金属金相组织发生变化，造成表层金属的强度和硬度降低，并产生残余应力，甚至会出现微观裂纹，这种现象称为磨削烧伤。

after machining, the residual stress in surface layer is compressive stress (−), and the residual stress in inner layer is tensile stress (+).

(2) Hot plastic deformation   Cutting or grinding heat results in thermal plastic deformation under conditions of high temperature. High temperature often causes the metal of the surface layer to expand. Here the surface layer presents thermal compression stress. After machining, the shrinkage of the thermal plastic deformation produced in the surface layer would be hindered by inner metal. Therefore, the residual stress in surface layer is tensile stress (+), and the residual stress in inner layer is compressive stress (−). Under grinding condition, the higher the grinding temperature is, the larger the extent of thermal plastic deformation, the larger the residual tensile stress, and even the cracks would occur sometimes.

(3) Variation of metallurgical structure   High machining temperature results in the variation of the metallurgical structure in the surface layer of the workpiece. Because different metallurgical structures have different specific volume, the variation of the metallurgical structure in the surface layer will lead to the change of the metal volume. When the hardened steel is ground, the residual tensile stress will be generated in surface layer, and the residual compressive stress will be generated in inner layer.

The actual residual stress as well its distribution in the machined surface layer is the comprehensive results of above three factors. Under certain conditions, one or two factors of the three factors may play the leading role.

5.5 Vibrations in machining processes

5.5.1 Introduction

In the metal cutting process, violent vibration and chatter often happen between the workpiece and the cutting tool. Vibration and chatter would cause chatter mark on machined surface, thus leading to poor surface finish. Sometimes, cutting variables have to be reduced in order to maintain regular machining. Vibration and chatter would cause the premature wear, chipping, and failure of the cutting tool, loosen the joints in machine tools and fixtures, thus lowering the precision and rigidity of machining facilities. Violent vibration would make objectionable noise which would hurt the physical and mental health of operators.

Vibration and chatter in machining are complex phenomena. In cutting operations there are three types of vibration: free vibration, forced vibration and self-excited vibration. As the free vibration can attenuate rapidly because of the action of damp, so it has less effect on the machining process. The forced vibration and self-excited vibration will be discussed only in this section.

5.5.2 Forced vibration in machining processes

1. Causes of forced vibration

Forced vibration, which will not be attenuated, is generally caused by an external periodic dis-

造成磨削烧伤的根本原因是磨削温度过高。因此避免和减轻磨削烧伤的基本途径是尽可能减少磨削热的产生并加快散热速度。具体措施有：①合理选择砂轮；②适当减小磨削深度和磨削速度，适当增加工件的转动速度和轴向进给量；③采用高效冷却方式，以有效降低磨削区温度。

3. 表面层金属残余应力

由于工件表面层金属的形状、或体积、或金相组织会发生改变，在表面层内就会保留有某种应力。残余应力形成的原因如下：

（1）冷态塑性变形　在加工中，工件表面层产生了剧烈的塑性变形。表层金属沿切削速度方向被拉长，而里层金属会阻止表层金属的变形。于是，加工后表层金属的残余应力是压应力（-），而里层金属则是拉应力（+）。

（2）热态塑性变形　切削热在高温条件下会引起热态塑性变形。高温往往使加工表面产生热膨胀，此时表层产生热压应力。加工后，表层已产生的热塑性变形收缩受到里层金属的阻碍。故加工后表面层残余应力为拉应力（+），里层则产生残余压应力（-）。在磨削时，磨削温度越高，热塑性变形越大，残余拉应力也越大，有时甚至会产生裂纹。

（3）金相组织变化　高的加工温度会引起工件表面层的金相组织发生相变。不同的金相组织有不同的密度，故相变会引起体积变化。磨削淬火钢时，表面层将产生残余拉应力，里层将产生残余压应力。

实际加工后表面层残余应力是上述三方面原因的综合结果。在一定条件下，其中一个或两个会起主要作用。

5.5　机械加工过程中的振动

5.5.1　概述

金属切削过程中，工件和刀具之间常发生强烈的振动。切削振动使得零件加工表面出现振纹，表面粗糙度值增大。为使加工继续，有时不得不降低切削用量。振动会引起刀具过早磨损、崩刃和失效，使机床、夹具等零件的连接部分松动，从而降低机床、夹具的刚度和精度。强烈的振动会发出刺耳的噪声，危害操作者的身心健康。

机械加工中的振动是一种复杂的现象。切削加工中产生的振动分为自由振动、强迫振动和自激振动三种类型。由于自由振动在阻尼的作用下会很快地衰减下去，因此对加工的影响不大。故本节只讨论强迫振动和自激振动。

5.5.2　机械加工过程中的强迫振动

1. 强迫振动产生的原因

强迫振动通常是由外界周期性干扰力（激振力）引起的不衰减振动。强迫振动的振源有机外振源与机内振源之分。机外振源的影响可通过隔振来消除。机内振源主要有：

（1）高速回转零件的不平衡　如电动机转子、带轮、联轴器、砂轮、齿轮等回转件，其转速越高，产生周期干扰力的幅值就越大。

（2）往复运动部件的换向冲击　如牛头刨床的滑枕、液压滑台等部件在改变运动方向

turbing forces (or exciting force). The vibration sources of the forced vibration have internal sources and external sources. The influence of external vibration sources can be eliminated easily by vibration isolation. Internal vibrating sources includes:

(1) **Imbalance of parts rotated at a high speed** The higher the rotating speed of such parts as motor's rotor, pulley, coupling, grinding wheel, gear etc., the larger the magnitude of periodic disturbance force produced by these revolving parts.

(2) **Impact of reciprocating parts** Such as the ram of the shaper, hydraulic slide and so on.

(3) **Manufacturing errors of transmission parts** Such as pitch error of gear, machining error of rolling bearing, thickness error of belt and so on.

(4) **Impact produced in cutting process** Such factors as the intermittent cutting, uneven machining allowance and hardness of the blank would cause the variation of the cutting force.

2. Characteristics of forced vibration

1) The frequency of the forced vibration is always equal to the frequency of the external disturbing force, and has nothing to do with the natural frequency of vibration system.

2) The forced vibration is caused by an external periodic disturbing force, and may not be attenuated by damp. The forced vibration does not in itself cause the variation of disturbing force.

3) The amplitude of the forced vibration has to do with disturbing force, rigidity of system and damp coefficient. The larger the disturbing force, the smaller the rigidity of the system and the damp coefficient, and the larger the amplitude of the forced vibration. When the frequency of the external disturbing force approaches or equals the natural frequency of vibration system, the amplitude of the forced vibration reaches the maximum, that is, resonant vibration occurs.

5.5.3 Self-excited vibration in machining processes

1. Self-excited vibration and its characteristics

Sometimes a machine tool would vibrate due to the cutting process itself instead of outside disturbing forces. This is because when the cutting stops, this kind of vibration will come to stop even if the machine tool keeps idle running. Obviously, this is not forced vibration. As this type of vibration is intensified and maintained by alternating cutting force generated in machining system, it is named as self-excited vibration. As it is a kind of undamped oscillation with high frequency and very large amplitude, it is generally called the chatter.

Self-excited vibration system in a machine tool is a closed system composed of a vibration system (technological system) and a regulation system (cutting process), as shown in Fig. 5-40. In machining process, accidental outside disturbances (such as blank error) would cause the variation of the cutting force. The variational cutting force exerting on machining system would make the system produce vibration, thus leading to the cyclic variation of the position of the cutter relative to the workpiece $y(t)$. The cyclic variation $y(t)$ would cause machining process to produce alternating cutting force $F(t)$, which again energized machining system to produce vibration. If the conditions of the self-excited vibration don't exist in machining system, the vibration caused by accidental outside disturbance would attenuate gradually because of damping. Once the vibration comes to a stop,

时产生的冲击。

（3）机床传动件的制造误差　如齿轮的齿距误差、滚动轴承制造误差、带厚度误差等。

（4）切削过程中的冲击　如断续切削、毛坯硬度和加工余量不均匀等均会引起切削力变化。

2. 强迫振动的特征

1）强迫振动的频率总与外界干扰力的频率相同，与系统的固有频率无关。这是强迫振动最本质的特征。

2）强迫振动是由周期性干扰力引起的，不会被阻尼衰减掉，振动本身并不能引起干扰力变化。

3）强迫振动振幅的大小与干扰力、系统刚度及阻尼系数有关。干扰力越大，系统刚度和阻尼系数越小，则振幅越大。当干扰力的频率与系统的固有频率相近或相等时，振幅达最大值，即出现"共振"现象。

5.5.3　机械加工过程中的自激振动

1. 自激振动及其特征

有时机床的振动是由切削过程本身引起的，而不是外界干扰力。这是因为切削停止后，即使机床还在空转，这种振动也没有出现，显然这不是强迫振动。由于这种振动是由加工系统本身产生的交变切削力加强和维持的，故称为自激振动。由于它是频率较高、振幅很大的不衰减振动，故又称颤振。

机床自激振动系统由振动系统（工艺系统）和调节系统（切削过程）组成，如图 5-40 所示。在加工过程中，偶然的外界干扰（如毛坯误差等）会引起切削力的变化。这个变化的切削力作用在加工系统上，就会使系统产生振动，于是引起工件和刀具间的相对位置发生周期性变化 $y(t)$，使加工过程产生交变切削力 $F(t)$，并再次激励加工系统产生振动。如果加工系统不存在自激振动的条件，这种偶然的外界干扰引起的振动将因阻尼作用而逐渐衰减。加工系统的振动一旦停止，交变切削力就随之消失。

自激振动能否产生以及振幅大小取决于每一个振动周期内系统所获得的能量 E^+ 与所消耗能量 E^- 的对比情况。如图 5-41 所示，当 $E^+ = E^-$ 时，振幅达到 A_0 值。当某瞬时的振幅为 A_1 时，由于 $E^+ > E^-$，多余的能量使振幅不断增大，直到 $E^+ = E^-$ 为止；当某瞬时的振幅为 A_2 时，由于 $E^+ < E^-$，则振幅将衰减，直到 $E^+ = E^-$ 为止。可见，只有当 $E^+ = E^-$ 时，系统将处于稳定状态。

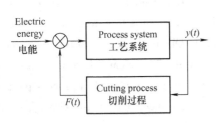

Fig. 5-40　Self-excited vibration system in machine tool 机床自激振动系统

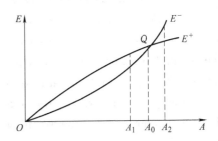

Fig. 5-41　Energy relation in a self-excited vibration system 自激振动系统的能量关系

the alternating force will disappear.

Whether self-excited vibration can take place or not and what the amplitude is depend on the contrast of the energy E^+ acquired to the energy E^- consumed by the system in a vibration period. As shown in Fig. 5-41, when $E^+ = E^-$, the vibration amplitude will be up to A_0. At the moment when amplitude $A_1 < A_0$, because $E^+ > E^-$, the surplus energy will increase the amplitude; at the moment when amplitude $A_2 > A_0$, because $E^+ < E^-$, the amplitude will be decreased to A_0. Therefore, only when $E^+ = E^-$, the system will be in a stable situation.

Self-excited vibration has the following characteristics: ① the self-excited vibration is a kind of undamped oscillation; ② the frequency of the self-excited vibration is equal to or close to the natural frequency of the vibration system; ③ whether self-excited vibration can take place or not and what the amplitude is depend on the contrast of the energy acquired to the energy consumed by the system in a vibration period.

2. Mechanism of self-excited vibration

Up to now, many scholars have given different theories to explain the mechanism of self-excited vibration in machining process. Two of them are introduced here.

(1) Regenerative chatter principle   As the feed and the minor cutting edge angle are relative small in cutting or grinding operations, the overlapping cutting would occur, i.e. there is an overlapping cutting area in the first feed and next feed. In cylindrical grinding shown in Fig. 5-42, the size of overlapping grinding area can be expressed by overlapping coefficient μ, that is

$$\mu = (B - f)/B \qquad (0 < \mu < 1)$$

where B stands for the width of the grinding wheel, and f stands for the longitudinal feed of the workpiece.

In cutting process, accidental outside disturbances would cause the vibration of the machining system and the chatter mark left on the machined surface. When the workpiece rotates to next revolution, the cutter is cutting the surface with chatter marks, leading to the variation of the cutting depth. The variation of the depth of cut must result in variations of the cutting force. This phenomenon is called regenerative effect. The vibration caused by regenerative effect is called the regenerative chatter. Under what conditions would the regenerative chatter happen in the machining system?

The principle of the regenerative chatter is shown in Fig. 5-43. It can be seen from Fig. 5-43 that the work surface y (dotted line) cut in subsequent revolution lags behind the surface y_0 (solid line) cut in previous revolution, the cutting action from A to B is called "cut-in", from B to C is called "cut-out". Because the average cutting thickness in cut-in area is less than that in cut-out area, the cutting force produced in cut-in area is also smaller. Obviously, the positive work done by cutting force is larger than the negative work in a vibration cycle. This means there is excess energy in the process system, as a result, regenerative chatter occurs in the system. If a process parameter (such as the rotation speed of the workpiece) is changed so as to keep y in phase with y_0, or make y advance a phase angle φ, the regenerative chatter can be avoided.

(2) Vibration mode coupling principle   There is no overlapping cutting in some machining

自激振动具有以下特性：
① 自激振动是一种不衰减的振动；② 自激振动的频率等于或接近于系统的固有频率；③ 自激振动能否产生及振幅的大小取决于振动系统在每一个周期内获得和消耗的能量对比情况。

2. 自激振动产生的机理

关于机械加工过程中自激振动产生的机理，许多学者已提出了许多不同的学说，下面介绍其中的两种学说。

(1) 再生颤振原理　在切削或磨削加工中，由于一般的进给量都不大，并且刀具的副偏角也较小，使得后一次进给和前一次进给的切削区必然会有重叠部分，即产生重叠切削。如图 5-42 所示的外圆磨削，重叠磨削区的大小可用重合度 μ 表示，即

$$\mu = (B - f)/B \qquad (0 < \mu < 1)$$

其中，B 代表砂轮宽度；f 代表工件的纵向进给量。

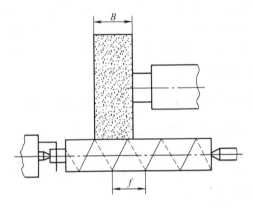

Fig. 5-42　Overlapping grinding 重叠磨削

在切削过程中，偶然的干扰会使加工系统产生振动并在加工表面上留下振纹。当工件转至下一转时，刀具在有振纹的表面上切削，使得切削深度发生变化，引起切削力作周期性变化。这种现象称为再生效应，由此产生的自激振动称为再生颤振。究竟系统在何种情况下会发生再生颤振呢？

图 5-43 所示为再生颤振原理图。由图看出，当后一转切削加工的工件表面 y（图中虚线）滞后于前一转切削的工件表面 y_0（图中实线）时，从 A 至 B 为切出，从 B 至 C 为切入。由于在切入工件的半个周期中的平均切削厚度比切出时的平均切削厚度小，切削力也小，则在一个振动周期中，切削力做的正功大于负功，说明有多余能量输入到系统中，因而系统产生了再生颤振。如果改变加工中的某项工艺参数（如工件转速），使 y 与 y_0 同相或超前一个相位角 φ，则可以避免再生颤振发生。

(2) 振型耦合原理　在一些加工中并不存在重叠切削，如车削矩形螺纹的外圆表面，但当切削深度增加到一定值时，也出现了自激振动。显然再生自激振动原理已不能解释其原因。而振型耦合原理可以对这种自激振动现象进行解释。它主要用于说明多自由度系统的自激振动现象。

processes, such as the cylindrical turning of the square thread, but when the depth of cut increases to a certain value, the self-excited vibration also appears. Obviously, this phenomenon can't be explained by regenerative chatter principle. Vibration mode coupling principle should be used to explain this phenomenon.

A cutter with mass m is hung on two springs with rigidity k_1, $k_2(k_1 < k_2)$ separately, as shown in Fig. 5-44. The dynamic cutting force F_d activizes simultaneously the two vibrations with vibration modal x_1, x_2 respectively. As $k_1 \neq k_2$, their resultant motion track in the (x_1, x_2) plane is a ellipse. Suppose the cutter is cutting into work along the track $A \rightarrow B \rightarrow C$, as the tool moving direction is opposite to the direction of the cutting force, the cutter does negative work; when the cutter cut away from work along the track $C \rightarrow D \rightarrow A$, the tool moving direction is the same as the direction of the cutting force, the cutter does positive work. Because the average cutting thickness in cut-out process is larger than that in cut-in process, the positive work done by cutting force is larger than the negative work in a vibration cycle. The redundant energy in the system can keep the vibration going on. If the moving track of the cutter relative to the work is in the direction opposite to the arrow, i.e. cut-in along $A \rightarrow D \rightarrow C$, cut-out along $C \rightarrow B \rightarrow A$, clearly, the negative work done by the cutting force is larger than the positive work. There is no redundant energy in the system to maintain vibration, and the former vibration would be attenuated continually.

5.5.4 Ways to control vibrations in machining processes

1. Ways to eliminate or decrease forced vibration

1) The most effective way to eliminate forced vibration is to find the vibration source and remove it. The frequency of vibration source after adjusting should be far away from the natural frequency of the weak mode of the machining system.

2) To use various vibration isolating and damping measures. The vibration isolating foundation and vibration isolators are used to isolate the machine tool or equipment from vibration source. Various damping devices can also be used.

3) To decrease exciting force. The high-speed rotating parts in machine tools, such as grinding wheel, chuck, rotor and so on, should be balanced dynamically. The defective parts of a machine should be replaced or repaired in time.

4) To enhance the vibration resistance of the technological system. A machine tool should be rigidly installed on its properly prepared foundation. The site where the machine tool is to be installed should be quite far away from the vibration sources, such as presses, forging machines, compressors, etc. To increase the rigidity of the technological system can improve the vibration resistance of the technological system effectively. To increase the damping capacity of the technological system can prevent and avoid vibration.

2. Ways to eliminate or decrease self-excited vibration

1) To choose machining variables properly. High speed machining can avoid self-excited vibration. The increase of the feed f is helpful for restraining regenerative chatter, but is restricted by surface roughness. The increase of the depth of cut a_p leads to the increase of cutting width a_w, causing

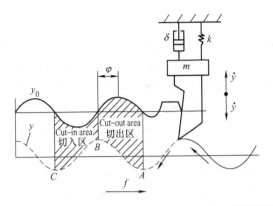

Fig. 5-43　Principle of regenerative chatter 再生颤振原理

如图 5-44 所示，质量为 m 的刀具悬挂在两个刚度为 k_1 和 k_2（$k_1 < k_2$）的弹簧上。动态切削力 F_d 以同一频率同时激起两个振型 x_1 和 x_2 的振动。因为 $k_1 \neq k_2$，它们的合成运动在（x_1，x_2）平面内的轨迹即为椭圆。假定刀尖沿由 $A \rightarrow B \rightarrow C$ 的轨迹切入工件，运动方向与切削力方向相反，刀具做负功；刀尖沿由 $C \rightarrow D \rightarrow A$ 的轨迹切出时，运动方向与切削力方向相同，刀具做正功。由于切出时的平均切削厚度大于切入时的平均切削厚度，在一个振动周期内，切削力所做的正功大于负功，因此有多余的能量输入振动系统，振动得以维持。如果刀具和工件的相对运动轨迹沿着与图中箭头相反的方向切入和切出，显然，切削力做的负功大

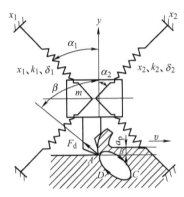

Fig. 5-44　Vibration mode coupling principle 振型耦合原理

于正功，振动就不能维持，原有的振动就会不断地衰减下去。

5.5.4　控制机械加工振动的方法

1. 消振或减小强迫振动的方法

1）消除强迫振动的最有效方法就是找出振源并消除之。也可通过调节振源频率，使其远离机床加工系统薄弱模态的固有频率。

2）采用各种隔振与减振措施。用隔振地基或隔振装置将需要防振的机床或部件与振源之间分开，还可采用各种减振装置。

3）减小激振力。对于机床上的高速回转件，如砂轮、卡盘、电动机转子等，应进行动平衡。应及时更换或修理机器中有缺陷的零件。

4）提高工艺系统抗振性。机床应牢固地装在预先准备好的地基上。机床的位置应尽量远离振源，如锻压设备、压缩机等。提高工艺系统刚度，可有效地改善工艺系统的抗振性。增大工艺系统的阻尼，能够有效地防止和消除振动。

2. 消除或减小自激振动的方法

1）合理选择切削用量。高速切削可以避免产生自激振动。进给量 f 增大有利于抑制再生颤振，但又受到表面粗糙度的制约。切削深度 a_p 增大会引起切削宽度 a_w 增大，容易产生颤振。

chatter easily.

2) To choose tool geometric parameters properly. The smaller the cutting edge angle κ_r, the larger the cutting width a_w, the easier the self-excited vibration. The larger the rake angle γ_o, the smaller the cutting force, the smaller the vibration amplitude. To decrease clearance angle α_o properly can increase the frictional damping between work and flank of cutter, which is helpful for improving the machining stability. But α_o can't be too small, the damping land can be machined on the flank, if necessary.

3) To enhance the vibration resistance of technological system.

4) To arrange the orientation of axis with low rigidity reasonably.

5) To use various vibration damping devices.

2）合理选用刀具的几何参数。主偏角 κ_r 越小，切削宽度越宽，因此越易产生振动。前角 γ_o 越大，切削力越小，振幅也越小。适当减小后角 α_o 可以加大工件和刀具后面之间的摩擦阻尼，有利于提高切削稳定性。但 α_o 一般不能太小，必要时可以在刀具后面上磨出一段消振棱。

3）提高工艺系统抗振性。

4）合理布置低刚度主轴的方位。

5）采用各种减振装置。

Chapter 6

Fundamentals of Machine Assembly Technology

第 6 章

机械装配技术基础

6.1 Introduction

Mechanical products are generally composed of many workparts and assemblies. General assembly is the last link in machine manufacturing. It mainly includes such jobs as assembling, adjusting, inspecting, and testing etc. The machine quality is finally ensured by assembling. Therefore, mechanical assembly plays a very important role in product manufacturing process.

6.1.1 Concept of assembly accuracy

1. Assembly

All machines are composed of workparts, combined members, subassemblies, and assemblies. According to the requirements for assembly accuracy, some related workparts, subassemblies and assemblies are put together by necessary connecting and fitting to compose a qualified machine product. This process is called assembly. In order to carry out assembling work successfully, a machine is generally divided into several independent parts which are called assembly unit. There are five classes of assembly units, i.e. workpart, combined part, subassembly, assembly, and machine.

A workpart is the most basic cell to compose a machine. It is made of a whole piece of metal or other material. Very a few workparts are direct mounted in the machine, but they are assembled into combined part, subassembly, and assembly in advance.

A combined part is the permanent connecting (welding, riveting, etc.) of some workparts or the combination of assembling one or two components on a datum part. The combined part may be still machined some times. Fig. 6-1 shows a gear assembly. For the sake of manufacturing technology, the gear consists of two parts. The member 3 is fitted to the datum part 1 and is fixed by the rivet.

A subassembly is composed of combined members and workparts, which are mounted on a datum part. Take the spindle in a headstock of a machine tool for example. Gears, sleeves, washers, keys and bearings all are assembled on a datum part—spindle, which is called the spindle subassembly. The assembling process is called the subassembly assembling.

An assembly is composed of subassemblies, combined parts and workparts, which are mounted on a datum part. The assembling process to assemble workparts, combined parts and subassemblies into assembly is called the assembly assembling. The headstock is an assembly using the box body as a datum part.

A machine is composed of assemblies, subassemblies, combined members and workparts, which are mounted on a datum part. The assembling process to assemble workparts, and assemblies into final product (machine) is called the final assembly. For example, a lathe is a machine using the bed as the datum part on which the headstock, feed gearbox, apron and other many kinds of parts are mounted.

6.1 概述

机械产品一般是由许多零件和部件装配而成的。总装配是机器制造中的最后一个阶段，它主要包括装配、调整、检验、试验等工作。机器的质量最终是通过装配保证的，装配质量在很大程度上决定机器的最终质量。因此，机械装配在产品制造过程中占有非常重要的地位。

6.1.1 装配精度的概念

1. 装配

任何机器都是由零件、套件、组件、部件等组成的。按照规定的技术要求，将若干个零件或部件进行必要的配合和连接，使之成为合格产品的过程，称为装配。对于结构比较复杂的产品，为保证装配工作顺利地进行，通常将机器划分为若干个能进行独立装配的部分，称为装配单元。装配单元一般分为零件、合件、组件、部件和机器五个等级。

零件是组成机器的最基本单元，它是由整块金属或其他材料制成的。零件一般都预先装成合件、组件、部件后才安装到机器上，直接装入机器的零件并不太多。

合件可以是若干零件永久连接（如焊接、铆接等）或者是在一个基准零件上，装上一个或若干个零件的组合。合件组合后，有可能还要加工。如图 6-1 所示的装配齿轮，由于制造工艺的原因，分成两个零件，在基准零件 1 上套装齿轮 3 并用铆钉 2 固定。

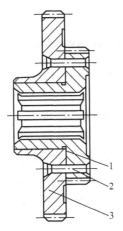

Fig. 6-1 Combined gear 齿轮合件
1—Datum part 基准零件　2—Rivet 铆钉　3—Gear 齿轮

组件是在一个基准零件上，装上一个或若干个合件及零件而成的。如机床主轴箱中的主轴，在基准轴件上装上齿轮、套、垫片、键及轴承的组合件称为组件。为此而进行的装配工作称为组装。

部件是在一个基准零件上，装上若干组件、合件和零件构成的。把零件装配成为部件的过程称为部装。如车床的主轴箱装配就是部装。主轴箱箱体为部装的基准零件。

在一个基准零件上，装上若干部件、组件、合件和零件就成为整个机器。把零件和部件

2. Assembly accuracy

(1) Relative position accuracy It means the distance accuracy, relative position accuracy between relevant parts and assemblies in a product, such as the center distance accuracy, coaxiality, parallelism, and perpendicularity between the shafts in a headstock.

(2) Relative motion accuracy It means the accuracies in moving direction and relative motion speed between movable parts in a product. For example, the parallelism, and perpendicularity between assemblies with relative motion belong to the accuracy in moving direction; transmission accuracy belongs to the accuracy in relative motion speed.

(3) Fit accuracy It includes fit quality and contact quality between mating surfaces. The fit quality would affect the fit character, and the contact quality would affect the contact rigidity and the stability of the fit character.

6.1.2 Factors affecting assembly accuracy

The factors of affecting assembly accuracy are as follows: ① Dimension error, form error and relative position error of mating parts; ② Poor finish of the mating surface; ③ Inaccurate location of assembled machine elements; ④ Poor fitting and adjusting of machine mating parts; ⑤ Non-observance of assembly process requirements (non-uniform tightening of screw fasteners, misalignment and deformation of parts in the pressing process and other operations, etc.); ⑥ Geometric errors of assembly equipment; ⑦ Thermal destortions of assembly system; ⑧ Poor adjustment for assembly equipment.

6.1.3 Relation of part accuracy to assembly accuracy

Since any workpart has a certain machining error, the error accumulation would affect the machine assembly accuracy. For example, the assembly accuracy requirement for the center height of both the headstock and the tailstock has to do with the manufacturing accuracies of the bed 4, headstock 1, tailstock 2 and other relevant parts, as shown in Fig. 6-2. If the accumulated errors exceed the allowance, unqualified products would appear. From the view of the assembling process, it is expected that the assembling process is just a simple connecting process—without any repair or adjustment, and the design tolerance of the machine still can be met. Therefore, higher assembly accuracy requires higher workpart accuracy. But the machining accuracy of a workpart is limited not only by the technological conditions, but by the cost. There are some machines with many components and very high assembly accuracy. Even giving no thought to the production cost in improving machining accuracy of workparts, the final assembly accuracy can not be reached yet.

Therefore, the assembly accuracy does not only depend on the accuracy of mating parts, but also on the assembly technology to some extent, especially for the machines with higher accuracy and small-lot production. In the long-term assembly practice, people have created many ingenious assembly methods according to the different machines, different production types and conditions. In the different assembly methods, there are different relations between the part accuracy and the assembly accuracy. In order to analyze this relation quantitatively, dimension chain theory is often

装配成最终产品的过程称为总装。例如，卧式车床就是以床身为基准件，装上主轴箱、进给箱、溜板箱等部件及其他组件、合件、零件所组成的。

2. 装配精度

产品的装配精度一般包括：

（1）相互位置精度　相互位置精度是指产品中相关零、部件间的距离精度和相互位置精度。如机床主轴箱中轴系之间中心距尺寸精度和同轴度、平行度、垂直度等。

（2）相对运动精度　相对运动精度是产品中有相对运动的零、部件之间在运动方向和相对速度上的精度。运动方向的精度常表现为部件间相对运动的平行度和垂直度。相对速度精度如传动精度。

（3）相互配合精度　相互配合精度包括配合表面间的配合质量和接触质量。配合质量影响配合的性质，而接触质量则会影响接触刚度和配合性质的稳定性。

6.1.2　影响装配精度的因素

影响装配精度的因素有：①配合零件的尺寸误差、形状误差和相互位置误差；②配合零件的表面粗糙度；③装配到机器上的零件的位置不准确；④机器配合件的装配和调整不良；⑤不遵守装配工艺要求（如紧固螺钉拧紧力大小不均匀，压力装配时零件未对准和零件变形等）；⑥装配设备的几何形状误差；⑦装配系统的热变形；⑧装配设备的调整不良。

6.1.3　零件精度与装配精度的关系

由于一般零件都有一定的加工误差，在装配时这些零件的加工误差累积就会影响装配精度。例如，车床主轴中心线和尾座中心线对床身导轨的等高要求，如图 6-2 所示，与床身 4、主轴箱 1、尾座 2 等零部件的加工精度有关。如果这些零件的累积误差超出装配精度指标所规定的范围，则将产生不合格品。从装配工艺角度考虑，当然希望这种累积误差不要超过装配精度指标所规定的允许范围，从而使装配工作只是简单的连接过程，不必进行任何的修配或调整就能满足装配精度要求。因此，一般装配精度要求高的，要求零件精度也要高。但零件的加工精度不但在工艺上受到加工条件的限制，而且受到经济上的制约。如有的机械产品的组成零件较多，而最终装配精度要求又较高时，即使把经济性置之度外，尽可能地提高零件的加工精度以降低累积误差，结果往往还是无济于事。

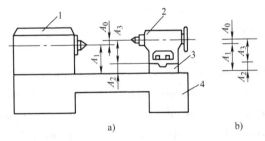

Fig. 6-2　Assembly dimension chain of spindle and tailstock with equal-height centerline
主轴与尾座具有等高中心线的装配尺寸链
1—Headstock 主轴箱　2—Tailstock 尾座　3—Baseplate 底板　4—Bed 床身

因此装配精度不仅取决于配合件的精度，而且在某种程度上也取决于装配技术。在装配

used in the assembly process. The dimension chain is set up firstly, the quantitative relation between the part accuracy and the assembly accuracy is determined through solving the dimension chain.

6.2 Assembly methods

As mentioned above, the accuracy of mating parts is the main factor of influencing machine assembly accuracy. The relations of workpart accuracy to assembly accuracy can be determined through establishing analyzing assembly dimension chain.

6.2.1 Assembly dimension chain

1. Concept of assembly dimension chain

In machine assembly relation, the dimension chain composed of relative part dimension or mutual position relation is called the assembly dimension chain. The closed link of assembly dimension chain is just the assembly accuracy or technical requirement to be ensured in assembly. The assembly accuracy (closed link) is the dimension and position relation formed after the parts and subassemblies are assembled. All the dimensions and position relations which have a direct influence on the assembly accuracy are component links.

Fig 6-3 shows the assembly relation between a shaft and a hole. The clearance A_0 between the shaft and the hole must be ensured after assembling. The clearance A_0 is the closing link of this assembly dimension chain. It is formed after fitting the shaft (dimension A_2) into the hole (dimension A_1). A_0 increases with the increase of A_1, and decreases with the increase of A_2. Therefore, A_1 is increasing link, and A_2 is decreasing link. The dimension chain equation is $A_0 = A_1 - A_2$.

2. Method to establish an assembly dimension chain

To find out the component of an assembly dimension chain correctly and establish the dimension chain are the basis of calculating the dimension chain.

(1) Method of finding out the assembly dimension chain Firstly, determine the closed link according to assembly accuracy; then start from any of parts at two ends of the closed link, march along the position direction required by assembly accuracy, take the assembly datum surface as a contact clue, search respectively the related parts affecting assembly accuracy until the same datum surface is found.

(2) Attentions in establishing an assembly dimension chain

1) Simplification of assembly dimension chain. Generally, as a machine has a complex structure, there are many factors affecting assembly accuracy. Under the precondition of ensuring assembly accuracy, those unimportant factors can be omitted so as to simplify assembly dimension chain.

2) Shortest route (least links) principle. It is known from the dimension chain theory that when the assembly accuracy (tolerance of closed link) is given, the less the number of component links, the larger the tolerance allocated to each component link. Then, the workpart can be machined more easily and economically. Therefore, only one dimension for one workpart is permitted to

精度要求高、生产批量较小时尤其如此。人们在长期的装配实践中,根据不同的机器、不同的生产类型和条件,创造了许多巧妙的装配方法。在不同的装配方法中,零件加工精度与装配精度间具有不同的相互关系。为了定量地分析这种关系,常将尺寸链的基本理论应用于装配过程中,即建立装配尺寸链,通过解算装配尺寸链,最后确定零件精度与装配精度之间的定量关系。

6.2 装配方法

如前所述,零件的精度是影响机器装配精度的最主要因素。通过建立、分析计算装配尺寸链,可以确定零件精度与装配精度之间的关系。

6.2.1 装配尺寸链

1. 装配尺寸链的基本概念

在机器的装配关系中,由相关零件的尺寸或相互位置关系所组成的尺寸链,称为装配尺寸链。装配尺寸链的封闭环就是装配所要保证的装配精度或技术要求。装配精度(封闭环)是零部件装配后最后形成的尺寸或位置关系。在装配关系中,对装配精度有直接影响的零部件的尺寸和位置关系,都是装配尺寸链的组成环。

图 6-3 所示为轴和孔的装配关系,要求轴孔装配后有一定的间隙。轴孔间的间隙 A_0 就是该尺寸链的封闭环,它是由孔尺寸 A_1 与轴尺寸 A_2 装配后形成的尺寸。在这里,孔尺寸 A_1 增大,间隙 A_0(封闭环)也随之增大,故 A_1 为增环。反之,轴尺寸 A_2 为减环。其尺寸链方程为 $A_0 = A_1 - A_2$。

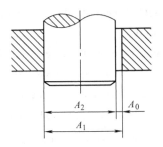

Fig. 6-3 Assembly relation between a shaft and a hole 轴和孔的装配关系

2. 装配尺寸链的查找方法

正确的查明装配尺寸链的组成,并建立尺寸链是进行尺寸链计算的基础。

(1) 装配尺寸链的查找方法 首先根据装配精度要求确定封闭环。再取封闭环两端的任意一个零件为起点,沿装配精度要求的位置方向,以装配基准面为查找的线索,分别找出影响装配精度要求的相关零件,直至找到同一基准表面为止。

(2) 查找装配尺寸链时应注意的问题

1) 装配尺寸链应进行必要的简化。机械产品的结构通常都比较复杂,对装配精度有影响的因素很多,在查找尺寸链时,在保证装配精度的前提下,可以不考虑那些较小的因素,

be a component link of the assembly dimension chain. That is, the positional dimension connecting two assembly datums is marked directly on the part drawing. In this way, the number of component links equals the number of related parts and assemblies, i. e. "one component one link". This is the shortest route (least links) principle of an assembly dimension chain.

The axial dimension chain formed in a gear assembly (Fig. 6-4) reflects the "one component one link" principle. Clearly, the marked dimensions shown in Fig. 6-5 are irrational, because it violates the "one component one link" principle.

3. Calculation method of an assembly dimension chain

The assembly method has a close relation to the calculation method of an assembly dimension chain. The calculation method of the assembly dimension chain varies with assembly method for the same assembly accuracy.

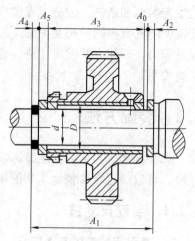

Fig. 6-4 Gear assembly dimension chain 齿轮装配尺寸链

6.2.2 Interchangeable assembly method

The essence of interchangeable assembly methods is to ensure the product assembly accuracy through controlling the machining errors of a workpart. Based on interchangeable extent, interchangeable methods are classified as strict interchangeable method and partial interchangeable method.

1. Strict interchangeable method

The mating parts are manufactured strictly following the design requirements on the print. Assembly accuracy still can be ensured by taking any workpart into assembly without any repair and adjustment in assembling operation. This method uses "Max-Min method" (Extreme value method) to solve the assembly dimension chain. Here the sum of machining tolerances of all parts can't be larger than the design tolerance T_0. It can be expressed as follows:

$$T_0 \geq \sum_{i=1}^{n-1} T_i \tag{6-1}$$

where, T_0—tolerance of the closed link (assembly accuracy);

T_i—tolerance of ith component link;

n—numbers of all links.

When counter-calculating an assembly dimension chain, "intermediate calculation method" ("interdependent dimension tolerance method") is often used. The method means that the tolerance of some component links, which are difficult to machine or whose tolerance are inappropriate to change (standard element), are predetermined, very a few links or one component link can be selected as "coordination link", which is easy to machine, or confined by quite small conditions in production, or measured by means of general-purpose measuring tool. The "coordination link" is al-

使装配尺寸链适当简化。

2）最短路线（最少环数）原则。由尺寸链理论可知，在装配精度一定时，组成环数越少，则各组成环所分配到的公差值就越大，零件加工越容易、越经济。因此在查找装配尺寸链时，每个相关的零、部件只应有一个尺寸作为组成环列入装配尺寸链，即将连接两个装配基准面的位置尺寸直接标注在零件图上。这样，组成环的数目就等于有关零、部件的数目，即"一件一环"，这就是装配尺寸链的最短路线（最少环数）原则。

图 6-4 所示齿轮装配后轴向间隙尺寸链就体现了"一件一环"的原则。如果把图 6-4 中的轴向尺寸标注成图 6-5 所示的两个尺寸，则违反了"一件一环"的原则，其装配尺寸链的构成显然不合理。

3. 装配尺寸链的计算方法

装配方法与装配尺寸链的解算方法密切相关。同一项装配精度，采用不同装配方法时，其装配尺寸链的解算方法也不相同。

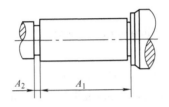

Fig. 6-5　Irrational dimensions
不合理的尺寸标注

6.2.2　互换装配法

互换法的实质就是用控制零件的加工误差来保证产品装配精度的一种方法。根据零件的互换程度不同，互换法又可分为完全互换法和不完全互换法。

1. 完全互换法

组成机器的每一个零件，装配时不需挑选、修配或调整，装配后即可达到规定的装配精度要求的装配方法称为完全互换法。采用完全互换法时，装配尺寸链采用极大极小法（极值法）解算，即尺寸链各组成环公差之和不能大于封闭环公差

$$T_0 \geq \sum_{i=1}^{n-1} T_i \tag{6-1}$$

式中　T_0——封闭环公差（装配精度）；

T_i——第 i 个组成环公差；

n——尺寸链总环数。

在进行装配尺寸链反计算时，通常采用中间计算法（或称"相依尺寸公差法"）。该方法是将一些比较难以加工和不宜改变其公差的组成环（如标准件）的公差预先确定下来，只将极少数或一个比较容易加工，或在生产上受限制较少和用通用量具容易测量的组成环定为协调环。这个环称为"相依尺寸"，意思是该环的尺寸相依于封闭环和其他组成环的尺寸和公差值。然后用公式计算相依尺寸的公差值和极限偏差。其计算过程如下：

so called "interdependent dimension", which means the demension is dependent on closed link and other component links. The tolerance of "interdependent dimension" should be calculated. Its calculation procedure is as follows:

(1) To establish assembly dimension chain and determine the "coordination link" Whether the basic dimension of the closed link is correct should be checked firstly. A standard part or public link can't be selected as a coordination link.

(2) To determine the tolerance of component links The average tolerance of each component link is firstly determined according to "equal tolerance" rule. Then the average tolerance of each component link can be regulated appropriately and determined according to its size and machining difficulty, but the tolerance of the "coordination link" must be calculated finally by the following formula.

$$T_{A_y} = T_{A_0} - \sum_{i=1}^{n-2} T_{A_i} \tag{6-2}$$

where, A_y—coordination link;

T_{A_y}、T_{A_0}、T_{A_i}—tolerance of the coordination link, closing link and other component links separately.

(3) To determine the limit deviation The limit deviation of component links except for the coordination link can be marked in "body-in" method, the deviation of the standard part remains the same as it is, then the limit deviation of the "coordination link" is calculated.

If the coordination link is the increasing link, then

$$ES_{A_y} = ES_{A_0} - \sum_{p=1}^{m-1} ES_{A_p} + \sum_{q=m+1}^{n-1} EI_{A_q} \tag{6-3}$$

$$EI_{A_y} = EI_{A_0} - \sum_{p=1}^{m-1} EI_{A_p} + \sum_{q=m+1}^{n-1} ES_{A_q} \tag{6-4}$$

If the coordination link is the decreasing link, then

$$ES_{A_y} = -EI_{A_0} + \sum_{p=1}^{m} EI_{A_p} - \sum_{q=m+1}^{n-2} ES_{A_q} \tag{6-5}$$

$$EI_{A_y} = -ES_{A_0} + \sum_{p=1}^{m} ES_{A_p} - \sum_{q=m+1}^{n-2} EI_{A_q} \tag{6-6}$$

where, A_p—increasing link;

A_q—decreasing link.

【Example 6-1】 Fig. 6-6a shows the assembly relation among the shaft, gear and washers. The axial clearance between a gear and a washer should be kept in 0.1 ~ 0.35mm. Given that $A_1 = 30$mm, $A_2 = 5$mm, $A_3 = 43$mm, $A_4 = 3_{-0.05}^{0}$mm (standard part), $A_5 = 5$mm. When an interchangeable assembly method is used, try to determine the tolerance and limit deviation of every component link.

Solution:

1) To establish an assembly dimension chain and verify the basic dimension of each link. Assembly dimension chain is shown in Fig. 6-6b, where $A_0 = 0_{+0.10}^{+0.35}$mm is closed link. The basic dimension of closed link is

(1) 建立装配尺寸链，确定"协调环" 首先要验算公称尺寸是否正确。不能选取标准件或公共环作为协调环。

(2) 确定组成环的公差 可先按"等公差"原则确定各组成环的平均公差值，然后根据各组成环尺寸大小和加工的难易程度再进行适当的调整，将其他组成环的公差值确定下来，最后利用公式求出协调环的公差值。即

$$T_{A_y} = T_{A_0} - \sum_{i=1}^{n-2} T_{A_i} \tag{6-2}$$

式中　　A_y——协调环；

T_{A_y}、T_{A_0}、T_{A_i}——协调环、封闭环和除协调环以外的其余组成环的公差值。

(3) 确定组成环的极限偏差 除协调环外的其余组成环的极限偏差，按"入体"原则标注，标准件按规定标注，然后计算"协调环"的极限偏差。

若协调环为增环，则

$$\text{ES}_{A_y} = \text{ES}_{A_0} - \sum_{p=1}^{m-1} \text{ES}_{A_p} + \sum_{q=m+1}^{n-1} \text{EI}_{A_q} \tag{6-3}$$

$$\text{EI}_{A_y} = \text{EI}_{A_0} - \sum_{p=1}^{m-1} \text{EI}_{A_p} + \sum_{q=m+1}^{n-1} \text{ES}_{A_q} \tag{6-4}$$

若协调环为减环，则

$$\text{ES}_{A_y} = -\text{EI}_{A_0} + \sum_{p=1}^{m} \text{EI}_{A_p} - \sum_{q=m+1}^{n-2} \text{ES}_{A_q} \tag{6-5}$$

$$\text{EI}_{A_y} = -\text{ES}_{A_0} + \sum_{p=1}^{m} \text{ES}_{A_p} - \sum_{q=m+1}^{n-2} \text{EI}_{A_q} \tag{6-6}$$

其中，A_p 为增环，A_q 为减环。

【例 6-1】 如图 6-6a 所示齿轮装配，轴固定，而齿轮空套在轴上回转，要求保证齿轮与挡圈的轴向间隙为 0.1~0.35mm，已知：$A_1 = 30$mm、$A_2 = 5$mm、$A_3 = 43$mm、$A_4 = 3_{-0.05}^{0}$mm（标准件）、$A_5 = 5$mm，现采用完全互换法装配，试确定各组成环公差值和极限偏差。

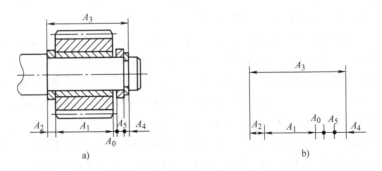

Fig. 6-6　Assembly relation between a gear and a shaft　齿轮与轴的装配关系

解 1) 建立装配尺寸链，验算各环的公称尺寸。装配尺寸链如图 6-6b 所示，其中 $A_0 =$

$$A_0 = A_3 - (A_1 + A_2 + A_4 + A_5) = 43\text{mm} - (30 + 5 + 3 + 5)\text{mm} = 0\text{mm}$$

As the washer A_5 is easy to machine and measure, A_5 is selected to be the coordination link.

2) To determine the tolerance and the limit deviation of component links. Based on the equal tolerance principle, the average tolerance of each link is:

$$T_{A_i} = \frac{T_{A_0}}{n-1} = \frac{0.25\text{mm}}{5} = 0.05\text{mm}$$

Shield ring A_4 is a standard part, $A_4 = 3_{-0.05}^{\ 0}\text{mm}$, $T_{A_4} = 0.05\text{mm}$. In the light of the size and machining difficulty, the tolerance of other component links is determined as following: $T_{A_1} = 0.06\text{mm}$, $T_{A_2} = 0.02\text{mm}$, $T_{A_3} = 0.1\text{mm}$, (The tolerance grade of each component link is about IT9). The limit deviations of A_1 and A_2 are determined following Basic Shaft System: $A_1 = 30_{-0.06}^{\ 0}\text{mm}$, $A_2 = 5_{-0.02}^{\ 0}\text{mm}$, and A_3 is determined following Basic Hole System: $A_3 = 43_{\ 0}^{+0.1}\text{mm}$.

3) To calculate the tolerance and the limit deviation of the coordination link.

Tolerance of A_5: $T_{A_5} = T_{A_0} - (T_{A_1} + T_{A_2} + T_{A_3} + T_{A_4}) = 0.25\text{mm} - (0.06 + 0.02 + 0.1 + 0.05)\text{mm} = 0.02\text{mm}$

Lower deviation of A_5: $\text{ES}_{A_0} = \text{ES}_{A_3} - (\text{EI}_{A_1} + \text{EI}_{A_2} + \text{EI}_{A_4} + \text{EI}_{A_5})$

$$0.35\text{mm} = 0.1\text{mm} - (-0.06\text{mm} - 0.02\text{mm} - 0.05\text{mm} + \text{EI}_{A_5})$$

$$\text{EI}_{A_5} = -0.12\text{mm}$$

Upper deviation of A_5: $\text{ES}_{A_5} = T_{A_5} + \text{EI}_{A_5} = 0.02\text{mm} + (-0.12)\text{mm} = -0.10\text{mm}$

Therefore, $A_5 = 5_{-0.12}^{-0.10}\text{mm}$

The dimensions of all component links are: $A_1 = 30_{-0.06}^{\ 0}\text{mm}$, $A_2 = 5_{-0.02}^{\ 0}\text{mm}$, $A_3 = 43_{\ 0}^{+0.1}\text{mm}$, $A_4 = 3_{-0.05}^{\ 0}\text{mm}$, $A_5 = 5_{-0.12}^{-0.10}\text{mm}$.

The advantages of an interchangeable assembly method are: simple assembly process and high efficiency; easy to organize the assembly work; easy to realize the specialization production of parts and assemblies; convenient to provide spare parts and so on. Therefore, no matter what the production type it is, if the economic machining accuracy of the workpart can be met, the interchangeable assembly method should be selected firstly. But when the assembly accuracy is very high, especially when the component link number is relatively large, it is difficult to machine the workpart economically. At that time, the partial interchangeable assembly method should be considered.

2. Partial interchangeable assembly method (Probability method)

For the most of machine products, even though their components are directly assembled into machines without being selected, repaired and adjusted, the assembly accuracy still can be met. But a small number of waste products may be produced. This method is called partial interchangeable assembly method. Its essence is to enlarge the tolerance of components. As the method is based theoretically on probability, it is also named as probability method.

When the component links are close to normal distribution, the closed link is also the approximate normal distribution. For the sake of calculation, the basic dimension of each component link is usually converted into average dimension. Then the average dimension A_{0M} of the closed link is

$$A_{0M} = \sum_{p=1}^{m} A_{pM} - \sum_{q=m+1}^{n-1} A_{qM} \tag{6-7}$$

$0^{+0.35}_{+0.10}$mm 为封闭环。封闭环公称尺寸为

$$A_0 = A_3 - (A_1 + A_2 + A_4 + A_5) = 43\text{mm} - (30 + 5 + 3 + 5)\text{mm} = 0\text{mm}$$

因为 A_5 是一个挡圈，易于加工和测量，故选为"协调环"。

2）确定各组成环公差值和极限偏差。各组成环按等公差值确定公差为

$$T_{A_i} = \frac{T_{A_0}}{n-1} = \frac{0.25\text{mm}}{5} = 0.05\text{mm}$$

挡圈 A_4 为标准件，$A_4 = 3^{\ 0}_{-0.05}$mm、$T_{A_4} = 0.05$mm。其余各组成环按其尺寸大小和加工难易程度选择公差为：$T_{A_1} = 0.06$mm、$T_{A_2} = 0.02$mm、$T_{A_3} = 0.1$mm，各组成环公差等级约为 IT9。A_1、A_2 按基轴制确定极限偏差：$A_1 = 30^{\ 0}_{-0.06}$mm，$A_2 = 5^{\ 0}_{-0.02}$mm，A_3 按基孔制确定其极限偏差：$A_3 = 43^{+0.1}_{\ 0}$mm。

3）计算协调环的公差值和极限偏差。

A_5 的公差值 $T_{A_5} = T_{A_0} - (T_{A_1} + T_{A_2} + T_{A_3} + T_{A_4}) = 0.25\text{mm} - (0.06 + 0.02 + 0.1 + 0.05)\text{mm} = 0.02\text{mm}$

A_5 的下极限偏差 $\text{ES}_{A_0} = \text{ES}_{A_3} - (\text{EI}_{A_1} + \text{EI}_{A_2} + \text{EI}_{A_4} + \text{EI}_{A_5})$

$$0.35\text{mm} = 0.1\text{mm} - (-0.06\text{mm} - 0.02\text{mm} - 0.05\text{mm} + \text{EI}_{A_5})$$

$$\text{EI}_{A_5} = -0.12\text{mm}$$

A_5 的上极限偏差 $\text{ES}_{A_5} = T_{A_5} + \text{EI}_{A_5} = 0.02\text{mm} + (-0.12)\text{mm} = -0.10\text{mm}$

所以协调环 A_5 的尺寸为 $A_5 = 5^{-0.10}_{-0.12}$mm

各组成环尺寸和极限偏差为：$A_1 = 30^{\ 0}_{-0.06}$mm，$A_2 = 5^{\ 0}_{-0.02}$mm，$A_3 = 43^{+0.1}_{\ 0}$mm，$A_4 = 3^{\ 0}_{-0.05}$mm，$A_5 = 5^{-0.10}_{-0.12}$mm。

完全互换装配方法的优点是：装配过程简单，生产率高；便于组织流水作业和自动化装配；易于实现零部件的专业协作与生产，备件供应方便。因此只要能满足零件经济精度要求，无论何种生产类型都应首先考虑采用完全互换法装配。但是，当装配精度要求较高，尤其是组成环数较多，零件难以按经济精度加工时，此时可考虑采用不完全互换法。

2. 不完全互换法（概率法）

大多数产品在装配时，各组成零件不需挑选、修配或调节，装配后即能达到装配精度的要求，但少数产品有可能出现废品，这种方法称为不完全互换法。其实质是将组成零件的公差值适当放大。这种方法是以概率论为理论依据，故又称为概率法。

当各组成环的尺寸分布均接近于正态分布时，封闭环尺寸也近似于正态分布。为便于计算，常把各环的公称尺寸换算成平均尺寸。于是封闭环的平均尺寸 $A_{0\text{M}}$ 为

$$A_{0\text{M}} = \sum_{p=1}^{m} A_{p\text{M}} - \sum_{q=m+1}^{n-1} A_{q\text{M}} \tag{6-7}$$

式中 $A_{p\text{M}}$——增环的平均尺寸；

where, A_{pM}—the average dimension of the increasing link;
A_{qM}—the average dimension of the decreasing link.

Tolerance of closed link is

$$T_{A_0} = \sqrt{\sum_{i=1}^{n-1} T_{A_i}^2} \tag{6-8}$$

where, T_{A_i}—the tolerance of each component link.

When the component links have the same non-normal distribution, and the difference among dispersion ranges of component links is not very large, if the number of component links is large enough ($m \geq 5$), then the closing link is always close to the normal distribution. The approximate formula to calculate the tolerance of the closed link can be derived as follows:

$$T_{A_{0E}} = k\sqrt{\sum_{i=1}^{n-1} T_{A_i}^2} \tag{6-9}$$

where, $T_{A_{0E}}$—the equivalent tolerance of the closed link;
k—the relative distribution coefficient, $k = 1.2 - 1.6$.

After the tolerance and average dimension of each component link are calculated, the tolerance of each component link should be marked in bilateral symmetric deviation for average dimension. Then, they are remarked in the forms of basic dimension and unilateral deviation, if necessary.

【Example 6-2】 As shown in Fig. 6-6a, the known conditions are the same as Example 6-1. Try to use a partial interchangeable assembly method to determine the tolerance and deviation of component links (suppose the component links are close to normal distribution).

Solution:

1) To establish an assembly dimension chain and verify the basic dimension of each link (see Example 1).

Considering that A_3 is difficult to machine, A_3 is selected to be the coordination link.

2) To determine the tolerance and the limit deviation of component links. As all component links are close to normal distribution ($k_i = 1$), the tolerance of component links is allocated following "equal tolerance" rule.

$$T_{A_{iM}} = \frac{T_{A_0}}{\sqrt{n-1}} = \frac{0.25\text{mm}}{\sqrt{5}} \approx 0.1\text{mm}$$

Based on the tolerance determined by the "equal tolerance" rule, considering comprehensively the machining difficulty of each part, the tolerance value of component links is adjusted properly.

A_4 is the dimension of a standard part, its tolerance has been determined. The tolerance of other component links are adjusted as follows: $T_{A_1} = 0.14\text{mm}$, $T_{A_2} = T_{A_5} = 0.05\text{mm}$. As A_1, A_2, A_5 all are external dimension, their limit deviation are determined according to "basic shaft system". Then $A_1 = 30_{-0.14}^{0}\text{mm}$, $A_2 = 5_{-0.05}^{0}\text{mm}$, $A_4 = 3_{-0.05}^{0}\text{mm}$, $A_5 = 5_{-0.05}^{0}\text{mm}$.

3) To find the tolerance and limit deviation of the coordination link:

$$T_{A_3} = \sqrt{T_{A_0}^2 - (T_{A_1}^2 + T_{A_2}^2 + T_{A_4}^2 + T_{A_5}^2)} = \sqrt{0.25^2 - (0.14^2 + 3 \times 0.05^2)}\text{mm} \approx 0.18\text{mm}$$

$A_{1M} = 29.93\text{mm}$, $A_{2M} = A_{5M} = 4.975\text{mm}$, $A_{4M} = 2.975\text{mm}$, $A_{0M} = 0.225\text{mm}$

A_{qM}——减环的平均尺寸。

封闭环的公差 T_{A_0} 为

$$T_{A_0} = \sqrt{\sum_{i=1}^{n-1} T_{A_i}^2} \qquad (6-8)$$

式中 T_{A_i}——各组成环的公差。

当各组成环具有相同的非正态分布,且各组成环分布范围相差又不太大时,只要组成环数足够多($m \geq 5$),封闭环总是接近正态分布的。此时,封闭环公差的近似计算公式为

$$T_{A_{0E}} = k\sqrt{\sum_{i=1}^{n-1} T_{A_i}^2} \qquad (6-9)$$

式中 $T_{A_{0E}}$——封闭环的当量公差;

k——相对分布系数,$k = 1.2 \sim 1.6$。

在计算出各环公差值以及平均尺寸后,各环的公差值对平均尺寸应标注成双向对称偏差。然后根据需要再改注成具有公称尺寸和相应上、下极限偏差的形式。

【例 6-2】 如图 6-6a 装配图,已知条件与例 6-1 相同。采用不完全互换法装配,试确定各组成环公差和极限偏差(设各组成零件加工接近正态分布)。

解 1)建立装配尺寸链,验算各环公称尺寸(与例 6-1 相同)。考虑到尺寸 A_3 较难加工,选它作为协调环,最后确定其尺寸和公差大小。

2)确定各组成环公差和极限偏差。因各组成环尺寸接近正态分布(即 $k_i = 1$),则按等公差原则分配各组成环公差为

$$T_{A_{iM}} = \frac{T_{A_0}}{\sqrt{n-1}} = \frac{0.25\text{mm}}{\sqrt{5}} \approx 0.1\text{mm}$$

以按等公差原则确定的公差值为基础,综合考虑各零件加工难易程度,对各组成环公差值进行合理调整:A_4 为标准件,其公差值已确定。其余各组成环公差调整如下:$T_{A_1} = 0.14\text{mm}$,$T_{A_2} = T_{A_5} = 0.05\text{mm}$。由于 A_1、A_2、A_5 皆为外尺寸,其极限偏差按基轴制确定,则 $A_1 = 30_{-0.14}^{0}\text{mm}$,$A_2 = 5_{-0.05}^{0}\text{mm}$,$A_4 = 3_{-0.05}^{0}\text{mm}$,$A_5 = 5_{-0.05}^{0}\text{mm}$。

3)计算协调环的公差和极限偏差。

$$T_{A_3} = \sqrt{T_{A_0}^2 - (T_{A_1}^2 + T_{A_2}^2 + T_{A_4}^2 + T_{A_5}^2)} = \sqrt{0.25^2 - (0.14^2 + 3 \times 0.05^2)}\,\text{mm} \approx 0.18\text{mm}$$

因为 $A_{1M} = 29.93\text{mm}$,$A_{2M} = A_{5M} = 4.975\text{mm}$,$A_{4M} = 2.975\text{mm}$,$A_{0M} = 0.225\text{mm}$

由式(6-7)得 $A_{0M} = A_{3M} - (A_{1M} + A_{2M} + A_{4M} + A_{5M})$

$$A_{3M} = 0.225\text{mm} + (29.93 + 4.975 + 2.975 + 4.975)\,\text{mm} = 43.08\text{mm}$$

即

$$A_3 = \left(43.08 \pm \frac{0.18}{2}\right)\text{mm} = 43_{-0.01}^{+0.17}\,\text{mm}$$

According to Eq. (6-7), there is

$$A_{0M} = A_{3M} - (A_{1M} + A_{2M} + A_{4M} + A_{5M})$$

$$A_{3M} = 0.225\text{mm} + (29.93 + 4.975 + 2.975 + 4.975)\text{mm} = 43.08\text{mm}$$

Therefore,
$$A_3 = \left(43.08 \pm \frac{0.18}{2}\right)\text{mm} = 43^{+0.17}_{-0.01}\text{mm}$$

By comparing the calculation result with that of Example 6-1, it is found that probability method can enlarge the manufacturing tolerance of the workpart, thus decreasing manufacturing cost.

6.2.3 Selective assembly method

The essence of this method is that the mating parts are manufactured in light of economic machining accuracy, that is, the tolerances of component links in the dimension chain are magnified to the economic and feasible degree. When assembling a proper workpart is selected to reach stipulated assembly accuracy. This method is used when design tolerances are smaller than production tolerances. A selective assembly involves direct selection of mating parts, sorting these into size group prior to assembly and a combination of the two techniques. Here the second assembly method—the grouping assembly method is mainly discussed.

The grouping assembly method is to magnify the manufacturing tolerances of component links into several (generally 3 – 4) times relative to the tolerance found by full interchangeable assembly method, thus making the component links be machined economically. Then the parts is grouped based on the measured size. Assembly is performed between the corresponding groups in order to meet the assembly requirements. As the parts in the same group can be interchangeable, this method is also named as the selective interchangeable method. For the assembly with fewer components (2 component links) and higher assembly accuracy, this method is often used.

【Example 6-3】 Fig. 6-7 shows the assembly relation between the piston pin and the pin hole. The basic size of the mating parts is $\phi28$mm. It is required that the min. interference $Y_{min} = 0.0025$ mm, and max. interference $Y_{max} = 0.0075$mm should be met in cool assembling. If the economic machining accuracy of the piston pin and the pin hole is 0.01mm, and grouping assembly method is used, try to find the grouping numbers and grouping dimensions.

Solution: The tolerance of the closed link $T_{A_0} = (0.0075-0.0025)\text{mm} = 0.0050\text{mm}$

When using full interchangeable assembly method, the average tolerance of the piston pin and the pin hole is only 0.0025mm. To machine such tolerance of the piston pin and the pin hole is difficult and costly. When using grouping assembly method, the tolerance of pin and hole can be magnified by four times in the same direction. That means the dimensions of pin and hole can be divided into 4 groups, as shown in Fig. 6-7b.

Fig. 6-7 shows the assembly relation between the piston pin and the pin hole. If the diameter of the piston pin is $\phi28^{\ 0}_{-0.010}$mm, and is divided into four groups, then the other diameters of corresponding holes can be found one by one. In this way, piton pins can be machined by means of centerless grinding machine, and the pin hole can be bored with a diamond boring machine. Then their diameters are measured by means of a precision measuring tool and divided into four groups in

比较以上两例的计算结果可以看出，在封闭环公差一定的情况下，用概率法可扩大零件的制造公差，从而降低零件的制造成本。

6.2.3 选择装配法

该方法的实质是把配合零件按经济精度制造，即将组成环的公差放大到经济可行的程度，然后选择合适的零件进行装配，以保证规定的装配精度要求。该方法主要用于产品的设计公差小于制造公差的场合。选择装配法有三种：直接选配法、分组装配法和复合选配法。这里仅讨论分组装配法。

分组装配法是将各组成环的制造公差相对完全互换法所求数值放大几倍（一般为 3~4 倍），使其尺寸能按经济精度加工，再按实测尺寸将零件分组，并按对应组进行装配以达到装配精度的要求。由于同组内零件可以互换，故这种方法又称为分组互换法。在大批大量生产中，对组成环数少而装配精度要求高的部件，常采用这种装配法。

【例 6-3】 图 6-7 所示为活塞销与活塞销孔的装配关系。活塞销直径 d 与销孔直径 D 的公称尺寸为 $\phi 28$mm，按装配要求，在冷态装配时应保证最小过盈量 $Y_{min}=0.0025$mm，最大过盈量 $Y_{max}=0.0075$mm。如果活塞销和销孔的经济加工精度为 0.01mm，拟采用分组装配法装配，试确定活塞销和销孔直径的分组数和分组尺寸。

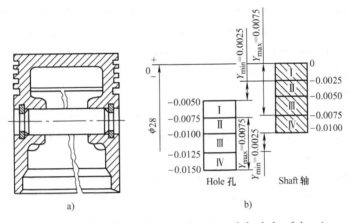

Fig. 6-7 Assembly relation between the pin and the hole of the piston
活塞销与活塞销孔的装配关系

解 封闭环的公差为 $T_{A_0}=(0.0075-0.0025)$mm $=0.0050$mm。

如果采用完全互换法装配，则销与孔的平均公差仅为 0.0025mm，制造这样精度的销轴与销孔既困难又不经济。而采用分组装配法，可将销轴与销孔的公差在相同方向上放大 4 倍，于是可得到分组数为 4，如图 6-7b 所示。

如果活塞销直径定为 $\phi 28_{-0.010}^{0}$mm，将其分为 4 组，则对应销孔的直径也可一一求出。这样，活塞销可用无心磨床加工，活塞销孔用金刚镗床加工，然后用精密量具测量其尺寸，并按实测尺寸大小分成 4 组，涂上不同颜色加以区别，分别装入不同容器内，以便进行分组装配。具体分组情况列于表 6-1 中。

the light of size. Different groups are coated with different colors and are put into different containers so as to perform grouping assembly conveniently. Specific grouping instances are listed in Tab. 6-1.

When grouping assembly method is used, the followings should be noticed.

1) The mating feature (clearance or interference) of each group after sorting must be the same as the original mating requirement. Therefore, the tolerances of mating parts (pin and hole) should be equal, i.e. $T_s = T_h$; Tolerances increase in the same direction, and the increscent times should be equal to the grouping numbers.

2) It is inappropriate to sort too many groups. Otherwise it would cause the production orgnization jobs complexity because the workloads of measuring, sorting and storing increase.

3) It is better to sort the mating parts into complete sets.

4) The surface roughness, form and position tolerances should still maintain the original requirements.

6.2.4 Individual fitting assembly method

In small-lot production, this method is often used for the product with higher assembly accuracy and more component links. Mating parts can be machined economically. When assembling, the dimension of the predetermined component link can be modified by filing, manual scraping, lapping to meet the assembly accuracy. The predetermined component link is named as "fitting link". This assembly method is called "individual fitting method".

1. Selection of the "fitting link"

Attentions should be paid as follows: ① The fitting link should not be the "public link"; ② Be easy to dismount and assemble; ③ The surface to be fitted should be simple and with small area.

2. To determine the limit dimension of the fitting link

As the component links in individual fitting method are machined economically, the sum of tolerances of all component links ($\sum_{i=1}^{n-1} T_{A_i}$) may exceeds the stated tolerance T_{A_0} of the closing link in assembling. The fitting link needs to be fitted so as to meet the assembly requirement. It is necessary to determine the limit dimension of the fitting link so that the allowance to be machined is enough but not overlarge.

The influence of the fitting link being fitted on the closed link is of two cases: the closed link becomes large after the fitting link is repaired; the closed link becomes small after the fitting link is repaired. To understand the varing trend of the closing link is the key to determine the limit dimension of the fitting link. Therefore, the analysis and calculation must be done according to different conditions.

(1) The closed link increases with the machining of the fitting link

Under this circumstance, in order to meet assembly accuracy by machining the fitting link, the max. dimension of the closing link $A'_{0\max}$ before fitting must be not larger than the stated max.

Table 6-1　Diameter groups of piston pin and hole 活塞销与活塞销孔直径分组

Group 组别	Sign color 标志颜色	Diameter of pin 销直径/mm $d = \phi 28_{-0.010}^{0}$	Diameter of hole 销孔直径/mm $D = \phi 28_{-0.015}^{-0.005}$	Fit 配合 Y_{min}/mm	Y_{max}/mm
I	Red 红	$\phi 28_{-0.0025}^{0}$	$\phi 28_{-0.0075}^{-0.0050}$	0.0025	0.0075
II	White 白	$\phi 28_{-0.0050}^{-0.0025}$	$\phi 28_{-0.0100}^{-0.0075}$		
III	Yellow 黄	$\phi 28_{-0.0075}^{-0.0050}$	$\phi 28_{-0.0125}^{-0.0100}$		
IV	Green 绿	$\phi 28_{-0.0100}^{-0.0075}$	$\phi 28_{-0.0150}^{-0.0125}$		

采用分组装配时应注意以下几点：

1）为保证分组后各组的配合性质及配合精度与原装配要求相同，应使配合件的公差相等；公差应同方向增大，且增大的倍数应等于分组数。

2）分组数不宜过多，否则就会因零件的测量、分类、保管工作量的增加而导致生产组织工作复杂。

3）最好能使分组的零件全部配套。

4）配合件的表面粗糙度、几何公差不能随尺寸公差的放大而放大，应保持原设计要求。

6.2.4　修配装配法

在单件小批生产中，对于那些装配精度要求高、组成环数又多的产品结构，常用修配法装配。修配法是将各组成环按经济精度制造，装配时通过手工锉、刮、研等方法修配尺寸链中某一组成环（称为修配环）的尺寸，使封闭环达到规定的装配精度要求。

1. 修配环的选择

采用修配法装配时应正确选择修配环，修配环一般应满足：①不是公共环；②便于装拆；③修配面形状比较简单，面积要小。

2. 修配环极限尺寸的确定

由于修配法中各组成环是按经济精度制造的，这样装配时就有可能使得各组成环的公差之和（$\sum_{i=1}^{n-1} T_{A_i}$）超过规定的封闭环公差 T_{A_0}。此时为了达到规定的装配精度要求，就需对修配环进行修配。为使修配环有足够而又不至于过大的修配量，就要确定修配环的极限尺寸。

修配环修配后对封闭环的影响不外乎有两种情况：修配环越修使封闭环越大；修配环越修使封闭环越小。明确修配环被修配后使封闭环变大还是变小，是确定修配环极限尺寸的关键。因此必须根据不同的情况分别进行分析计算。

（1）修配环越修使封闭环越大的情况　在这种情况下，为使修配时能通过修配修配环来满足装配精度，就必须使修配前封闭环的最大尺寸 $A'_{0\max}$ 在任何情况下都不能大于封闭环规定的最大尺寸 $A_{0\max}$，即

$$A'_{0\max} \leq A_{0\max} \tag{6-10}$$

式（6-10）可用公差带图来描述，如图6-8所示。

为使修配的劳动量最小，应使 $A'_{0\max} = A_{0\max}$。此时，修配环无须修配，就能达到 $A_{0\max}$ 的要求，即最小修配量 $Z_{\min} = 0$。由极值法解尺寸链计算公式得

dimension $A_{0\max}$ of the closing link in any case. That is

$$A'_{0\max} \leq A_{0\max} \tag{6-10}$$

Eq. (6-10) can be illustrated by the tolerance band shown in Fig. 6-8.

In order to reduce the fitting workload to the minimum, let $A'_{0\max} = A_{0\max}$. Under this circumstance, $A_{0\max}$ can be met without machining the fitting link. That means the min. fitting allowance $Z_{\min} = 0$. According to the formula of the calculating dimension chain in extremum method, we can get

$$A'_{0\max} = \sum_{p=1}^{m} A_{p\max} - \sum_{q=m+1}^{n-1} A_{q\min} \tag{6-11}$$

One limit dimension can be found by Eq. (6-11). The other can also be found easily according to the economic machining accuracy of the fitting link.

【**Example 6-4**】 Fig. 6-9 shows the assembly sketch of carriage and guideway. It is required to ensure the assembly clearance $A_0 = 0 - 0.6$mm. It is determined to adopt individual fitting method to ensure assembly accuracy and to choose the pressing plate 3 as the fitting link. Given that $A_1 = 30_{-0.15}^{0}$mm, $A_2 = 20_{0}^{+0.25}$mm, $T_{A_3} = 0.10$mm, when fitting M surface, under the circumstance of $Z_{\min} = 0.1$mm, find the maxmum. fitting allowance $Z_{\max}$.

Solution: 1) To draw the dimension chain (Fig. 6-9).

2) To calculate the limit dimension of the fitting link. According to Eq. (6-11),

$$A_{3\max} = (0.06 + 29.85 - 20.25)\text{mm} = 9.66\text{mm}$$

$$A_{3\min} = A_{3\max} - T_{A_3} = (9.66 - 0.1)\text{mm} = 9.56\text{mm}$$

Therefore, $A_3 = 10_{-0.44}^{-0.34}$mm.

When $Z_{\min} = 0.1$mm, $A_3 = (10_{-0.44}^{-0.34} - 0.1)$mm $= 10_{-0.54}^{-0.44}$mm.

3) To calculate the maxmum fitting allowance $Z_{\max}$

It can be seen from the tolerance band illustration shown in Fig. 6-8b that

$$Z_{\max} = T_{A'_0} - T_{A_0} + Z_{\min} = (0.15 + 0.25 + 0.1 - 0.06 + 0.1)\text{mm} = 0.54\text{mm}$$

(2) The closed link decreases with the machining of the fitting link Under this circumstance, in order to meet assembly accuracy by machining the fitting link, the min. dimension of the closing link $A'_{0\min}$ before fitting must be not less than the stated min. dimension $A_{0\min}$ of the closing link in any case. That is

$$A'_{0\min} \geq A_{0\min} \tag{6-12}$$

Eq. (6-12) can also be illustrated by a tolerance band diagram, as shown in Fig. 6-10. Fig. 6-10a shows that when $A'_{0\min} = A_{0\min}$, the fitting allowance is the minimum, i.e. $Z_{\min} = 0$. Then

$$A'_{0\min} = \sum_{p=1}^{m} A_{p\min} - \sum_{q=m+1}^{n-1} A_{q\max} \tag{6-13}$$

$$A'_{0\max} = \sum_{p=1}^{m} A_{p\max} - \sum_{q=m+1}^{n-1} A_{q\min} \qquad (6\text{-}11)$$

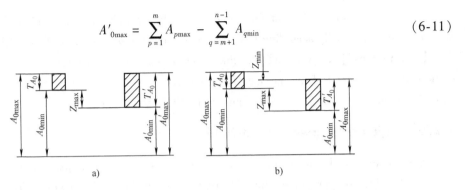

Fig. 6-8 Relative position of actual value A'_0 to prescribed value A_0
封闭环实际值 A'_0 与规定值 A_0 的相对位置
a) $A'_{0\max} = A_{0\max}$ b) $A'_{0\max} < A_{0\max}$

由式（6-11）可求出修配环的一个极限尺寸。再根据修配环的经济加工精度，另一个极限尺寸也可方便地求出。

【例6-4】 图6-9所示为机床溜板与导轨装配简图，要求保证间隙 $A_0 = 0 \sim 0.6\text{mm}$。现采用修配法来保证装配精度，选择压板3为修配环。已知：$A_1 = 30_{-0.15}^{0}\text{mm}$，$A_2 = 20_{0}^{+0.25}\text{mm}$，$T_{A_3} = 0.10\text{mm}$。试求在最小修配量 $Z_{\min} = 0.1\text{mm}$ 情况下，修配 M 面时 A_3 的尺寸及其极限偏差，并计算最大修配量 $Z_{\max}$。

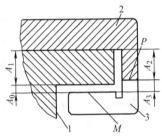

Fig. 6-9 Assembly sketch of the carriage 2 and the guideway 1
机床溜板和导轨的装配简图
1—Guideway 导轨 2—Carriage 溜板 3—Pressing plate 压板

解： 1）画出尺寸链图（图6-9）。
2）计算修配环的极限尺寸。由式（6-11）得

$$A_{3\max} = (0.06 + 29.85 - 20.25)\text{mm} = 9.66\text{mm}$$

$$A_{3\min} = A_{3\max} - T_{A_3} = (9.66 - 0.1)\text{mm} = 9.56\text{mm}$$

即 $A_3 = 10_{-0.44}^{-0.34}\text{mm}$。

当 $Z_{\min} = 0.1\text{mm}$ 时，$A_3 = (10_{-0.44}^{-0.34} - 0.1)\text{mm} = 10_{-0.54}^{-0.44}\text{mm}$。

3）计算最大修配量 $Z_{\max}$。由图6-8b所示的公差带关系图，可以得出

$$Z_{\max} = T_{A'_0} - T_{A_0} + Z_{\min} = (0.15 + 0.25 + 0.1 - 0.06 + 0.1)\text{mm} = 0.54\text{mm}$$

（2）修配环越修使封闭环越小的情况 在这种情况下，为保证装配要求，必须使装配后封闭环的最小尺寸 $A'_{0\min}$ 在任何情况下都不小于封闭环规定的最小尺寸 $A_{0\min}$，即

In the similar way, one limit dimension of the fitting link can be found by Eq. (6-13). The other can be found easily according to the economic machining accuracy of the fitting link.

【Example 6-5】 In Example 6-4, other conditions are not changed, only the surface to be fitted is the surface P instead of the surface M (see Fig. 6-9). Try to find A_3 and its deviation and the max. fitting allowance Z_{max}. (Let students do by themselves).

6.2.5 Adjustment assembly method

Adjustment assembly method means changing the position of an adjustable part in product structure, or choosing an appropriate adjustable part to achieve assembly accuracy. These two adjustable parts play a role of compensating accumulated errors, so they are called compensating member. Washers, shims, adjusting screws, thread bushings, and wedges, etc. can be used as compensating member. This method is advantageous, for it allows the use of wide-tolerance components, a simple assembly process with high accuracy is easily achieved. Commonly used adjusting assembly methods are stationary adjusting method, movable adjusting method, and error counteraction adjusting method.

1. Stationary adjusting assembly method

In an assembly dimension chain, a workpart is selected to be an adjusting member. This member is a set of special-purpose parts which are manufactured in a certain dimension interval (such as bushing, washer, and shim, etc.). The adjusting parts with different sizes can be changed according to the accumulated errors formed by component links so as to reach the assembly requirements. As the position of an adjusting part is stationary, this method is named as the stationary adjusting assembly method.

【Example 6-6】 In the assembly relation of the gear and shaft shown in Fig. 6-6a, the known conditions are the same as Example 6-1. When adopting the stationary adjusting assembly method, try to find the dimension deviations of all component links, and determine the grouping numbers as well as the dimension of each component.

Solution:

1) To establish the assembly dimension chain (the same as Example 6-1).

2) To select the adjusting part. As A_5 is a washer easy to machine, dismantle and assemble, it is selected to be the adjusting part.

3) To determine the tolerance of component links. Except for A_4 (standard part), other component links are machined in the light of economic machining accuracy. Set
$$T_{A_1} = T_{A_3} = 0.20 \text{mm}, \quad T_{A_2} = T_{A_5} = 0.10 \text{mm}$$
The deviation of each link is marked according to the body-in rule, then
$A_1 = 30_{-0.20}^{0}$ mm, $A_2 = 5_{-0.10}^{0}$ mm, $A_3 = 43_{0}^{+0.20}$ mm, and A_4 remains unchangeable.
According to the dimension chain calculating formula, there is
$$A_5 = 5_{0}^{+0.10} \text{mm}$$

4) To determine the adjusting range of the adjusting link A_5. As all component links are machined in the light of economic machining accuracy, its accumulated tolerance $T_{A_{0\Sigma}}$ must be larger

$$A'_{0\min} \geqslant A_{0\min} \tag{6-12}$$

式（6-12）也可用公差带图来描述，如图 6-10 所示。显然，当 $A'_{0\min} = A_{0\min}$ 时，如图 6-10a 所示，修配量最小，即 $Z_{\min} = 0$，于是有

$$A'_{0\min} = \sum_{p=1}^{m} A_{p\min} - \sum_{q=m+1}^{n-1} A_{q\max} \tag{6-13}$$

同理，利用式（6-13）可求出修配环的一个极限尺寸，再根据给定的经济加工精度确定修配环的另一极限尺寸。

【例 6-5】 在例 6-4 中，将修配 M 面改为修配 P 面（图 6-9），其他条件不变，求修配环 A_3 的尺寸及其极限偏差，并计算最大修配量 $Z_{\max}$。（此例留给读者自行完成）

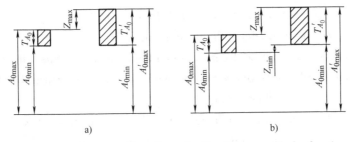

Fig. 6-10　Relative position of actual value A'_0 to prescribed value A_0
封闭环实际值 A'_0 与规定值 A_0 的相对位置
a) $A'_{0\max} = A_{0\max}$　b) $A'_{0\max} > A_{0\max}$

6.2.5　调整装配法

该方法是在装配时用改变可调整件在产品结构中的相对位置或选用合适的调整件以达到装配精度的方法。这两种零件都起到补偿装配累积误差的作用，故称为补偿件。垫圈、垫片、调整螺钉、螺纹套筒、楔等零件都可以用作补偿件。该方法优势明显，因为它允许使用公差大的零件，使得装配过程简单易行，且精度高。常见的调整方法有固定调整法、可动调整法、误差抵消调整法三种。

1. 固定调整法

在装配尺寸链中，选择某一零件为调整件，该零件是按一定尺寸间隔分级制造的一套专用件（如轴套、垫片、垫圈等）。根据各组成环形成的累积误差的大小来更换不同尺寸的调整件，以达到装配精度要求的方法称为固定调整法。

【例 6-6】 如图 6-6a 所示的齿轮与轴的装配中，已知条件与例 6-1 相同。现采用固定调整法，试确定各组成环的尺寸偏差，并求调整件的分组数和分组尺寸。

解 1）建立装配尺寸链（同例 6-1）。

2）选择调整件。因 A_5 为一垫圈，加工容易，装拆方便，故选其为调整件。

3）确定组成环的公差。除 A_4（标准件）外，其余各组成环均按经济精度制造。取 $T_{A_1} = T_{A_3} = 0.20\text{mm}$，$T_{A_2} = T_{A_5} = 0.10\text{mm}$。

各环按入体原则标注，则 $A_1 = 30_{-0.20}^{\ 0}\text{mm}$，$A_2 = 5_{-0.10}^{\ 0}\text{mm}$，$A_3 = 43_{\ 0}^{+0.20}\text{mm}$，而 A_4 不变。

根据尺寸链计算公式，可得 A_5 的极限尺寸为

$$A_{5\max} = 5.10\text{mm} \qquad A_{5\min} = 5\text{mm}$$

than the stated tolerance T_{A_0}. The difference between $T_{A_{0\Sigma}}$ and T_{A_0} is just the adjusting range of the adjusting link.

$$T_{A_{0\Sigma}} = T_{A_1} + T_{A_2} + T_{A_3} + T_{A_4} + T_{A_5} = (0.20 + 0.10 + 0.20 + 0.05 + 0.10)\,\text{mm} = 0.65\,\text{mm}$$

Then the adjusting range R is

$$R = T_{A_{0\Sigma}} - T_{A_0} = (0.65 - 0.25)\,\text{mm} = 0.40\,\text{mm}$$

5) To determine the grouping number N. Take the difference between the closing link tolerance and the adjusting link tolerance as the size grouping interval Δ of the adjusting link, that is

$$\Delta = T_{A_0} - T_{A_5} = (0.25 - 0.10)\,\text{mm} = 0.15\,\text{mm}$$

Then the grouping number N of the adjusting link is

$$N = R/\Delta + 1 = 0.40/0.15 + 1 = 3.66 \approx 4$$

Several explanations on determining the grouping number:

① Grouping numbers can not be decimal. When the difference between the calculational value and the integral value is relative large, N can be approximated to the integer by changing the tolerance of component links or the adjusting link.

② Grouping numbers can not be overlarge, generally 3-4 groups are OK. To reduce the tolerance of the adjusting part is helpful to decrease the grouping numbers.

6) To determine the size of each group adjusting part. There are many methods to determine the size of each group adjusting part. Here a fundamental to determine the size of the adjusting part is introduced.

① When N is the odd number, the dimension in middle group is firstly determined. The dimensions of other groups are determined by plus or minus a Δ correspondingly.

② When N is the even number, taking a predetermined adjusting part size as the symmetrical center, each group of size is determined according to the size difference Δ.

In this example, $N = 4$. Taking $A_5 = 5^{+0.10}_{0}$ mm as the symmetrical center and $\Delta = 0.15$ mm as the size grouping interval, each group size is determined as follows: $5^{-0.125}_{-0.225}$ mm, $5^{+0.025}_{-0.075}$ mm, $5^{+0.175}_{+0.075}$ mm, $5^{+0.325}_{+0.225}$ mm.

After putting A_1, A_2, A_3, A_4 into assembly, the axial clearance is firstly measured. Then taking A_4 down, you can choose one adjusting part with proper thickness from a set of adjusting parts and put it into assembly. The assembly accuracy can be met after reassembling A_4.

2. Movable adjusting assembly method

The assembly accuracy is ensured by changing the relative position of an adjusting part in product structure. This method is named as the movable adjusting assembly method. Fig. 6-11 illustrates the movable adjusting method. The screw in Fig. 6-11a is used for adjusting the axial position of end cover in order to obtain the proper bearing clearance. The screw in Fig. 6-11b is used for adjusting the axial position of the wedge in order to ensure the fit clearance of guideways.

3. Error counteraction adjusting method

In product assembly, the machining errors can be partly counteracted by adjusting the mutual position of related parts so as to improve the assembly accuracy. This method is called an error counteraction adjusting method. The method is more widely used in machine tool assembly. For exam-

即 $A_5 = 5^{+0.10}_{\ 0}$ mm。

4）确定调整环 A_5 的调整范围。由于各环均按经济精度制造，其累积公差值 $T_{A_{0\Sigma}}$ 必然大于规定的公差值 T_{A_0}。这两者之差即为调整环的调整范围。

因为 $T_{A_{0\Sigma}} = T_{A_1} + T_{A_2} + T_{A_3} + T_{A_4} + T_{A_5} = (0.20 + 0.10 + 0.20 + 0.05 + 0.10)\text{mm} = 0.65\text{mm}$

则调整范围 R 为

$$R = T_{A_{0\Sigma}} - T_{A_0} = (0.65 - 0.25)\text{mm} = 0.40\text{mm}$$

5）确定调整环的分组数 N。取调整封闭环公差与调整环制造公差之差，作为调整环尺寸分组间隔 Δ，即

$$\Delta = T_{A_0} - T_{A_5} = (0.25 - 0.10)\text{mm} = 0.15\text{mm}$$

则调整环的分组数 N 为

$$N = R/\Delta + 1 = 0.40/0.15 + 1 = 3.66 \approx 4$$

关于确定分组数的几点说明：

① 分组数不能为小数。当计算的值和圆整后的值相差较大时，可以通过改变各组成环的公差或调整环的公差，使 N 值近似为整数。

② 分组数不能过多，一般以 3~4 组为宜。调整件公差的减小有助于减少分组数。

6）确定各组调整件的尺寸。确定各组调整件尺寸的方法有多种，这里介绍一种确定原则。

① 当 N 为奇数时，首先确定中间一组的尺寸，其余各组尺寸相应的加上或减去一个 Δ 值。

② 当 N 为偶数时，以预先确定的调整件尺寸为对称中心，再根据尺寸差 Δ 确定各组尺寸。本例中 $N = 4$，故以 $A_5 = 5^{+0.10}_{\ 0}$ mm 为对称中心，并以尺寸差 $\Delta = 0.15$ mm 的间隔确定各组尺寸分别为：$5^{-0.125}_{-0.225}$ mm，$5^{+0.025}_{-0.075}$ mm，$5^{+0.175}_{+0.075}$ mm，$5^{+0.325}_{+0.225}$ mm。

待 A_1、A_2、A_3、A_4 装配后，测量其轴向间隙值，然后取下 A_4，从一组调整件中选择一个适当厚度的 A_5 装入，再重新装上 A_4，即可保证所需的装配精度。

2. 可动调整法

用改变调整件在产品结构中的相对位置来保证装配精度的方法称为可动调整法。图 6-11 所示为可动调整法的应用。图 6-11a 是主轴箱用螺钉来调整端盖的轴向位置，最后达到调整轴承间隙的目的；图 6-11b 表示小刀架上通过调整螺钉来调节镶条的位置来保证导轨副的配合间隙。

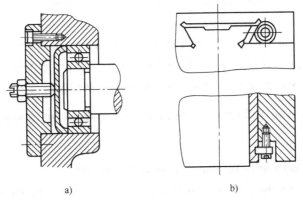

a)　　　　　　　　b)

Fig. 6-11　Applications of movable adjustment methods　可动调整法的应用

ple, when assembling the lathe spindle, the radial runout of the spindle is controlled by regulating the radial runout direction of the front and back bearings.

The advantages of the adjusting assembly method lie in that the workparts can be machined economically and assembled conveniently, and that higher assembly accuracy can be achieved. The disadvantage of the adjusting assembly method is that it needs to add a set of adjusting devices and requires good adjusting technique. Because the adjusting assembly method has outstanding advantages, it has got wide application.

Every assembly method mentioned above has its own characteristics. That which of assembly methods is adopted has been determined generally in design stage. Only in this way, the technical requirements of all parts and assemblies can be determined rationally by calculating the dimension chain. As the same product would be often produced in different production conditions, it may adopt different assembly methods. The general rule to select assembly method is: to give the priority to the selection of the strict interchangeable method; to adopt the partial interchangeable method when there are more component links in large volume production; if the assembly accuracy is quite high in mass production, the grouping interchangeable method should be used when there are less component links, and the adjusting method should be used when there are more component links; when other assembly methods can not be used in job and small-batch production, fitting assembly method can be used.

6.3 Assembly process planning

The rational assembly process is specified in the forms of document expressed by charts, tables and words. This document is called assembly process regulation. Assembly process regulation is a major technological document to direct assembly production. To prepare assembly process regulation is one of the main tasks in production preparation.

6.3.1 Major contents of assembly process regulation

The major contents include: ① to analyze the product and its assembly requirements, to determine assembly method and organizational form, and to plot assembly unit; ② to draw up assembly sequence and plot assembly operation, to draw assembly process chart and prepare assembly process regulation; ③ to select or design the tool, fixture and equipment used in the product assembly process; ④ to determine assembly technological requirement, quality inspecting method and tools for every assembly operation; ⑤ to calculate assembly time standard.

6.3.2 Principles and original information required in assembly process planning

1. Fundamental principles in planning assembly process

1) To ensure and improve product assembly quality so as to prolong the service life of product;
2) Through arranging assembly operation rationally, to reduce the amount of manual labor as

3. 误差抵消调整法

在产品装配时,通过调整有关零件的相互位置,使其加工误差相互抵消一部分,以提高装配的精度,这种方法称为误差抵消调整法。这种方法在机床装配中应用较多。如在车床主轴装配中,通过调整前后轴承的径向跳动方向来控制主轴的径向跳动。

调整装配法的优点在于不仅零件能按经济精度加工,而且装配方便,可以获得比较高的装配精度。缺点是要另外增加一套调整装置并要求较高的调整技术。但由于调整法优点突出,因而使用较为广泛。

上述各种装配方法各有其特点。一种产品究竟采用何种装配方法来保证装配精度,通常在产品设计阶段就应确定下来。只有这样,才能通过尺寸链计算合理确定各个零部件在加工和装配中的技术要求。但是,同一产品往往会在不同的生产类型和生产条件下生产,因而就可能采用不同的装配方法。选择装配方法的一般原则是:优先选择完全互换法;在生产批量较大,组成环数又较多时,应考虑采用不完全互换法;大量生产中,在封闭环精度较高,组成环数较少时可考虑采用分组互换法,环数较多时采用调整法;在装配精度要求很高,又不宜选择其他方法,或在单件小批生产中,可采用修配法。

6.3 装配工艺规程的制订

将合理的装配工艺过程用文件的形式规定下来就是装配工艺规程。它是指导装配生产的主要技术文件,制订装配工艺规程是生产技术准备工作的主要内容之一。

6.3.1 装配工艺规程的主要内容

装配工艺规程的主要内容包括:
1) 分析产品及其装配技术条件,确定装配方法与组织形式,划分装配单元。
2) 拟订装配顺序,划分装配工序,绘制装配工艺系统图,编制装配工艺规程。
3) 选择和设计产品装配过程中要使用的工具、夹具和设备。
4) 确定各工序装配技术要求、质量检查方法和检查工具。
5) 计算装配时间定额。

6.3.2 制订装配工艺规程的基本原则及所需要的原始资料

1. 装配工艺规程设计的基本原则
1) 保证产品装配质量,力求提高质量,以延长产品的使用寿命。
2) 合理安排装配工序,尽量减少钳工手工劳动量,缩短装配周期,提高装配效率。
3) 尽量减少装配占地面积,提高单位面积的生产率。
4) 要尽量减少装配工作所占用的成本。

2. 制订装配工艺规程所需的原始资料
1) 产品的总装图和部件装配图。装配图应清楚地表示出零、部件间相互连接情况及其联系尺寸;装配的技术要求;零件的明细表等。
2) 产品验收的技术标准。

much as possible so as to shorten assembly period and increase assembly efficiency;

3) To reduce the occupying area for assembly as much as possible;

4) To reduce the cost occupied by assembly work as much as possible.

2. Original information required in planning assembly process

1) General assembly drawing of a product and assembly drawing of an assembly. Assembly drawing should show clearly the joint situation and connection dimension among workparts and assemblies, technical requirements and detail list of workparts.

2) Technical standards for production acceptance.

3) Production program of a product. Production program of a product decides its production type. Different production types will lead to much difference in assembly organizational forms, assembly methods, the division of the technological process, the specialization degree of the facility and technological equipment and the manual work ratio.

4) Existing production conditions. Existing production conditions include the available assembling facilities and technological equipment, workshop area and the technical level of workers etc.

6.3.3 Approaches and Procedures to assembly process planning

1. To analyze the assembly drawing of a product and technical acceptance condition

This work includes learning the specific structure of a product, analyzing structural technology of a product, verifying the assembling technical specifications and acceptance standard, analyzing and calculating assembly dimension chain.

2. To determine assembly method and organizational form

The assembly method and organizational form of a product depend on the structural feature and production program of this product, and is also related to the existing production condition and facilities. To select the rational assembly method is the key to ensure assembly accuracy. The assembly method should be determined according to the dimension chain theory and specific production condition.

There are two organizational forms in assembly process: stationary type and movable type. Stationary type assembly means all assembly tasks are completed in a fixed site. It can be centralized assembly or distributed assembly. The former means all assembly tasks of the whole machine are completed by one worker or a group of workers in a fixed site. The latter means the assembling of parts and general assembling are completed separately by several workers or several groups of workers in different working sites. Stationary type assembly is usually used in job and small volume production, or the batch production of heavy, bulky machinery. Movable type assembly means that the parts and assemblies are transported from one assembling place to another by conveyors in assembling sequence, and part of the assembling work is completed in each assembling place. The movable type can be free movement or compulsory movement. The former is suitable for assembling small-size and light products or assemblies in large volume production. The latter has two forms: continuous and intermittent. The movable type assembly is usually used for the product in mass or large volume production in the form of transfer line.

3) 产品的生产规程。生产规程决定了产品的生产方法。生产方法不同，致使装配的生产组织形式、装配方法、工艺过程的划分、设备与工艺装备的专业化水平、手工作业量的比例均有很大不同。

4) 现有的生产条件。包括现有装配设备和工艺装备、车间面积和工人技术水平等。

6.3.3 制订装配工艺规程的方法与步骤

1. 研究产品的装配图及验收技术条件

了解产品及部件的具体结构；分析产品的结构工艺性；审核产品装配的技术要求和验收标准；分析与计算产品装配尺寸链。

2. 确定装配方法与组织形式

装配的方法和组织形式主要取决于产品的结构特点和生产规程，并考虑现有的生产技术条件和设备。选择合理的装配方法是保证装配精度的关键。应结合具体的生产条件，从机械加工和装配的角度出发应用尺寸链理论，确定装配方法。

装配的组织形式主要分为固定式和移动式两种。固定式装配是全部装配工作在固定的地点完成。固定式装配又有集中式和分散式之分。所谓集中式是指整台机器的所有装配工作都由一个或一组工人在一个工作地完成，而分散式是把整台机器分为部装和总装，分别由几个或几组工人同时在不同的工作地分散完成。固定式装配多用于单件小批生产，或质量大、体积大的批量生产中。移动式装配是将零、部件用输送带或输送小车按装配顺序从一个装配地点有节奏地运送到另一个装配地点，各个装配地点分别完成一部分装配工作。移动式装配分为自由移动式和强制移动式两种。自由移动式装配适合于在大批量生产中装配那些尺寸和质量都不大的产品或部件，而强制移动式装配又有连续移动和间歇移动两种形式。移动式装配常用于产品的大批大量生产中，以组成流水作业线和自动作业线。

3. 划分装配单元，确定装配顺序，绘制装配工艺系统图

划分装配单元是制订装配工艺规程中最重要的一个步骤。在确定装配顺序时，首先选择装配的基准件。装配基准件通常应是产品的基体或主干零、部件。基准件应有较大的体积和质量，有足够的支承面，以满足陆续装入零、部件时的作业需求。例如，床身零件是床身组件的装配基准零件；床身组件是床身部件的装配基准组件；床身部件是机床产品的装配基准部件。

确定装配顺序的一般原则是先难后易、先内后外、先下后上、先重大后轻小、先精密后一般。另外还应注意，处于同一基准件方位的装配工作尽可能集中安排；使用相同安装设备的装配应集中安排；电线、气路和油路的安装应和相应的工序同时进行；具有危险性的零部件安装应放在最后。

为了清晰地表示装配顺序，常用装配工艺系统图（图6-12）来表示。它是表示产品装配流程以及零、部件间相互装配关系的示意图。每一个装配单元（零件、合件、组件、部件）可用一个长方格来表示。在方框内写入每个装配单元的名称、编号和数量。有时在图上还要加注一些工艺说明，如焊接、钻孔、冷压和检验等内容。

3. To plot assembly unit, determine assembling sequence and draw assembly process chart

To plot assembly unit is the most important step in assembly process planning. The datum part should be selected firstly in determining assembly sequence. The datum part for assembling is usually the base or major part. It should be large, heavy, and have large enough support surface so as to enclose the parts and assemblies in succession. For instance, the lathe bed is the datum part of the bed sub-assembly, the bed sub-assembly is the datum part of bed assembly, and further, the bed assembly is also the datum part of the whole lathe.

The general principle for the determining assembly sequence is: difficult firstly and quite easy afterwards, inside firstly then outside, heavy part firstly and light afterwards, precise part firstly and ordinary afterwards. Besides, it should also be noticed that the assembling tasks which are located in the same datum part, or which will be done by using the same assembling equipment should be arranged concentratedly. The installation of electric circuit, pneumatic circuit and hydraulic circuit should be done together with relative operations. The parts with risks should be assembled lastly.

In order to show the assembly sequence clearly, the assembly process chart is often used, as shown in Fig. 6-12. It is a schematic diagram to show the assembling process and the assembly relation between workparts and assemblies. Each assembly unit (workpart, combined part, subassembly, assembly) is represented by a rectangular block. The name, code and number of each assembly unit are filled in the block. Sometimes, some process descriptions, such as welding, drilling, cold pressing, inspection and so on, are supplemented in the diagram.

4. To plot and design assembly operations

Main contents include: ①to plot assembly operation and decide operation contents; ②to determine the equipment and tools used in each operation; ③to lay down operating specification for each operation, such as the magnitude of pressure in interference fit, assembling temperature, rating torque for fastening bolt and so on; ④to specify the assembly quality and its inspection way; ⑤to determine the time standard for each operation and balance the operation time.

5. To compile assembly process file

The way to compile assembly process file is similar to that of the machining process.

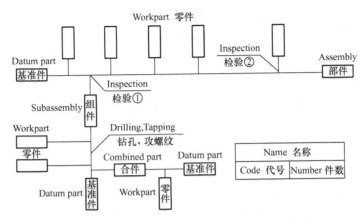

Fig. 6-12　Assembly process chart 装配工艺系统图

4. 划分装配工序，进行装配工序设计

主要内容有：①划分装配工序，确定工序内容；②确定各工序所需的设备和工具；③制订各工序装配操作规范，如过盈配合压入力大小、装配温度以及紧固螺栓联接的额定力矩等；④制订各工序装配质量要求与检测方法；⑤确定工序时间定额，平衡各工序节拍。

5. 编制装配工艺文件

其编写方法与机械加工工艺文件基本相同。

Chapter 7

Brief Introduction to Advanced Manufacturing Technology

第7章

先进制造技术简介

7.1 Introduction

7.1.1 Definition of advanced manufacturing technology

Manufacturing technology is a general name of all production technologies which are used in manufacturing industries to produce various necessities for national economic construction and people's life. It refers to a group of technologies which are used to change raw materials and other production elements economically and rationally into finished or semi-finished products which can be directly used with higher added value.

Advanced manufacturing technology (AMT) is a general name of manufacturing technologies, which keep on adopting the achievements in information technology and modern management technology, and utilize them synthetically to the whole manufacturing processes including product design, planning, production, inspection, management, marketing and recycling. Its purpose is to realize the production with high quality, high productivity, low cost, cleanness, and flexibility, and to enhance the adaptability and competitively.

7.1.2 Characteristics of advanced manufacturing technology

Compared with the traditional manufacturing technology, AMT has the characteristics as follows:

(1) Practicality. AMT does not pursue the advanced and new technology, but pay attention to the best practical effect. It centers on raising benefit and aims at increasing the competitivity of enterprise.

(2) Extensiveness. AMT is not limited to the machining technology, but covers the whole process from product design, planning to production, marketing and even recycling.

(3) Dynamic characteristic. AMT has not been a fixed mode, but is developing dynamically. It has different characteristics, emphases, targets and contents in different times and different countries or areas.

(4) Integration. AMT stresses on inter-permeating and inter-merging among various disciplines or majors. The border line is generally dimmed and disappeared. Technologies involved tend to be systematic and integrated.

(5) Systematicness. AMT has become a systematic engineering which can control material-flow, energy-flow, and information-flow in manufacturing processes.

(6) Emphasis on the production with high quality, high efficiency, low cost, cleanness and flexibility.

7.2 Advanced machining technology

7.2.1 Ultra-precision machining technology

1. Ultra-precision machining and its classification

Ultra-precision machining is a general name of all machining methods used to obtain a certain level of machining quality. Ultra-precision machining technology is the key technology to develop high-tech products and to realize equipment modernization. The development of ultra-precision machining technology represents the development level of a country's manufacturing industry.

Precision machining and ultra-precision machining represent the different development periods

7.1 概述

7.1.1 先进制造技术的定义

制造技术是制造业为了进行国民经济建设和人民生活而生产各类必需物资所使用的一切生产技术的总称，是将原材料和其他生产要素经济合理地转化为可直接使用的、具有较高附加值的成品/半成品和技术服务的技术群。

先进制造技术是指那些不断吸取信息技术和现代管理技术的成就，并将其综合应用于包括产品设计、规划、生产、检验、管理、销售和产品回收利用等整个制造过程的所有制造技术的总称。其目的是实现高质、高效、低耗、清洁、柔性生产，同时提高生产的适应性和市场竞争能力。

7.1.2 先进制造技术的特点

与传统制造技术相比，先进制造技术的特点如下：

（1）实用性　先进制造技术不是一味追求先进或新颖，而是注重技术的应用效果，注重提高效益和企业的市场竞争力。

（2）广泛性　先进制造技术并不仅局限于加工技术，而是包括从产品设计、规划，到生产、销售甚至循环再利用的整个过程。

（3）动态性　先进制造技术没有固定模式而是动态发展的技术。在不同时期、不同国家或地区先进制造技术的特点、重心、目标和内容有所不同。

（4）集成性　先进制造技术强调各学科、各专业的相互渗透和融合。学科间的界限通常很模糊，技术趋于系统化和集成化。

（5）系统性　先进制造技术是一项系统工程，可控制制造过程中的物料、能量及信息流。

（6）强调生产的优质、高效、低耗、清洁和灵活性。

7.2 先进加工技术

7.2.1 超精密加工技术

1. 超精密加工及其分类

超精密加工是指加工质量能够达到某一等级的所有加工方法的总称。超精密加工技术是开发高科技产品、实现装备现代化不可或缺的关键技术。它的发展代表一个国家制造业的发展水平。

精密加工和超精密加工代表了加工技术发展的不同阶段，按加工误差大小，加工可分为普通加工、精密加工和超精密加工。从目前发展水平看，加工误差为 $0.1 \sim 1\mu m$，表面粗糙度为 $Ra0.02 \sim 0.1\mu m$ 的加工方法称为精密加工；加工误差小于 $0.1\mu m$，表面粗糙度小于 $Ra0.02\mu m$ 的加工方法称为超精密加工。

of machining technology. Based on the magnitude of the machining error, machining can be divided into ordinary machining, precision machining and ultra-precision machining. In terms of the development level of machining technology, the machining method whose machining error is $0.1 - 1 \mu m$, and surface roughness is $Ra0.02 - 0.1 \mu m$ is called precision machining; the machining method whose machining error is less than $0.1 \mu m$, and surface roughness is less than $Ra0.02 \mu m$ is called ultra-precision machining.

According to the mechanisms and characteristics of machining, ultra-precision machining can be further divided into ultra-precision cutting, ultra-precision grinding, ultra-precision lapping, ultra-precision non-traditional machining, and combined machining, etc. The ultra-precision cutting and ultra-precision grinding are mainly introduced in this section.

2. Ultra-precision cutting

Ultra-precision cutting is characterized by using sharp diamond cutting tools to perform turning and milling operations. Diamond cutting tools have a low affinity with non-ferrous metals, higher hardness, and better heat conductivity. Furthermore, cutting tool can be ground to a very sharp edge (the edge radius is less than $0.01 \mu m$) which enables diamond cutters to machine the surfaces with low roughness (smaller than $Ra0.01 \mu m$). Two ultra-precision cutting methods are introduced here.

(1) Mirror turning of magnetic disc substrate   Magnetic disc storage is one of the main peripheral devices of computers. With the rapid development of computer technology, the data storage density of magnetic discs increases continuously, and the floating height of magnetic head decreases rapidly. In order to maintain a gap of less than $0.3 \mu m$ between the floating head and the magnetic disc rotating at 3600 r/min so that the head can read/write data rapidly and accurately, the disc surface should have very high precision. For example, if the flying height with $0.33 \mu m$ is required, the surface roughness of the disc must be $Ra0.015 \mu m$ or less. In recent years, the overseas researchers have developed a new running method, i.e. running in skipping way, in order to increase the data storage density of discs by a factor of at least 20. The minimum interval skipped by magnetic head in the world is 3nm. This means there is no air layer between the head and the disc, that is to say, the gap between them approaches to zero. The light-duty micro-miniature magnetic head has been developed in order to alleviate the mechanical damage of the discs resulted from the direct contact. The subminiaturization of the head requires both the head and the disc should be machined with higher precision. Therefore, high-precision machining of disc substrates is a vital subject in the development of magnetic disc storages.

The blank of magnetic disc substrates is disc-type aluminum alloy with higher surface flatness. Both sides of the blank are turned with a diamond cutter before annealing, and then surface polishing is carried out. The magnetic disc substrate can be machined on a high-precision lathe by means of diamond cutter sharpened carefully.

(2) Mirror milling of aircraft organic glass   Windows in modern larger passenger planes are made of organic glass. During the planes take off and land on the ground, windows are frequently suffered from the hitting of gritty dust carried by the atmosphere. After a certain number times of take-off and landing, the surfaces of the windows would become very coarse. This influences on the view of passengers and pilots. Therefore, the windows should be renovated by polishing. The time

根据加工方法的机理和特点，超精密加工可以分为超精密切削、超精密磨削、超精密研磨、超精密特种加工和复合加工等。本节主要介绍超精密切削和超精密磨削方法。

2. 超精密切削加工

超精密切削的特点是借助锋利的金刚石刀具对工件进行车削和铣削。金刚石刀具与非铁（有色）金属亲和力小，其硬度、耐磨性以及导热性都非常优越，且能刃磨得非常锋利（刃口圆弧半径小于 $0.01\mu m$），从而可加工出表面粗糙度小于 $Ra0.01\mu m$ 的表面。

（1）磁盘基片的镜面车削　磁盘存储器是计算机的主要外部设备之一。随着计算机技术的飞速发展，磁盘单位面积的存储密度也在不断提高，磁头在磁盘上的浮动高度急剧减小。要使磁头与以 3600 r/min 旋转的磁盘间稳定地保持 $0.3\mu m$ 以下的间隙，磁头快速、准确无误地存取信息，这就要求磁盘表面具有很高的精度。例如，如果要求浮动高度为 $0.33\mu m$，则磁盘表面粗糙度就要在 $Ra0.015\mu m$ 以下。近年来，国外为了把磁盘记录密度再提高 20 倍以上，正在开发新的磁头走行方式，即磁头跳跃走行方式。目前世界上磁头跳跃的最小间隔为 3nm，使磁头与磁盘之间没有空气层，间隙近于零。为了减轻直接接触磨损等机械损伤，又研究开发了负荷轻的超小型磁头。由于磁头超小型化，对磁头和盘片平面质量的加工要求更高了。因此，磁盘基片的高精度加工是磁盘存储器开发中的重要课题。

磁盘基片使用平面度好的铝合金圆盘做毛坯，铝板两面用金刚石车刀车削后进行退火处理，再进行表面抛光。磁盘基片可在高精度磁盘车床上采用仔细刃磨过的金刚石车刀加工。

（2）飞机玻璃的镜面铣削　现代大型客机窗户是用有机玻璃制成的。飞机起飞与降落时，玻璃屡遭大气中夹带沙尘碰撞，飞行一定起降次数后，窗户玻璃表面就变得十分粗糙，直接影响飞行员和乘客的视野，这就需要对玻璃进行重新抛光修复。采用传统抛光方法，修复一块玻璃通常需要 1h 左右，当玻璃有较深零星刻痕时，加工时间更长。若采用镜面铣削方法，所需时间不到抛光时间的一半，从而大大缩短飞机维修时间。此方法已被许多大飞机维修中心广为采用。

镜面铣削切削速度通常为 30m/s 左右。镜面铣削平面度可达 $0.1\mu m$。粗糙度除取决于机床、刀具因素外，还与工件材料本身特性有关。为了能加工出完美工件，主轴在换刀后必须进行动平衡，以尽量减少动不平衡对工件表面造成的波纹。

3. 超精密磨削加工

超精密磨削技术是在一般精密磨削基础上发展起来的一种亚微米级加工技术。其目的不仅要提供镜面级的表面粗糙度，还要保证获得精确的几何形状和尺寸。超精密磨削一般多采用金刚石、立方氮化硼等超硬磨料砂轮，其加工误差可达 $0.1\mu m$ 以下，表面粗糙度小于 $Ra0.025\mu m$。镜面磨削是一种超精密磨削方法，其加工表面粗糙度可达 $Ra0.02 \sim 0.01\mu m$。之所以称为镜面磨削是因为这种磨削方法可以获得光泽如镜的表面。要想实现镜面磨削，磨粒的粒度必须尽可能小，如 $2\mu m$，甚至 $0.2\mu m$。因此必须使用金刚石微粉砂轮。由于微粉砂轮极易堵塞，就必须经常进行修整。由于金刚石非常硬，用常规的砂轮修整方法很难修整，因此就得采用去除金刚石磨粒周围的结合剂的方法使磨粒裸露。如果结合剂去除的过多，金刚石磨粒容易脱落；如果去除的太少，就可能不能形成切削刃和足够的容屑空间。所以，金刚石微粉砂轮的修整是超精密磨削的关键技术

needed to renovate a piece of glass is about an hour if traditional polishing process is adopted. The repairing time would be even longer if there are deeper nicks scattered on the surfaces of these glasses. However, if mirror milling is used instead of traditional polishing, the machining time would be reduced by at least 50%, thus shortening the maintenance time of the aircraft window greatly. Mirror milling has been widely used in many maintenance centers of larger aircrafts.

Generally, the cutting speed of mirror milling is about 30m/s, and the flatness obtained can reach up to 0.1μm. The surface roughness of machined glass depends on machine tool, cutting tool as well as characteristics of workpiece materials. In order to obtain perfect machining effects, after a new cutting tool is mounted on the spindle, dynamic balance must be performed for the spindle to reduce the ripples on the workpiece surface resulted from dynamic unbalance of the spindle.

3. Ultra-precision grinding

Ultra-precision grinding technique is one kind of sub-micron machining techniques developed on the basis of precision grinding. The purpose of ultra-precision grinding is not only to achieve mirror-like surface finish, but also to obtain exact geometric shape and accurate dimension. Grinding wheels made of super-hard materials, such as CBN wheels and diamond wheels, are often used in ultra-precision grinding. Its machining error is 0.1μm, and surface roughness is less than Ra0.025μm. Mirror grinding is a kind of ultra-precision grinding process in which surface roughness of Ra0.02~0.01μm can be achieved. The "mirror grinding" means the grinding method can produce a glassy surface as a mirror. In order to realize mirror grinding, the grain size should be as small as possible, for example 2μm, or even 0.2μm. Therefore, the grinding wheel made of diamond powder must be used. As the wheel is easy to jam, it should be often dressed. As the diamond is very hard, it is very difficult to cut it by means of ordinary dressing methods. The way to remove the bonding materials around diamond grains must be found so as to expose the grains. If too much bonding material is removed, the diamond grains would fall off easily; too little bonding material is removed, the cutting edge, or enough chip space can't possibly be formed. Therefore, the dressing technology of a diamond powder wheel is one of the key technologies in ultra-precision grinding. Currently, there are two advanced methods to dress the grinding wheels with super-hard grains: one is electrolytic dressing, the other is electrical discharge dressing.

The electrolytic in-process dressing (EIPD) is shown in Fig. 7-1. The power supply of EIPD is used to control the current, voltage, and pulse width in dressing process. When the electric brush contacts the wheel shaft smoothly, the wheel is connected to the positive pole, and the electrode is the negative pole. In the small clearance between the positive and the negative pole (0.1 to 0.3mm), electrolysis occurs through supplying electrolyte (grinding fluid) and electrical current.

The grinding practices on different samples (such as optical glass, silicon, ceramics, etc.) show that the surface roughness obtained by EIPD grinding is Ra0.02 - 0.01μm, reaching to the mirror finish.

7.2.2 Nanofabrication technology

1. Introduction

Nanofabrication evolves from microfabrication. Since the transistor appeared in 1947, microe-

之一。目前有两种先进的超硬磨料砂轮的修整方法。一种是电解修整法，另一种是电火花修整法。

图 7-1 所示为电解在线修整法（Electrolytic In-Process Dressing，EIPD）。电解在线修整系统中的电源用于控制修整电流、电压及脉冲持续时间。当电刷与砂轮轴平稳接触时，金属结合剂砂轮成为阳极，而电极成为阴极。在阴极和阳极间的细小间隙（0.1 ~ 0.3mm）内通以磨削液和电流，间隙内将发生电解反应。

不同试件（光学玻璃平面、硅片平面和陶瓷内孔）的磨削实践表明，EIPD 磨削的表面粗糙度为 $Ra0.02 ~ 0.01\mu m$，达到了镜面水平。

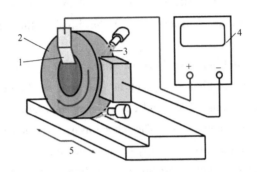

Fig. 7-1　Electrolytic in-process dressing（EIPD）电解在线修整法
1—Brush 电刷　2—Diamond wheel 金刚石砂轮
3—Electrolyte 电解液　4—Power supply 电源　5—Feed 进给

7.2.2　纳米加工技术

1. 概述

纳米加工是由微细加工发展而来的。自从 1947 年晶体管面世以来，微电子和集成电路行业成为不断将加工技术推向新尺度的主要推动力。大规模生产中，实际可实现的集成电路的最小特征尺寸是 65nm。而批量生产 45nm 尺度的集成电路也已在研究之中。根据纳米技术中提到的 100nm 的尺度标准，目前新一代集成电路已进入纳米领域。

纳米加工技术是在纳米尺度范围（0.1 ~ 100nm）内对原子、分子等进行操纵和加工的技术。它是一门涉及多个学科领域的交叉学科，是在现代物理学、化学和先进工程技术相结合的基础上诞生的，是一门与高新技术紧密结合的新科技。纳米加工的目的是低成本、大批量地构造出可作为元件、装置或系统的纳米级结构。纳米加工对于所有的纳米技术领域都很重要，尤其对于将工程和科学相结合的传统纳米技术领域。

纳米加工技术的发展有两条途径：一是将传统的超精加工技术，如机械加工、电化学加工、离子束蚀刻、激光加工等向极限精度逼近，使其达到纳米加工能力；二是开拓新效应的加工方法，如采用扫描隧道显微术、原子力显微术等操纵技术的加工方法是一种非常有前景

lectronics and Integrated Circuit (IC) industry has been the main driving force to continuously push fabrication technologies to their new dimensional limit. The actually achieved minimum circuit feature in mass production is 65nm. And volume manufacturing of 45nm ICs is already on the horizon. According to the 100nm dimensional mark, the current generation of IC is already in the nano-technology regime.

Nanofabrication is the design and manufacture of devices with dimensions measured in nanometer scale (0.1 – 100nm) by manipulating and restructuring atoms or molecules. It is a crossing discipline developed on the basis of the combination of modern physics, chemistry and advanced engineering technologies. It is a developing technology that has close connections with high technologies. Nanofabrication aims at building nanoscale structures, which can act as components, devices, or systems, in large quantities at potentially low cost. Nanofabrication is vital to all nanotechnology fields, especially for the realization of nanotechnology that involves the traditional areas across engineering and science.

The development of nano-fabrication follows two ways. One is to improve the precision of traditional ultra-precision machining technologies, such as precision machining, Electrochemical Machining (ECM), Ion Beaming Etching (IBM) and Laser Beam Machining (LBM) etc., to an extreme level that approaches to the nanometer scale. The other way is to exploit new techniques. For example, nanofabrication using manipulative techniques is a promising way of producing nano-based electronic components using processes such as Scanning Tunneling Microscopy (STM) and Atomic Force Microscopy (AFM) etc.

The reasons why so much attention is given to nano-devices are their unique performances, unprecedented functions as well as the particular phenomena resulted from interactions among several kinds of energy, which contribute to high energy efficiency of materials, improved built-in intelligence and properties of the devices. However, the fabrication techniques are comparatively complex and costly. Currently, methods used to build nanoscale structures and nano-structured materials are commonly characterized as "top-down" and "bottom-up".

The top-down strategies have basically evolved from conventional lithographic techniques, in which nanoscale structures are fabricated from a bulk material by gradually removing or subtracting bits of the material in series. The top-down approach has been proven to be critical tool for the sustained evolution of the electronic, computer, optical and micro-electro-mechanical system industries. Examples of this approach are Electron Beam Lithography (EBL) and Ion Beam Lithography (IBL) etc. However, there continue to be many obstacles and challenges that confront top-down techniques as these techniques approach their fundamental size limits. Also, this method can only be used for patterning features typically in two-dimensions (2D).

Bottom-up fabrication strategies involve manipulation or synthetic methods of biochemistry in directly assembling sub-nanoscale building blocks, such as atomic, and molecular etc. into required nanoscale patterns of which bio-medical, chemical, and physical sensors and actuators are obvious applications. The typical bottom-up techniques are nanofabrication using manipulative techniques or Scanning Probe Microscopy (SPM), Self Assembly (SA) and epitaxy techniques (Molecular Beam

的电子元件纳米加工法。

纳米器件之所以得到广泛的关注，是因为它们具有独特的性能、前所未有的功能、在于多种形式的能量相互作用时所呈现的奇特现象，进而带来材料的高能量效率、内置式的智能和性能改善等。但纳米器件的制造方法相当复杂，制作成本很高。目前，纳米器件和纳米结构材料的制备方法有"自上而下"和"自下而上"两类。

"自上而下"法是由传统的光刻技术发展而来的，它通过各种方法从块状材料中不断地去除少量材料来形成最终的纳米级结构。实践证明"自上而下"法对于电子、计算机、光学和微机电系统行业的持续发展非常关键。"自上而下"法的代表工艺有电子束光刻加工和离子束光刻加工等。然而，"自上而下"法面临许多障碍和挑战，因为依靠这种工艺来进一步减小电子器件尺寸变得越来越困难，而且这些技术大多只能用于制作二维图形。

"自下而上"法是指通过将亚纳米级结构单元如原子、分子等直接组装成所需要的纳米级构造的操纵和生物化学合成技术。这些纳米构造可用于生物医学、化学、传感器和驱动器等。采用操纵技术的纳米加工或扫描探针显微加工、自组装、外延生长技术（分子束外延生长、液相外延生长、气相外延生长等）都属于"自下而上"的纳米加工法。由于不涉及材料去除，因此"自下而上"加工法不会浪费原材料。

2. 典型纳米加工技术

（1）传统纳米加工　传统纳米加工仅指"自上而下"的纳米加工法。该方法的加工精度为纳米级，而不是产品尺寸为纳米级。因此传统纳米加工可定义为产品尺寸精度为 1~100nm 的去材料加工工艺。传统纳米加工可分为四类：

1）确定性纳米级机械加工。该加工方式采用固定形状刀具或受控制系统控制的刀具，依靠刀具形状或刀具运动路径加工出所需三维零件。且切削量可以很小，约为几十纳米。典型工艺有金刚石车削和纳米级磨削。

2）游离磨料纳米加工。该方法采用游离磨粒去除少量材料，包括抛光、研磨和珩磨。

3）非机械纳米加工。包括聚焦离子束加工、电火花微加工和准分子激光加工。

4）光刻加工。该工艺采用掩膜获得需要形状的产品，一般用于生产二维图案，在三维产品生产方面有较大局限性。主要包括 X 射线光刻、光刻—电铸—模铸复合成形加工、电子束光刻。

（2）采用操纵技术的纳米加工　采用操纵技术的纳米加工是一种极具前景的加工方式，它借助于扫描隧道显微镜、原子力显微镜等实现纳米级电子元件的加工。

扫描隧道显微技术的原理是：与悬臂梁相连的细小针尖与导电的工件表面接触，在超真空环境下，使细小金属针尖与导电材料表面保持很小的距离（小于 1nm），当在针尖和零件接触处施以偏置电压时，在针尖和工件间隙处将产生隧道电流（图 7-2）。通过对电流进行监测可获得反馈信息。反馈电流通常在 10pA 到 10nA 之间。所施加的电压应能降低能量势垒使电子可在间隙中形成隧道。探针针尖需经过化学抛光或研磨，通常材料为钨、铱或铂铱合金。

Epitaxy, Liquid Phase Epitaxy, Vapor Phase Epitaxy etc.). There is no raw material waste in bottom-up fabrication processes because they do not remove materials.

2. Typical nanofabrication techniques

(1) Traditional nanomachining Traditional nanomachining refers only to the "top-down" nanofabrication approach. It is more concerned with the precision rather than the characteristic size of the product. So nanometric machining is defined as the material removal process in which the dimensional accuracy of a product can be achieved is 100nm or better, even towards 1nm level. Nanometric machining can be classified into four categories:

1) Deterministic mechanical nanomachining. This method is to utilize fixed and controlled tools, which can specify the profiles of 3-D components by a well defined tool surface and path. The method can remove materials in amounts as small as tens of nanometers. It includes typical diamond turning and nanogrinding etc.

2) Loose abrasive nanomachining. This method uses loose abrasive grits to remove small amount of materials. It consists of polishing, lapping and honing etc.

3) Non-mechanical nanomachining. It comprises Focused Ion Beam Machining, micro-EDM, and Excimer Laser Machining.

4) Lithographic method. It employs masks to specify the shape of the product. Two dimensional shapes are the main outcome. Severe limitations occur when 3-D products are attempted. It mainly includes X-ray lithography, LIGA, Electron Beam Lithography (EBL).

(2) Nanofabrication of using manipulative techniques Nanofabrication of using manipulative techniques is a promising way of producing nano-based electronic components using processes such as Scanning Tunneling Microscopy (STM), Atomic Force Microscopy (AFM), etc.

STM is a process that relies on a very sharp tip connected to a cantilever beam to touch a surface composed of atoms that is electrically conductive. It is a process that is conducted in an ultra-high vacuum where a sharp metal tip is brought into extremely close contact (less than 1nm) with a conducting surface (Fig.7-2). A bias voltage is applied to the tip and the sample junction where electrons tunnel quantum-mechanically across the gap. A feedback current is monitored to provide feedback and is usually in the range between 10pA and 10nA. The applied voltage is such that the energy barrier is lowered so that electrons can tunnel through the air gap. The tip is chemically polished or ground, and is made of materials such as tungsten, iridium, or platinum-iridium.

STM can be used as not only a measuring tool but also as the tool to make nano-structures and nano-materials. However, since STM is controlled by detecting a tunneling current between the probe and the sample, the measured sample is limited to conductive materials. To overcome this problem, Atomic Force Microscopy (AFM) was developed, which is a powerful technique that can be used to fabricate structures at the nanoscale.

AFM detects minute forces between the probe and the sample instead of tunneling current. As shown in Fig.7-3, a sharp tip connected to a cantilever beam is brought into contact with the surface of the sample. The surface is scanned, causing the beam to deflect, which is monitored by a scanning laser beam. The tip is micro-machined from materials such as silicon, tungsten, diamond, i-

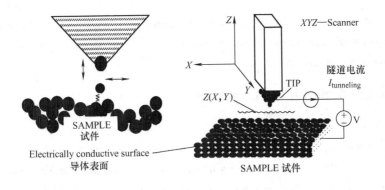

Fig. 7-2 Principle of Scanning Tunneling Microscopy (STM)
扫描隧道显微镜原理

扫描隧道显微镜不仅是一种测量工具，而且也是制备纳米结构和材料的工具。然而，扫描隧道显微术是利用探针和工件间的隧道电流来工作的，因此被测工件仅限于导体材料。为了解决该问题，出现了原子力显微镜。原子力显微术是一种用于生产纳米级结构的强有力方法。

与扫描隧道显微镜利用隧道电流工作不同，原子力显微镜是利用探针和工件间微弱的力工作的。如图 7-3 所示，探针的细小尖端与悬臂梁相连，尖端和工件表面接触。探针尖端在工件表面运动时，由于探针和工件间微弱力的作用使得悬臂梁偏斜。扫描激光束可用于检测梁的偏斜程度。探针尖端由硅、钨、金刚石、铁、钴、钐、铱或钐钴永磁体等材料制成。原子力显微镜的优点在于它可在标准气压下测量和加工任何类型的材料。

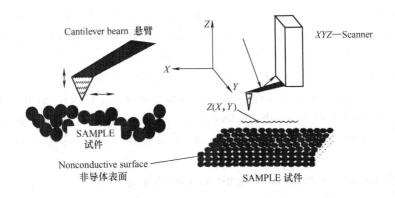

Fig. 7-3 Atomic force microscopy (AFM) 原子力显微镜

ron, cobalt, samarium, iridium, or cobalt-samarium permanent magnets. The advantage of AFM is that any type of materials can be measured and machined and the operation can be conducted under normal atmospheric conditions.

7.2.3 High-speed machining technology

1. Introduction

High-speed machining (HSM) or High-speed cutting is an advanced production technology with great potential. Most of the time, any process which employs a spindle that can operate at high rotation speed is labeled HSM, without any relation to the benefits produced. In fact, like the introduction of NC and later CNC, HSM is a revolutionary process that will change the way metal removal is conceived.

HSM was first proposed by Carl. J. Salomon in 1931. Based on his metal cutting studies on steel, non-ferrous and light metals at cutting speeds of 440 m/min (steel), 1,600 m/min (bronze), 2,840 m/min (copper) and up to 16,500 m/min (aluminum), he assumed that under conventional machining conditions, machining temperatures rise with the increase of cutting speed, while from a certain cutting speed upward machining temperatures start dropping again. Fig. 7-4 shows the schematic diagram of high-speed machining hypothesis put forward by Salomon.

Salomon's fundamental research showed that for any material there is a certain range of cutting speeds where machining cannot be made due to excessively high temperatures because the temperatures exceed melting points of any cutting tool material (called "the death valley"). For this reason, HSM can also be termed as cutting speeds beyond that limit.

HSM is a relative term from the viewpoint of the materials being machined because of the vastly different speeds at which different materials can be machined with acceptable tool life. For example, it is easier to machine aluminum at 1,800m/min than titanium at 180m/min. And a cutting speed of 500m/min is considered high-speed machining for cutting alloy steel whereas this speed is considered conventional in cutting aluminum. At present, the common-used two definitions of HSM were given by the International Academy for Production Engineering (CIRP) and the Institute of production Engineering and Machine Tool (PTW) at the Darmstadt University of Technology Darmstadt. The CIRP termed machining processes performed at the speed of 500 – 7000m/min as HSM. The PTW defines HSM as being such that conventional cutting speeds are exceeded by a factor of 5 to 10. In Ultra-high speed machining, the cutting speeds should be about 10 times higher than those used in conventional processes. The high cutting speed ranges in machining various materials can be seen in Fig. 7-5.

2. Advantages and limitations of high-speed machining (HSM)

HSM is characterized by very high cutting speed, small depths of cut and high feed rates which give HSM many advantages when compared with conventional machining processes.

1) In HSM, cutting tool and workpiece temperature are kept low which gives a prolonged tool life in many cases. On the other hand, the cuts are shallow and the engagement time for the cutting edge is extremely short. Research shows 90% – 95% cutting heat produced in HSM would be taken

7.2.3 高速加工技术

1. 引言

高速加工或高速切削是一种极具潜力的先进加工技术。很多情况下，人们不管产生的效益如何，把任何主轴高速回转的切削均称为高速切削。事实上，正如数控和计算机数控的出现，高速切削也是一项不断发展的技术，它将改变金属切削加工在人们心目中的印象。

高速切削最早由 Carl Salomon 于 1931 年提出。在分别对钢、非铁金属以及轻合金在不同切削速度下（钢：440m/min；青铜：1600m/min；黄铜：2840m/min；铝：16500m/min）进行大量切削实验的基础上，Salomon 指出：在常规切削范围内，切削温度随切削速度的增大而提高，但当切削速度增大到某一数值以上，切削温度会随切削速度的增大而下降。图 7-4 为 Salomon 高速切削理论示意图。

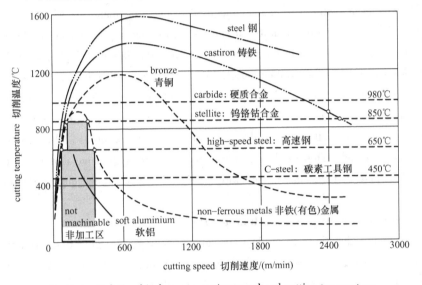

Fig. 7-4　Relationship between cutting speed and cutting temperature
切削速度与切削温度的关系

Salomon 的研究还表明，对于任何材料存在一个速度范围。在这一范围内，切削温度超过了任何刀具材料的熔点，造成加工无法进行（该范围被称为"死谷"）。因此高速切削也可定义为切削速度大于该极限值的切削加工。

从加工材料的观点出发，高速切削是一个相对的概念，因为不同材料要保证刀具有足够寿命所采用的切削速度大不相同。例如，以 1800m/min 的线速度加工金属铝和以 180m/min 的线速度加工金属钛，两者相比之下，前者更容易实现。而 500m/min 的切削速度对于合金钢加工来讲属于高速切削，但对于铝材加工来讲则属于传统加工范畴。目前沿用的两种高速切削定义分别是由国际生产工程协会（CIRP）和达姆施塔特工业大学的生产工程与机床研究所（PTW）提出的。CIRP 提出：以线速度 500～7000m/min 进行的切削为高速切削。PTW 则提出：以高于 5～10 倍普通切削速度进行的切削称为高速切削，而超高速切削则是以比常规切削速度高 10 倍左右的速度进行的切削。各种材料高速切削的速度范围如图 7-5 所示。

away by chips flowing at high rates. And heat accumulated in the workpiece is not so much. Therefore, HSM is suitable to machine workpieces that are prone to deform in high temperature.

2) In HSM, the low cutting force gives a small and consistent tool deflection. This is one of the prerequisites for a highly productive and safe process. As the depths of cut are typically shallow in HSM, the radial forces on the tool and spindle are low. This saves spindle bearings, guide-ways and ball screws.

3) Since both cutting speeds and feed rates in HSM are 5 – 10 times higher than traditional machining, material removal rates can be improved by 3 – 6 times higher and processing time of a workpiece can be reduced to 1/3 correspondingly. In addition, owing to the high surface quality produced in HSM it is possible in many cases to eliminate subsequent finish machining entirely or in part.

4) HSM provides a better way to machine hard-to-machine materials with high productivity, less tool wear and better machining quality. It can be used to accomplish plastic working of hard and brittle materials. Also, it is suitable for machining ductile materials with high surface integrity.

The problems existed in the application of HSM have to do with the material and the geometry of the workpiece to be machined. The common disadvantages are claimed to be the following aspects. As the high acceleration and deceleration, rapid start and stop of the spindle would speed up the wear of guideways, ball screws and spindle bearings, the spindles and controllers with high performances are required. This kind of machine tool is quite expensive and has higher maintenance costs. Excessive tool wear associated with HSM calls for advanced cutting tool materials and coatings.

3. Applications of high-speed machining (HSM)

HSM, with many advantages in productivity and efficiency, currently finds its way into almost every field of machining. Let us compare the kinematics background of turning and milling procedures. In turning, where the kinematic mechanism is based on a rotating workpiece, it is much more difficult to cope with the huge, quickly rotating weights and safe workpiece chucking at high speeds, and thus the conditions for the use of HSM are considerably less appropriate than in milling in general. Consequently, high speed turning does not have wide industrial application yet.

Unlike milling, in which the short chips are beneficial for high speed machining, drilling at high cutting speeds produces long chips which have to be taken out of the hole. It is impossible to quickly discharge the chip, as it is in milling. During drilling, the heat is absorbed by the drill and the wall of the drilled hole, so high speed drilling is impossible without internal cooling of the solid carbide drill that allows the cutting fluid to go directly to the contact positions between the cutting edge and the material machined. Drilling is defined as a high speed procedure if the cutting parameters exceed the conventional ones by a factor of at least 2.

Consequently, the main focus of HSM is on the milling processes. The following considerations are dedicated to milling technology. HSM will only succeed if there is a perfect interaction among machine tool, tool, workpiece and tool clamping techniques, cutting fluids, cutting parameters, such as spindle speeds, cutting speeds and feeds.

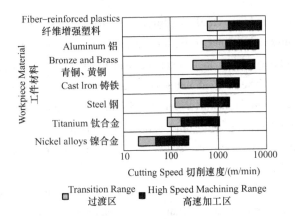

Fig. 7-5 High cutting speed ranges in machining various materials 各种材料切削速度范围

2. 高速切削的优缺点

高速切削的切削速度高，切削深度小，进给速度高，因而具有许多传统加工所不具有的优点。

1）多数情况下，高速切削时刀具和工件的温度较低，因而可以延长刀具寿命。另一方面，高速切削的切削深度小，切削刃切削时间短。有研究表明，高速加工中90%到95%的热量可被高速流走的切屑带走，工件内部累积的热量很少。因此，高速切削适合于加工易于热变形的零件。

2）高速切削中切削力小，刀具变形一致，变形量也小。这是实现高效、安全生产的必要条件。另外，由于高速切削的切削深度浅，作用在刀具和主轴上的径向力小，对主轴轴承、导轨和滚珠丝杠非常有利。

3）由于高速切削中切削速度和进给速度比传统切削参数高5~10倍，因此材料去除率可提高3~6倍，加工时间也可相应缩减到原来的1/3。此外，加工出的工件表面质量高，因此很多情况下可完全或部分消除后续的精加工。

4）高速切削可用于难加工材料的切削，且生产率高，刀具磨损小，零件加工质量高。采用高速切削可以实现硬、脆材料的塑性去除加工，以及韧性材料的高表面完整性加工。

高速切削存在的问题与所要加工工件的材料和几何形状有关。常见的缺点有几个方面。加工中由于加速度大以及主轴的快速起停加速了导轨、滚珠丝杠和主轴轴承的磨损，需要高性能的机床主轴及控制器，因此机床价格昂贵，且维修成本高。高速切削刀具会发生过度磨损，因此要求更优的刀具材料和刀具涂层。

3. 高速切削的应用

高速切削在生产率和切削效率方面具有很大优势，目前几乎在各个领域均有应用。下面对车削和铣削的运动学原理进行比较。车削加工要求工件旋转。若高速车削大型零件，要做到工件高速旋转很困难，而且很难保证工件的安全装夹。为此，与铣削相比，车削一般不太

The main application fields of HSM are three industry sectors. The first category is industry which deals with machining light alloys to produce automotive components, small computer parts or medical devices. This industry needs fast metal removal because the technological process involves many machining operations. The second category is aircraft industry which involves machining of long aluminum parts, often with thin walls. The third industry sector is the die and mould industry. Besides, Milling of electrodes in graphite and copper is another excellent area for HSM. Graphite can be machined in a productive way with Ti (C, N) or diamond coated solid carbide end mills.

4. High-speed machining machines

HSM is only possible if all elements of the system of machine tool, tools, workpiece are optimally matched. The strict kinematic and dynamic requirements that must be fulfilled by the corresponding machine design demand modular design approaches with innovative solutions in machine manufacturing, e. g. frequently granite castings for the frames, and advanced drive and control technology.

For this reason, the implementation of high speed milling technology in industry resulted in a wide variety of high speed machining centers and machines, as required for different machining tasks, such as light alloy machining in the aerospace industry, with cutting requiring a great expenditure of energy, or finishing of hardened steel dies in the tool-and-die-making industry. Some of the significant assemblies and components of a high speed machine system will be discussed briefly.

Motorized spindles with ball bearing have been proven effective as main spindles because of their good dynamic performance. Among the feed drives, the electromechanical servo linear motors are dominant, but the linear direct drives, which enable much higher feed rates (100m/min) and acceleration values of 5 to $10g$ (50-100m/s^2) are only at the experimental stage. For small-and medium-sized HSM machines, the machine frames are made of granite, whose damping capacity is 6 to 10 times higher than grey cast iron, and whose thermal expansion coefficient is only 1/3 to 1/5 of steel. For large-sized machines, the great stiffness required has to be achieved by means of appropriate welded steel constructions. In this domain, an innovative designs based on parallel kinematic mechanisms is carried out and the so called non-Cartesian axis concepts (hexapods, tripods) are introduced, both in terms of structural stiffness and thermal stability.

Concerning the axis allocation, the three major axes X, Y and Z are, as a rule, dimensioned as Cartesian linear axes. In addition, circular and swiveling axes are implemented for the transition from 3-to-5-axis milling in different variants. Non-Cartesian axis concepts (e. g. hexapods) are particularly suitable for 5-axis milling due to the corresponding control technology, extreme feed rates (100m/min) and acceleration values up to $3g$.

5. High-speed cutting tools

HSM demands special qualification in the cutting tool material, tool design and dimensioning. As a matter of fact, successful HSM depends most of all on the selection of the adequate cutting tool material. Thus, the use of PCD and diamond coated cemented carbide is standard for aluminum machining. For the machining of cast iron and partially hardened steel, CBN cutting material is used, and so are newer developments in fine grain and superfine grain cemented carbides and cermets,

适合于高速切削。因而，高速车削仍然没能得到广泛的应用。

与高速铣削产生的短切屑相比，高速钻削产生的切屑较长，这些切屑需要从加工孔中排出。对于钻削，切屑不能像在铣削中一样可方便、及时地排出。因此，高速钻削需要对硬质合金钻头进行内部冷却，使切削液直接流向切削刃和加工材料的接触部位。高速钻削是切削参数增加到常规钻削加工参数两倍以上的钻削加工。

基于上述几点，高速切削主要应用于铣削加工。以下均专门针对铣削加工而言。高速铣削只有在机床、刀具、工件和刀具装夹、切削液、切削参数（如主轴转速、切削速度和进给量）等各方面相互协调的情况下才能顺利进行。

高速切削主要应用于三大领域。一是轻合金的加工，可加工汽车零件、计算机用小零件或医疗装置用小零件。这些零件的加工工序较多，需要较高的加工速度。二是航空工业，用于加工尺寸较长的薄壁铝件。三是模具行业。此外，石墨或铜电极也非常适合采用高速铣削。借助于覆有TiC（N）复合粉或金刚石涂层的整体硬质合金立铣刀可实现高生产率的石墨材料高速铣削。

4. 高速切削机床

实现高速切削需要优化配置由机床、刀具、工件组成的加工系统的各个模块。因此，为实现不同设计所提出的严格的运动学和动力学要求，需要在机床生产中采用具有创新方案的模块化设计方法，如采用花岗岩浇注的床身，引入先进的驱动和控制技术等。

高速铣削技术在工业中的应用促使了多种类型的高速加工中心和设备的出现。它们是为了满足各种加工任务而研发的。如为实现航空工业能量消耗较大的轻合金的加工以及工、模具行业淬硬钢模具的精加工出现了相应的高速加工设备。以下将介绍高速加工设备的一些重要的功能部件。

实践证明，装有球轴承的电主轴以其优良的动力学性能成为高速加工设备有效的主轴系统。而在进给驱动方面，直线电动机的优越性最为显著。直线电动机驱动可实现很高的进给速度（100m/min）和加速度（$5\sim10g$），但目前仍处于实验阶段。对于中小型的高速加工机床，机床床身由花岗岩制成。花岗岩床身的吸振能力是灰铸铁床身的$6\sim10$倍，而热膨胀系数仅为钢制床身的$1/3\sim1/5$。对于大型机床，为满足所需刚度要求，必须采用合适的焊接钢结构。对于这类机床，为达到结构刚度和热稳定性要求，引入了基于并联运动机构的创新设计方案以及所谓的非直角坐标系运动轴（六自由度并联，三自由度并联结构）的概念。

在轴的分配方面，通常X、Y、Z为直线坐标轴。此外，三到五轴铣床还有不同数目的回转轴。由于其相应的控制技术、极高的进给速度和高达$3g$的加速度，非直角坐标系运动轴的概念（如六自由度并联）非常适合于五轴铣削。

5. 高速切削刀具

高速切削要求刀具材料、结构与尺寸都满足要求。事实上，实现高速切削最重要的任务就是选择恰当的刀具材料。聚晶金刚石刀具和金刚石涂层硬质合金刀具适于铝件的加工。而加工铸铁和部分淬硬钢则采用立方氮化硼刀具以及新研制的细晶粒和超细晶粒硬质合金刀具和耐高温的晶须增强陶瓷刀具。

设计高速切削刀具时，两个问题非常关键：刀具的动平衡性和容许离心力。如果刀具不平衡，在高速情况下产生的离心力将超过切削力。然而，刀具不平衡的后果仅在不平衡度很

each with appropriate coatings, as well as highly heat resistant whisker-reinforced ceramic cutting materials.

Concerning tool design, in HSM, the following two issues are crucial: tool imbalance and acceptable centrifugal forces. Tool imbalances may generate forces at high speeds that exceed the cutting forces of the cutting procedure. However, definite consequences can only be found out at very high imbalance values. The expected negative effect on tool life and the load of the spindle bearing have to be considered when specifying the imbalance quality levels. Milling cutters for HSM have to be designed in such a way that the tool body or the chucking elements do not break and even rotate at speeds in the upper range of the speed limit. Consequently, newly developed mills for HSM are tested on centrifugal test benches.

During all centrifugal tests, the fracture of a clamp screw for an indexable insert was the reason for the breakdown. However, according to the assessment by the tool manufacturer, this is considered less dangerous than the tool body bursting. In any case, the passive safety of the HSM machines should be high, since operating errors may also result in tool fracture.

7.2.4 Modern non-traditional machining technology

1. Introduction

Machining processes can be mainly divided into two groups, conventional machining processes and non-traditional (or unconventional) machining (NTM) processes. The former type involves cutting with single-point or multipoint cutting tools, each with a clearly defined geometry, and abrasive processes, such as grinding. Conventional machining techniques dominated the manufacturing industry until the mid-1900s. Their total dominance, however, has been reduced with the introduction of numerous new commercial (nontraditional) manufacturing techniques since the 1950s, ranging from ultrasonic machining of metal dies to the nano-scale fabrication of optoelectronic components using a variety of lasers.

NTM processes were invented under the following situations where traditional machining processes are unsatisfactory or uneconomical.

1) Innovative materials such as super-alloys, composites, ceramics and many other advanced materials which are difficult to or cannot be processed by traditional machining methods require new manufacturing technologies. These materials are difficult to be fabricated probably because they are too hard, too brittle, or too flexible to resist cutting forces.

2) Production of parts with complex internal or external profiles or small holes is another impetus of NTM.

3) Demanded requirements for surface finish and tolerances can not be satisfied because of undesirable or unacceptable temperature rise, residual stresses and other factors produced in conventional machining process.

NTM refers to processes in which nontraditional energy transfer mechanism and/or non-traditional media for energy transfer are involved. Compared with traditional machining processes, NTM processes show the following features:

大的情况下才能被发现。在确定刀具平衡精度等级时要考虑到刀具不平衡度对刀具寿命和主轴轴承所受载荷的影响。为了确保铣刀刀体和刀片夹紧装置以极限速度高速旋转时不会发生破碎，需要对新研制的高速铣刀在离心实验台上进行测试。

在所有离心实验中，可转位刀片紧固螺钉的断裂是实验停止的原因。然而，据刀具生产厂家的评估，这种情况与刀体的爆裂相比，危险程度较低。无论如何，高速切削机床要保证高的被动安全性，因为误操作也会造成刀具的破碎。

7.2.4 现代特种加工技术

1. 引言

加工工艺可分为两大类：传统加工和特种加工。传统加工包括采用具有特定几何形状的单刃或多刃刀具进行的加工以及采用磨料的加工工艺（如磨削加工）。直到20世纪中期，传统加工工艺在制造业中一直占主导地位。从20世纪50年代以来，这种主导地位已经随着各种新的特种加工技术的出现而有所下降。特种加工技术包括金属型的超声波加工以及使用各类激光进行的光电元件的纳米加工等。

特种加工的出现是由于传统加工不能满足或不能经济地满足以下条件：

1) 传统加工方法不能或难以加工的新材料的应用需要新的加工技术。这些材料包括超级合金、复合材料、陶瓷以及其他高性能材料。难以加工的原因在于材料硬度高、脆性高或太过柔韧而难以承受加工中产生的切削力。

2) 具有复杂内、外部轮廓的零件以及小孔的加工是推动特种加工发展的另一个因素。

3) 传统加工受加工中产生的高温、残余应力以及其他因素的影响难以获得高的表面质量和尺寸精度。

特种加工是指那些采用非常规的能量转换机理或非常规介质进行能量转换的加工工艺。与传统加工相比，特种加工具有以下特点：

1) 去除材料时有切屑或无切屑。例如，磨料喷射加工中产生微切屑，而电解加工则是通过原子尺度的电化学溶解作用去除材料的。

2) 加工中可不需要刀具。如激光束加工中，加工由激光完成。

3) 刀具材料可不必比工件材料硬度高。

4) 特种加工中机械能不是必要的，各种能量均可用于加工。

特种加工通常根据采用的能量类型分类，包括采用机械能的、电能的、热能的和化学能的特种加工工艺。超声波加工（Ultrasonic Machining, USM）、磨料喷射加工（Abrasive Jet Machining, AJM）、水射流加工（Water Jet Machining, WJM）和磨料水射流加工（Abrasive Water Jet Machining, AWJM）属于机械能加工工艺。电解加工（Electrochemical Machining, ECM）是采用电化学能的代表工艺。电解磨削（Electro Chemical Grinding, ECG）和电解钻孔（Electro Jet Drilling, EJD）是另外两种电化学能加工工艺。电火花加工（Electro-Discharge Machining, EDM）、激光束加工（Laser Beam Machining, LBM）和电子束加工（Electron Beam Machining, EBM）是三种主要的热能加工工艺。化学加工工艺包括化学铣削（Chemical Milling, CHM）和光化学铣削（Photochemical Milling, PCM），它们通过化学反应从金属表面去除材料。

1) Material removal may occur with chip formation or even no chip formation may take place. For example, in Abrasive Jet Machining (AJM), chips are of microscopic size and in case of Electrochemical Machining (ECM) material removal occurs due to electrochemical dissolution at the atomic level.

2) There may not be a physical tool present. For example in Laser Jet Machining (LJM), machining is carried out by laser beam.

3) The tools need not be harder than the workpiece material.

4) Mostly, NTM processes do not necessarily use mechanical energy to provide material removal. They use different energy domains to provide machining.

NTM techniques have often been classified according to the principal type of energy utilized to remove material—mechanical, electrical, thermal, and chemical. Ultrasonic Machining (USM), Abrasive Jet Machining (AJM), Water Jet Machining (WJM), and Abrasive Water Jet Machining (AWJM) belong to the mechanical processes. Electrochemical Machining (ECM) is the primary representative of Electrochemical processes. Electro Chemical Grinding (ECG) and Electro Jet Drilling (EJD) are another two ECM processes. Electro-discharge Machining (EDM), Laser Beam Machining (LBM) and Electron Beam Machining (EBM) are the three primary thermal energy-based processes. Chemical processes, including Chemical Milling (CHM) and Photochemical Milling (PCM), etc. refer to the removal of material from metal surfaces through purely chemical reactions.

2. Modern non-traditional machining processes

In the following section we will review several modern NTM processes.

(1) Ultrasonic Machining (USM) USM is a mechanical material removal process which is commonly used to produce holes and cavities in hard and brittle workpieces. In USM process, hard, brittle particles contained in a slurry are accelerated toward the surface of the workpiece by a machining tool oscillating at ultrasonic frequency. Through repeated abrasions, the tool machines a cavity of a cross section identical to its own, as shown in Fig. 7-6.

The process of USM can be expressed as follow: a tool which is shaped converse to that of the desired hole or cavity and made of brass, copper, or stainless steel is positioned so that a small cutting gap is formed. The tool is vibrated at high frequency (about 20kHz) with an amplitude of around $15-50\mu m$ over the workpiece in an abrasive slurry. Material removal occurs when the abrasive particles suspended in the slurry-filled gap are struck and propelled into the workpiece by rapidly vibrating tool. Tiny crack initiation will be formed at each abrasive particle impact site followed by crake propagation and brittle fracture of the material. The slurry will carry away the debris from the cutting area. The tool is gradually moved down maintaining a constant gap between the tool and the workpiece surface. The process continues to finish machining of the workpiece.

The basic mechanical structure of an USM, as shown in Fig. 7-7, is very similar to a drill press. The major components of Ultrasonic machines are slurry delivery and return system, feed mechanism, transducer, and horn or concentrator.

USM is primarily used for machining hard and brittle work material (dielectric or conductive) such as metallic alloys, glass, semiconductors, ceramics, carbides, precious stones etc. A standard

2. 现代特种加工工艺

以下简单介绍几种现代特种加工工艺。

（1）超声波加工（USM） 超声波加工利用机械能去除材料，常用于在硬、脆材料上进行孔和型腔的加工。超声波加工借助于超声频振动的工具使磨料悬浮液高速撞击、抛磨被加工表面而去除材料。借助于不断地撞击与抛磨可加工出与工具自身形状一致的型腔，如图7-6所示。

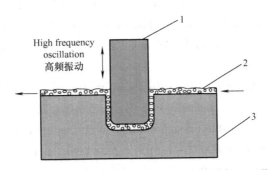

Fig. 7-6 Ultrasonic machining process 超声波加工工艺
1—Tool 工具　2—Slurry 磨料悬浮液　3—Workpiece 工件

超声波加工时，首先定位工具，使工具和工件间形成很小的间隙。工具由铜或不锈钢制成，其形状与被加工孔或型腔的形状一致。工具置于磨料悬浮液中，并在工件上方以超声频（约20kHz）进行小幅（15~50μm）振动。由于工具的高速振动，工作液中的磨料颗粒撞击工件表面，并在撞击点处产生微小裂纹源。随后，裂纹不断扩展并最终使工件材料发生脆性断裂。从工件上脱落的材料微粒被工作液冲走。之后，工具逐渐向下运动以保持工具与工件表面间的间隙。不断重复以上过程直至完成零件的加工。

超声波加工设备的结构与钻床相似。图7-7所示为其基本结构，包括磨料工作液循环系统、进给机构、超声振动系统和变幅杆。

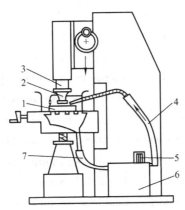

Fig. 7-7 Typical ultrasonic machine 典型超声波机床
1—Workpiece 工件　2—Horn 变幅杆　3—Transducer 换能器
4—Slurry to machining zone 流向加工区的磨料悬浮液　5—Slurry pump 悬浮液泵
6—Slurry tank 磨料悬浮液箱　7—Return slurry 返回的磨料悬浮液

machine process will cause brittle material to crack and break. Ultrasonic machining uses thousands of tiny impacts and very little pressure to move material. This rarely results in a break, even in very brittle materials.

The tool used in ultrasonic machining is different from the ones used in a standard process. It must be highly wear resistant, usually made of tough, strong and ductile materials like low-carbon steels and other ductile metallic alloys. This allows the abrasives to impact the tool, but not damage it.

The abrasives used in the slurry are the same as those used in most grinding wheels. Boron carbide (B_4C), silicon carbide (SiC) and aluminum oxide (Al_2O_3) are widely used mostly due to their hardness and low cost. Occasionally, diamond and cubic boron nitride are used to work the hardest materials.

USM competes with traditional processes based on its strength of machining hard and brittle materials as well as on the workpiece geometry complexity. Surface finish in USM can be an order of magnitude better than that achievable through milling. With fine abrasives, the tolerance of 0.0125 mm or better and surface finish of $Ra0.2-1.6 \mu m$ can be held. In addition, USM produces little or no sub-surface damage and no heat-affected zone. The quality of an ultrasonic cut provides reduced stress and a lower likelihood of fractures that might lead to device or application failure over the life of the product.

Its main limitations are low material removal rate (typically $0.8 cm^3/min$), rapid tool wears and limited machining area and depth. However, various investigations have shown that higher material removal rates (up to 6mm/min) can be achieved with increased grain size (up to an optimal diameter) and concentration of abrasives in the slurry, and increased amplitude and frequency of the oscillations of the tool.

(2) Abrasive Jet Machining (AJM) AJM removes material through the mechanical action of a focused stream of abrasive-laden gas. In AJM, abrasive particles are made to impinge on the work material at a high velocity. The jet of abrasive particles is carried by dry air, nitrogen (N_2) or carbon dioxide (CO_2). The high velocity stream of abrasive is generated by converting the pressure energy of the carrier gas or air to its kinetic energy and hence high velocity jet. The nozzle directs the abrasive jet in a controlled manner onto the work material, so that the distance between the nozzle and the workpiece and the impingement angle can be set desirably. The high velocity abrasive particles produce sufficient force to remove the material by micro-cutting action as well as brittle fracture of the work material. Fig. 7-8 schematically shows the material removal process.

AJM is different from standard shot blasting, as in AJM, smaller-diameter abrasives are used and the parameters can be controlled more effectively providing better control over product quality. In AJM, generally, the abrasive particles of around $50 \mu m$ grit size would impinge on the work material at velocity of 200m/s from a nozzle of I.D. of 0.5 mm with a stand off distance of around 2mm. The kinetic energy of the abrasive particles would be sufficient to provide material removal.

Fig. 7-9 shows the components of an AJM system. Carrier gas is issued from a gas cylinder or air compressor and passes through a pressure regulator to obtain the desired working pressure. The gas is then passed through an air dryer and a series of filters to remove any residual water vapor and

超声波加工主要用于加工硬而脆的材料（导体或非导体），如合金、玻璃、半导体、陶瓷、碳化物、宝石等。一般的加工方法通常会使脆性材料产生裂纹甚至发生断裂，而超声波加工采用无数小冲击和非常小的压力来去除材料，即使加工非常脆的材料也几乎不会产生断裂和裂纹。

与普通工艺中使用的工具不同，超声波加工中的工具必须具有高的耐磨性。工具一般由韧性好、强度高、延性好的材料，如低碳钢和其他韧性金属合金。这样，磨料虽对工具有冲击，但不会损坏工具。

悬浮液中的磨料和普通砂轮中所用材料相同。常用磨料为硬度高而价格低的碳化硼 B_4C、碳化硅 SiC 和氧化铝 Al_2O_3。加工特硬材料有时也会使用金刚石和立方氮化硼。

与传统加工相比，超声波加工在加工脆性材料和复杂形状工件方面有较大优势。加工工件的表面粗糙度可比铣削加工好一个量级。用小粒度的磨料，加工公差可达 0.0125mm 以下，表面粗糙度可达 Ra0.2~1.6μm。此外，超声波加工不会或很少会对材料内部产生破坏，也不会产生热影响区。且加工中产生的应力小，不易产生裂纹。而应力和裂纹通常是导致产品在使用中失效的主要原因。

其不足之处在于材料去除速度较慢（一般为 $0.8cm^3/min$）、工具有磨损以及加工区域与深度受限制。然而，很多研究表明采用较大磨料粒度（达到最优直径）和磨料浓度的磨料悬浮液，并提高工具的振幅和频率也可获得较高的材料去除速度（6mm/min 以上）。

（2）磨料喷射加工（AJM） 磨料喷射加工利用磨粒射流高速冲击工件表面，通过机械作用去除材料。磨粒射流由磨粒和干燥的空气、氮气或二氧化碳气体混合形成。高压气体的压力能转变为动能形成高速磨粒流，又称高速射流。喷嘴将磨粒射流以一定的喷射角度喷射在工件上，同时使喷嘴与工件间保持合适的距离。高速运动的磨粒在工件上产生足够大的力，在显微切削作用下去除材料或使工件材料发生脆性断裂而被去除。图 7-8 所示为磨料喷射加工简图。

与喷丸加工不同，磨料喷射加工采用的磨料粒度小，而且加工参数可进行有效控制，因而可获得更好的加工质量。一般来讲，粒度为 50μm 左右的磨粒以 200m/s 的速度从内径为 0.5mm 的喷嘴中喷出，撞击 2mm 距离处的工件进行加工。此时磨粒的动能足以达到去除材料的目的。

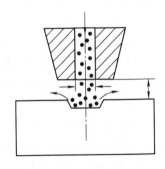

Fig. 7-8　Abrasive Jet Machining（AJM）磨料喷射加工

oil vapor or particulate contaminant. Next, the carrier gas enters a closed chamber known as the mixing chamber. The abrasive particles enter the chamber from a hopper through a metallic sieve. The sieve is constantly vibrated by an electromagnetic shaker. The mass flow rate of abrasive entering the chamber depends on the amplitude of vibration of the sieve and its frequency. The abrasive particles are then carried by the carrier gas to the machining chamber. The machining is carried out as high-velocity abrasive particles are issued from the nozzle onto a workpiece traversing under the jet.

The most commonly used abrasives are aluminum oxide (Al_2O_3) and silicon carbide (SiC) in sizes from $10\mu m$ to $50\mu m$. Occasionally, glass beads and sodium bicarbonate are used for surface finishing and surface cleaning respectively. Their shape can be either irregular or spherical. Nozzles are made of tungsten carbide or sapphire, which is very expensive. The internal diameter of nozzles generally ranges from 0.2 to 0.8 mm.

The important machining characteristics in AJM are material removal rate (MRR), machining accuracy and life of the nozzle. The MRR of AJM is affected by four major process variables, nozzle-to-workpiece distance, abrasive flow rate, gas pressure and the type of abrasive being used.

Flexibility, low heat production and ability to machine hard and brittle materials are the main advantages of AJM. Its flexibility owes from its ability to use hoses to transport the gas and abrasive to any part of the workpiece. AJM does not heat the workpiece material. So it can be used for machining heat sensitive materials without worrying about machining stresses and thermal damages.

Limitations of AJM are: MRR is rather low; abrasive particles tend to get embedded particularly if the work material is ductile; tapering occurs due to flaring of the jet; environmental load is rather high.

AJM is commonly used for surface cleaning of castings and forgings, removing rust and film for steel plates, pre-treating for painting and plating, drilling and cutting small sections of glass, ceramics or hardened metals, deburring and deflashing, and cutting small holes or slots etc.

(3) Laser Beam Machining (LBM) Laser Beam Machining (LBM) is a thermal material removal process that utilizes a high-energy coherent light beam to melt and vaporize particles on the surfaces of metallic and nonmetallic workpieces. All laser machining applications make use of a laser beam after it has passed through a lens, as illustrated in Fig. 7-10. This causes the beam to converge to a focal point at a precise distance from the lens. When focused and concentrated in this manner, a laser beam containing less than 100W power is capable of vaporizing or melting any known material.

The mechanism of material removal during LBM includes different stages such as melting, vaporization, and chemical degradation (chemical bonds are broken which causes the materials to degrade). When a high energy density laser bream is focused on the workpiece surface, the thermal energy is absorbed which heats and transforms the work volume into a molten, vaporized or chemically changed state that can easily be removed by flow of high pressure assist gas jet.

图 7-9 所示为磨料喷射加工系统的组成。来自气缸或空气压缩机的气体，经压力阀 2 调压获得必要的工作压力后，流经干燥器 3 及一系列过滤器 4 以去除气体中残余的水分、油分或颗粒状杂质。然后，气体进入密闭的混料室 13。同时，磨料经由金属筛从料斗进入混料室。金属筛在振动器作用下不断振动。磨料喂入量由振动器的振幅和频率控制。这样，混合后的磨料流进工作室，经喷嘴 8 喷出后用在横向运动的工件 9 上进行加工。

常用磨料是粒径为 10 ~ 50μm 的刚玉和碳化硅。有时也会用到玻璃珠或碳酸氢钠以进行表面抛光或表面清理。磨料形状可为球状或不规则形状。喷嘴材料为碳化钨或昂贵的蓝宝石，内径约为 0.2 ~ 0.8mm。

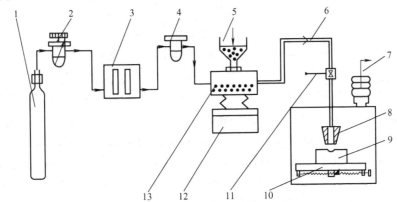

Fig. 7-9　Components of an AJM system 磨料喷射加工系统组成
1—Gas supply 气源　2—Pressure control valve 压力阀　3—Drier 干燥器　4—Filter 过滤器
5—Abrasive feeder 喂料斗　6—Abrasive mixed with carrier gas 磨料混合气　7—Exhaust 废气
8—Nozzle 喷嘴　9—Workpiece 工件　10—Table 工作台　11—Electro-magnetic on-off valve 电磁阀
12—Electro-magnetic shaker 电磁振动器　13—Mixing chamber 混料室

磨料喷射加工的主要加工参数是材料去除率、加工精度和喷嘴寿命。材料去除率受喷嘴与工件间距离、磨料速度、气体压力和磨料类型四种工艺参数的影响。

磨料喷射加工的主要优势体现在加工的灵活性、冷切削和适合加工脆硬材料三个方面。由于使用软管运送磨料，因此可将气体和磨料运送至工件的任何部位，从而具有好的加工灵活性。磨料喷射加工属于冷切削，加工时不会使工件材料发热，因此可用于加工热敏材料，而无需担心加工应力和热损伤对工件的影响。

其不足之处在于材料去除率低；磨粒易于嵌入工件材料内部，尤其当工件材料为塑性材料时；由于高速射流呈喇叭状撞击工件表面，加工时易产生锥度；环境负荷较重。

磨料喷射加工主要加工范围包括：铸、锻件表面清理；钢板除锈、去涂层；油漆、电镀表面的预加工；玻璃、陶瓷或硬金属材料的小范围钻孔、切割；去毛刺、飞边；小孔、浅槽的加工等。

(3) 激光束加工（LBM）　激光束加工采用高能、相干光束熔融、气化金属或非金属工件表面材料，是一种靠光热效应去除材料的加工工艺。所有的激光束加工使用的都是通过聚焦物镜汇聚后的激光束，如图 7-10 所示。光束经聚焦物镜，在距离物镜某个确定距离处汇聚于焦点。当激光以这种方式集中、聚焦后，功率不超过 100W 的激光也能够气化、熔融任何材料。

激光束加工中，材料去除包括几个阶段：熔融、气化和化学降解（化学键断裂从而使

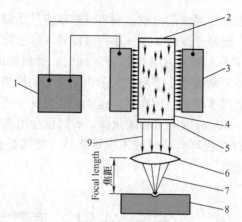

Fig. 7-10 Schematic diagram of LBM process 激光束加工示意图
1—Power supply 电源   2—100% reflecting mirror 全反射镜   3—Flash lamp 闪光灯
4—Laser-discharge tube 激光管   5—Partially reflecting mirror 部分反射镜   6—Len 透镜
7—Focused light 聚焦激光   8—Workpiece 工件   9—Single wavelength in-phase light 单波长同相激光

Laser beams are widely used for cutting, drilling, marking, welding, sintering and heat treatment. The common-used two types of LBM processes are Laser Beam Drilling and Laser Beam Cutting.

Nd:YAG lasers are best suited for operation in a pulsed mode for maximum energy output. That is, energy is stored until a threshold value is reached and then rapidly discharged at frequencies of up to 100 kHz. Owing to their preferred pulsed mode operation at high-energy outputs, Nd:YAG lasers are best suited for drilling operations (besides welding and soldering). In drilling, energy transferred into the workpiece melts the material at the point of contact, which subsequently changes into plasma and leaves the region (Fig. 7-11). A gas jet (typically, oxygen) can further facilitate this phase transformation and the departure of material.

Laser beam drilling can be utilized for all materials, though it should be targeted for hard materials and hole geometries that are difficult to achieve with other methods. Furthermore, laser drilling can also be used for superfast hole drilling, at rates above 100 holes per second, of diameters of as low as 0.02mm, when using a fast-moving mirrors/optics arrangement. Large holes can be achieved by a "trepanning" approach, where the laser starts at the center of the hole and follows a spiral path that eventually cuts the circumference of the circle in a continuous wave mode.

Typical past examples of laser drilling have included small nickel-alloy cooling frames for land-based turbines with almost 200 inclined holes, gas turbine combustion liners for aircraft engines with up to 30,000 inclined holes, ceramic distributor plates for fluidized bed heat exchangers, and plastic aerosol nozzles. Some more recent cases of laser drilling include bleeder holes for fuel-pump covers and lubrication holes in transmission hubs in the automotive industry, and fuel-injector caps and gas filters in the aerospace industry.

The term laser cutting is equivalent to (continuous) contour cutting in milling: a laser spot reflected onto the surface of a workpiece travels along a prescribed trajectory and cuts into the material. Multiple lasers can work in a synchronized manner to cut complex geometries. Continuous wave gas lasers are suitable for laser cutting. They provide high average power and yield high material removal rates and smooth cutting surfaces, in contrast to pulsed-mode lasers that create periodic sur-

材料降解）。当高能激光束聚焦于工件表面时，吸收的热能使加工区域熔融、气化或发生化学状态改变，然后可以很方便地通过高压气体将加工区域的材料吹走。

激光束广泛应用于切割、打孔、打标、焊接、烧结和热处理等工艺。常用的两种激光加工工艺是激光打孔和激光切割。

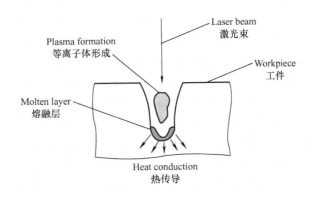

Fig. 7-11　Laser beam drilling process 激光打孔

Nd：YAG 激光器非常适合以脉冲方式工作的、高输出功率的加工。也就是说，当能量积聚至极限点后，以高达 100kHz 的频率快速放电。由于其理想的、高输出功率的脉冲工作方式，Nd：YAG 激光器非常适于打孔加工（除了焊接和钎焊外）。打孔时，传递至工件的能量使接触点处的材料熔融，并随后转变为等离子体脱离加工区域，如图 7-11 所示。气体喷流（通常为氧气）可进一步加速状态的转变和材料的去除。

激光打孔可用于加工所有材料，但最主要的应用是加工其他加工方式难以完成的高硬度材料和复杂几何形状的孔。此外，借助于快速移动的光学装置，激光打孔能够以每秒加工 100 个孔的速度完成直径低至 0.02mm 的孔的加工。激光加工大尺寸孔时，可采用套钻法。激光首先位于孔中心，然后沿螺旋轨迹运动，最终以连续波方式加工孔周边。

激光打孔曾用于固定式叶轮机中有 200 个斜孔的镍合金冷却框架的加工，飞机燃气涡轮发动机中有 30000 个斜孔的燃烧室衬套的加工，流化床换热器的陶瓷多孔分布板以及塑料气雾喷嘴的加工。最近激光打孔也被用于加工汽车燃油泵盖的溢流孔，传动器齿毂的润滑油孔以及航空工业中的喷油嘴帽和气体滤清器。

激光切割相当于连续路径轮廓铣削。激光光斑照射在工件表面，并沿预定轨迹运动来切割材料。加工复杂形状工件时，可采用多头激光切割，使多束激光同时参与加工。连续波气体激光器适于进行激光切割。与脉冲式激光器相比，它们不会生成周期性分布的粗糙表面，具有平均输出功率高，材料去除率高，切割表面光滑的优点。自 20 世纪 70 年代以来，CO_2 激光器在激光切割行业一直占主导地位。

激光切割的材料去除原理与激光打孔相似。在稳态加工中，输入能量主要传导至加工部

face roughness. The CO_2 laser has dominated the laser cutting industry since the early 1970s.

The material removal mechanisms in laser cutting are similar to those in laser drilling: in steady-state operation, the input energy is balanced primarily by the conduction energy that melts and vaporizes the material. As the light beam moves forward, continuous molten (erosion) front forms because of high temperature gradients (Fig. 7-12). The kerf (narrow slot) left behind has parallel walls. For thin-walled metal workpieces the kerf width is typically less than 0.5mm, so that there is very little material waste. In a large number of cases, fast-flowing gas (e.g., oxygen) streams are utilized to assist laser cutting. They remove material and keep the focusing lens clean and cool.

Although most metals can be cut by lasers, materials with high reflectivity (e.g., copper, tungsten) can pose a challenge and necessitate the application of an absorbent coating layer on the workpiece surface. Furthermore, for most metals, the effective cutting speed exponentially decreases with increasing depth of cut. CO_2 and Nd:YAG lasers can also be utilized in the cutting of ceramics and plastics/composites. Overall, typical industrial applications of laser cutting include removing flash from turbine blades, cutting die boards, and profiling of complex geometry blanks.

(4) Electron-beam machining (EBM)　EBM is a process that utilizes the impact energy of high-velocity electrons. It is similar to LBM except laser beam is replaced by high velocity electrons. In EBM process, high-velocity electrons concentrated into a narrow beam with high energy density ($10^6 - 10^9 W/cm^2$) are directed toward the workpiece, creating large amount of heat in short time (usually a fraction of a microsecond), heating, melting and vaporizing portions of the workpiece material. The process usually takes place inside a vacuum, thus protecting the metal from the outside atmosphere.

Fig. 7-13 shows the schematic diagram of EBM process. A stream of electrons is started by a voltage differential at the cathode. The concave shape of the cathode grid concentrates the stream through the anode. The anode applies a potential field that accelerates the electrons. The electron stream is then forced through the valve, the magnetic lens and the deflection coils. Magnetic lenses are used to focus the electron beam to the surface of the workpiece. And by means of the deflection system the beam is positioned as needed, usually by means of a computer. Finally, the beam is focused into a narrow beam and directed onto the surface of the workpiece, heating, melting, and vaporizing the material.

Electron beam can be used to perform different operations by controlling the energy density and energy interval of the beam in machining. Heating the selected area of the workpiece surface rapidly can perform electron beam surface treating. Melting the selected area of the workpiece can accomplish electron beam welding. Utilizing electron beam with high power density to melt and vaporize the workpiece material can realize electron beam drilling and cutting operation. And electron beam lithography can be performed by using the chemical changes occurred when polymer materials are impinged by electron beam with low energy density.

EBM has the following features:

1) Electrons can be focused into a very narrow beam ($1 - 0.1 \mu m$), so EBM is suitable for micro-manufacturing.

位，使材料熔融、气化。随着激光束向前移动，由于温度梯度高，会不断形成熔融锋面，如图 7-12 所示。切割留下的切缝侧壁平行，且当切割薄金属件时，切缝宽度一般不超过 0.5mm，材料损失小。多数情况使用辅助高速气流吹走熔融材料，同时冷却聚焦物镜并使其保持清洁。

尽管激光可切割绝大多数金属，加工高反射率材料（如铜和钨）却相当有挑战性，一般需要在工件表面镀上一层吸收率高的薄膜层。此外，对于多数金属，激光切割的有效切割速度随切割厚度的增加而呈指数递减。CO_2 和 Nd：YAG 激光器均可用于切割陶瓷、塑料及复合材料。激光切割在工业中的典型应用有涡轮叶片去毛刺、模板切割和复杂几何形状板料的成形切割。

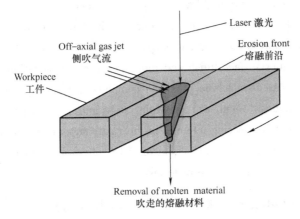

Fig. 7-12　Laser beam cutting process 激光切割过程

（4）电子束加工（EBM）　电子束加工利用高速电子的冲击动能进行加工。电子束加工与激光束加工相似，不同之处在于电子束取代了激光束。加工时，高速电子聚焦成能量密度很高（$10^6 \sim 10^9 W/cm^2$）的细电子束施加到材料表面，在短时间内（通常几分之一微秒）产生大量热量，对工件材料局部加热，使材料发生熔融和气化，从而去除材料。电子束加工通常在真空环境下进行，以防止外部大气对加工工件造成影响。

图 7-13 所示为电子束加工示意图。阴极加电压后发射电子，凹状控制栅极使电子穿过带高电位的阳极 6，经过加速之后的电子束依次通过光阑 8、电磁透镜 10 和偏转线圈 11，汇聚成细束冲击工件材料表面。工件材料被加热后发生熔融、气化而被去除。电磁透镜可将电子束聚焦于工件表面。偏转器通常由计算机控制，可将电子束定位至加工位置。

根据加工中功率密度和能量注入时间的不同，电子束可实现多种加工目的。将工件材料局部加热可进行电子束热处理；使材料局部熔化可进行电子束焊接；采用高能量密度的电子束使材料熔融、气化，可进行电子束打孔、切割等加工；而利用低能量密度电子束轰击高分子材料时产生的化学变化，可进行电子束光刻加工。

电子束加工具有以下特点：

1) 电子束可聚焦成 $1 \sim 0.1 \mu m$ 的细束，可用于实现微细加工。

2) 电子束加工时，材料因气化而被去除，机械作用力很小，不易出现变形和应力，可用于加工各种力学性能的导体、半导体和非导体材料。

2) In EBM, materials are removed by vaporization instead of large cutting forces, which leads to less deformation and stress in workpieces. Thus, EBM is capable of machining conductors, semiconductors and insulators with various mechanical properties.

3) Materials oxidized easily and semiconductor materials with higher purity can be machined by EBM process because the process is performed in vacuum.

4) EBM has higher productivity because of the higher power density of the electron beam. Fifty holes, 0.4mm in diameter, of a steel sheet with 2.5mm thickness can be drilled in one second by electron beam.

5) The main limitation of EBM might be that the initial investment of an EBM system is quite higher since both specialized facility and vacuum system are necessary.

All above-mentioned modern processes are characterized by the following common features: higher power consumption and lower material removal (or additive) rates than traditional fabrication processes, but yielding better surface finish and integrity (i.e., less residual stress and fewer microcracks). NTM processes can machine precision components from sophisticated materials. They offer the advantages of a reduced number of machining steps and a higher product quality. Interest in NTM has been on rise due to increasing interests in the vastly superior properties of innovative materials such as superalloys, composites and ceramics. NTM offers an attractive alternative and often the only choice for the processing of these materials.

7.2.5 Rapid prototyping & manufacturing technology

1. Introduction

Manufacturing community is facing to two important challenges. One is the substantial reduction of product development time. The other is the improvement on flexibility for manufacturing low-volume and multi-variety products. CAD and CAM have significantly improved the traditional product design and manufacturing processes. However, following problems still exist.

1) Rapid creation of 3D models and prototypes. New designs often have unexpected problems. A prototype is often used as part of the product design process to allow designers the ability to explore design alternatives, test theories and confirm performance prior to starting production of a new product. The product-development cycle time pressures demand to create these prototypes rapidly.

2) Cost-effective production of patterns and moulds with complex configuration. Traditionally, the development of molds is performed by machining and heat treating. It requires substantial calendar time and has significant associated costs. Therefore, it is of interest during product development to be able to quickly produce patterns and moulds with complex configurations. The rapid production of molds or tools can be accomplished through the use of the rapid prototyping processes followed by some subsequent processes.

2. Rapid Prototyping and Manufacturing (RP&M)

Rapid Prototyping and Manufacturing (RP&M) refers to a set of technologies that can automatically construct physical models from CAD data. Using the methods of RP, models or prototypes normally can be built in a matter of hours or a few days, depending on the number of models being pro-

3）电子束加工在真空中进行，工件表面不会出现氧化和污染，适合于加工易于氧化的材料和纯度要求很高的半导体材料。

4）电子束能量密度高，因此生产效率高。例如，电子束可在 1 秒内，在厚度为 2.5mm 的钢板上加工出 50 个直径为 0.4mm 的孔。

5）电子束加工的初始投资成本高，因为除了整套专用设备外，还需要真空系统。

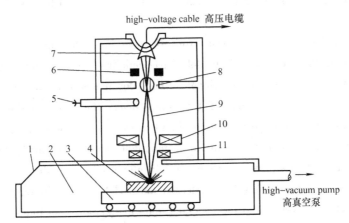

Fig. 7-13 Schematic diagram of EBM process 电子束加工示意图
1—Viewing port 观察孔　2—Vacuum chamber 真空室　3—Worktable 工作台　4—Workpiece 工件
5—Optical viewing system 光学观测系统　6—Anode 阳极　7—Cathode grid 阴极栅　8—Valve 光阑
9—Electron stream 电子束　10—Magnetic lens 电磁透镜　11—Deflection coils 偏转线圈

与传统加工工艺相比，上述现代特种加工方法功率消耗大，材料去除速度低，但具有较好的表面粗糙度和完整性（残余应力小，微裂纹少）。特种加工可实现高性能材料的精密加工，具有加工步骤少，产品质量高的优点。由于应用中需要具有优越性能的超级合金、复合材料和陶瓷等新材料，特种加工已经引起人们的关注，成为加工这些材料的极具吸引力的方法之一，有时甚至是唯一可用的方法。

7.2.5　快速原型制造技术

1. 引言

制造业面临两大挑战，一是大大缩短产品开发时间，二是提高小批量、多品种产品生产的灵活性。CAD/CAM 技术已经显著改善了传统的产品设计与制造过程，但问题依然存在：

1）三维模型和原型的快速创建。新设计总会有不可预知的问题。因此，样机生产成为设计过程的一部分，以便于设计者探求设计方案、测试设计理论并在新产品投入生产之前验证产品的性能。为了缩短产品开发周期，要求能够快速获得产品的原型样机。

2）复杂结构模具的经济、有效生产。一般来讲，模具生产需经过加工和热处理过程，时间长，耗费成本高。因此，人们对于在产品开发阶段就能快速加工出复杂结构的模具非常感兴趣。工、模具的生产可通过快速原型技术以及一些后续加工来实现。

2. 快速原型制造

快速原型制造是指根据 CAD 模型数据自动生成物理模型的一类技术的总称。采用快速原型技术，可在几小时或几天内获得产品的模型或样机。具体加工时间取决于同时加工的零

duced at the same time, the size and the structure complexity of the model.

Different from conventional machining, RP&M creates a physical model by means of RP machine based on CAD model. RP&M belongs to material adding processes instead of material removing processes. In RP processes, the part is fabricated by depositing layers in light of the part contour in x-y plane, and then the single-layer material is stacked up layer by layer discontinuously in z direction. Therefore, the prototypes are very exact on the x-y plane but have stair-stepping effect in z-direction.

As shown in Fig. 7-14, the process starts with 3D modeling of the product and then STL file is exported by tessellating the geometric 3D model. In tessellation, various surfaces of a CAD model are piecewise approximated by a series of triangles, and coordinates of vertices of triangles and their surface normal directions are listed. These STL files are checked for defects and are repaired if faulty is found. Defect free STL files are used as an input to various slicing software. At this stage the choice of the part deposition orientation is the most important factor since part building time, surface quality, amount of support structures, cost etc. are all influenced. Once the part deposition orientation is decided and slice thickness is selected, tessellated model is sliced by parallel planes to create a series of equal-thickness layers. And the generated data in standard data formats like SLC (stereolithography contour) or CLI (common layer interface) is stored. This information is used to generate physical model. The software that operates RP systems generates laser-scanning paths or material deposition paths depending on the basic deposition principle used in RP machine. Information computed here is used to deposit the part layer-by-layer on RP system platform. The final step is post-processing. At this stage, generally some manual operations are necessary. Sometimes the surface of the model is finished by sanding, polishing or painting for better surface finish or aesthetic appearance.

According to the energy type used in RP processes, there are laser beam RP processes, heat energy RP processes and mechanical energy RP processes etc. Examples of the laser beam RP processes are Stereolithography Apparatus (SLA), Selective Laser Sintering (SLS) and Laminated Object Manufacturing (LOM) etc. 3D Printing and Fused Deposition Modeling (FDM) belong to the mechanical energy RP processes.

At present, there are more than thirty kinds of RP technologies. Each of them has its own features. However, compared with traditional processes, RP processes, no matter what kind of technologies are used, have the common features: high production flexibility; short product development cycle time; fully digitized production; broad range of building materials.

3. Typical rapid prototyping processes

There are currently many methods of RP processes, with four being relatively common. The major differences among these methods are materials used and part building techniques. Here, these four important RP processes namely Stereolithography (SL), Selective Laser Sintering (SLS), Fused Deposition Modeling (FDM) and Laminated Object Manufacturing (LOM) are described.

(1) Stereolithography (SL)   As shown in Fig. 7-15, the basic principle of SLA process lies that small molecules are polymerized into large molecules under the initiation of photons. In the process, photosensitive liquid resin is cured selectively into solid polymer because of polymerization

件数目、模型的尺寸以及模型结构的复杂程度。

与传统加工不同，快速原型制造根据 CAD 模型，使用快速原型机生成物理模型。快速原型工艺属于材料添加工艺而不是材料去除工艺。在快速原型工艺中，通过在 x-y 平面依照零件轮廓堆积成层，然后在 z 方向上非连续地将单层材料层层堆积，最后得到零件原型。因此快速原型件在 x-y 面内形状非常准确，但在 z 方向上会呈现出阶梯状外观。

如图 7-14 所示，首先构建产品的三维模型，然后将模型分片并以 STL 格式输出。分片时，CAD 模型的各表面用一系列三角面片分段表示，同时给出每一面片的顶点坐标及法向。输出的 STL 文件需经过检查以判断是否有缺陷，如有，则需进行修补。检查无误的 STL 文件输入到各种分层软件中。该阶段的首要任务就是选择成型方向，因为成型方向会影响零件的成型时间、成型质量、支撑结构以及加工成本。当成型方向和片层厚度确定以后，以小三角平面表示的模型被平行平面分割成等厚的一系列片层。分层后的数据存储成 slc 或 cli 等标准格式，用于生成物理模型。软件驱动快速原型机生成与所采用的材料堆积原理有关的激光扫描路径或材料堆积路径。产生的信息用于在快速原型机工作台上逐层堆积材料形成零件。最后进行后处理，一般需要手工完成。为了获得更好的表面质量或外观，有时还需对原型件进行喷砂、抛光或喷漆等处理。之后即可对模型进行测试和验证。

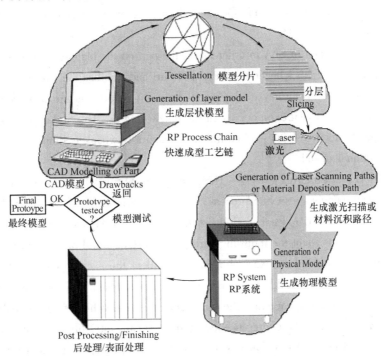

Fig. 7-14　Principle of Rapid Prototyping　快速原型原理

根据能量形式，快速原型工艺分为激光成型、热成型和机械成型。激光成型工艺有立体光刻、选区激光烧结和叠层制造等。而三维打印和熔融沉积造型则属于机械成型工艺。

目前，有三十多种快速原型技术，每一种都有其自身的特点。然而和传统工艺相比，不管采用的技术如何，快速原型工艺具有生产灵活性高；产品开发周期短；完全数字化生产和成型材料范围广的特点。

when exposed to ultraviolet light. The ultraviolet laser beam is moved along a path defined by a computer model to create a thin layer of solid resin. Then the resin bath is lowered, causing a layer of liquid resin to cover the cured layer. And the process is repeated, resulting in a 3D layered object.

In SLA process, the software firstly interprets and pre-processes the CAD data and slices it into a series of thin horizontal layers and converted to machine specified control data files based on the part, building and recoating parameters. The machine control data is then downloaded into the equipment for part building. At the beginning of the process, the platform is rested just below the liquid surface whose depth is the light absorption limit. The laser beam is deflected horizontally in X and Y axes by galvanometer-driven mirrors so that it moves across the surface of the resin to produce a solid pattern. After the first layer is built, the platform is lowered equal to one slice thickness. The total thickness of the layer is depending on the scanning speed of the laser and the depth penetration factor of the resin. A blade spread viscous liquid resin on the previously solidified layer as the blade traverses the vat for building next layer. This ensures smoother surface and reduced recoating time. The new slice is then scanned and solidified. Once the complete part is deposited, it is removed from the vat and then excess resin is drained. The part is post-cured in an UV oven after removing support structures and then ready for use, although some machining may be necessary. The SLA process is shown in Fig. 7-16.

Support is usually needed in SLA in order to put the part on the platform. Also, support is needed for overhanging and tilted surfaces hence minimize part curl and stabilize the part while being built.

The Surface finish and accuracy of the model built in SL process are usually very good, but sometimes stepping effect is visible on curves and ramped sections. The SLA method provides a good combination of speed, accuracy and surface finish. Its main drawback is limited selection of materials for use as functional models.

(2) Selective Laser Sintering (SLS)　　The principle of SLS is similar to that of SLA. The main difference is the materials used. In SLS process, powder materials are utilized instead of photo-curable resin. SLS process uses CO_2 laser to sinter or fuse successive layers of powder material.

An SLS machine consists of the following parts: feed piston, build piston, spreading apparatus, CO_2 laser, scanning device and an excess bin used to collect the excess powder. All these components are kept in nitrogen rich build chamber to reduce the explosion risk associated with fine powders and to prevent oxidization reaction. The build chamber is also heated to just below the melting point of the material to reduce the energy needed by the laser to sinter the materials. The feed piston is used to measure and feed powder that is spread over the build piston by a spreading apparatus.

In SLS process, fine powder (20 to 100μm in diameter) is spread on the substrate by a powder-leveling roller. The powder material is preheated to a temperature slightly below melting point in order to minimize thermal distortion and facilitate fusion to the previous layer. Laser beam controlled by a scanning device traces cross section on the surface to heat up to sintering temperature so that the scanned powder is bonded. The powder that is not scanned will remain in place to serve as the support to the next layer of powder. After the first cross section is completed the feed elevator raises one layer thickness and the build chamber lowers one layer thickness. The roller then spreads the next

3. 典型快速原型工艺

目前有多种快速原型工艺，其中常用的有四种。这些方法主要的区别在于所用的材料和模型的构建方法不同。在此介绍立体光刻、选区激光烧结、熔融沉积造型和叠层制造这四种工艺。

（1）立体光刻　如图7-15所示，立体光刻的基本原理是小分子在光子的作用下聚合成大分子结构。加工时，由于紫外线的照射，聚合作用将液态光敏树脂有选择性地固化为固态聚合物。紫外线激光束在计算机模型数据的控制下沿特定路径移动，生成一薄层截面；然后工作台下降，使得一层液态树脂铺在刚刚固化的树脂表面。重复该过程，最终生成三维层状模型。

立体光刻时，软件首先对CAD数据进行解译和预处理，并将CAD模型分割成一系列平行的薄层。然后根据零件、成型及涂层参数将数据转换成适于具体设备的控制文件，并传输至设备以创建原型。开始加工时，工作台位于液面以下光吸收极限深度处，激光束在检流计驱动镜的控制下在 X 和 Y 方向上水平偏斜，从而在树脂表面移动，生成固体模型。第一层固化完成后，工作台下降一个片层厚度的距离。片层厚度与激光扫描速度和树脂的深度穿透系数有关。刮刀横向移动将粘稠的树脂铺覆在前一固化层上以便于成型下一层。以此不仅可使成型表面光滑，而且可以缩短铺覆树脂时间。之后即可扫描并固化新层。一旦整个零件沉积完成，可从桶中取出，去除多余的树脂和支撑结构，并在紫外线炉中进行后固化处理。之后，原型件可被立即使用，有时也需再进行其他加工。图7-16为SLA示意图。

Fig. 7-15　Basic components of SLA
立体光刻基本组成

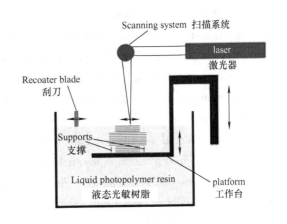

Fig. 7-16　Working principle of SLA
立体光刻基本原理

立体光刻通常需要支撑，主要是为了将零件放在工作台上。另一方面，当成型悬臂或斜面结构时，支撑可减少零件翘曲并使成型时零件结构稳定。

立体光刻成型的原型件具有表面质量高，精度好的特点，但曲面和斜面有时会呈阶梯状。立体光刻在成型速度、精度和表面质量方面的综合性能较好，其主要不足是用于生产功能模型的材料有限。

（2）选区激光烧结　选区激光烧结的基本原理与立体光刻相似，主要不同在于所用的

layer of powder evenly over the first cross section. The next cross section is then sintered. This process is continued until the part is completed. Fig. 7-17 shows the working principle of SLS.

Unlike SLA, SLS does not need the support structures because the excess powder on the build piston acts as a support for any overhanging features. Materials used in the process can be metal powder, ceramic powder, polymeric powder like polystyrene, polycarbonate or polyamide etc. However, product produced may suffer shrinkage and warpage due to sintering and cooling. And surface finish of the product is relatively rough. Also, the end product can be porous, and therefore may require infiltration to improve strength and surface finish.

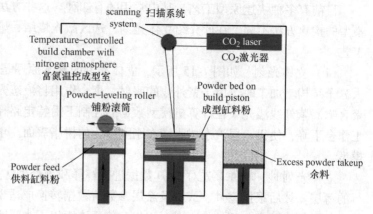

Fig. 7-17 Working principle of SLS 选区激光烧结工作原理

(3) Fused Deposition Modeling (FDM) An FDM machine is composed of heated extrusion nozzles, filament feed devices, a build platform, and a nozzle control apparatus. The whole system is contained within a heated environment to reduce the amount of energy needed to melt the filament in nozzle.

In FDM process, as shown in Fig. 7-18, two spools of thermoplastic filament (one spool of building material and one spool of supporting material) feed in the heated, movable extrusion head. The movement of the head is controlled by computer. The material inside the head is heated slightly above (approximately 0.5℃ or 1℃) its melting temperature so that it solidifies within a very short time (approximately 0.1s) after extrusion and cold-welds to the previous layer. The head traces the exact outline of each cross-section layer of the part. As the head moves in X and Y axes, molten thermoplastic material is extruded out of the nozzles by a precision pump. After the first layer is completed, the build platform is lowered one layer thickness for building next layer. This process is continued until the part is completed. Then the part can be taken out and any support structures can be removed manually or, when water soluble supports are employed, they can be simply dissolved by put into particular chemical solution.

Thermoplastic material is often used as molding material in FDM process, which includes Acrylonitrate Butadiene Styrene (ABS), Polycarbonate (PC), Polyphenylsulfone (PPSF) and investment casting wax. The main advantage of FDM process is the very durable parts that can be made using various engineering plastics. And the drawback might be that it takes much longer time to build a product, and the layering is clearly visible because of the extrusion type process.

(4) Laminated Object Manufacturing (LOM) In LOM process as shown in Fig. 7-19, a special type of adhesive-coated sheet material is laid down on the platform. A heated laminating roller passes over the sheet material and the adhesive bonds it to the surface beneath. The laser head

材料。SLS 使用的是粉末状材料而非光敏树脂。该工艺采用 CO_2 激光将一层层的粉末状材料熔融或烧结成型。

SLS 装置由进给缸、成型缸、铺粉装置、CO_2 激光器、扫描装置和用于收集多余料粉的料粉收集桶组成。所有组成部分均处于富氮成型室中,以防止出现由细微粉末引起的爆炸和氧化现象。成型室被加热到恰好是材料的熔点以下,这样可降低激光烧结材料所需能量。进给活塞用于测量和送料,送至成型缸的粉料由铺粉装置铺在成型活塞之上。

加工时,铺粉滚筒将粒径为 20~100μm 的细粉铺在底板上。粉料被预热至稍低于熔点温度,以减小热变形并加快上一层的熔融。扫描装置控制激光在截面上移动,对材料进行加热、烧结,从而将激光扫描过的粉料烧结在一起。未被扫描的粉料将保留在原有位置,支撑下一层料粉。第一层截面形成后,供料缸升降台上升一个层厚高度,成型缸升降台下降一个层厚高度,铺粉滚筒在第一层截面上均匀地再铺上一层料粉,进行下一层截面的烧结。重复上述过程,直到完成整个零件。图 7-17 为 SLS 工作原理图。

与 SLA 不同,SLS 不需要支撑结构,因为成型缸中的多余粉料可为任何悬臂结构提供支撑。成型材料可以是金属粉、陶瓷粉以及聚苯乙烯、聚碳酸酯或聚酰胺等聚合物粉。然而,由于烧结及之后的冷却,成型件会发生翘曲和收缩。成型件表面相当粗糙且多孔,因此需要进行浸渗以提高强度和表面质量。

(3) 熔融沉积造型(FDM) FDM 设备由加热喷头、送丝机构、成型工作台以及喷头控制装置组成。整个系统处于一定温度下以减少喷头处熔化料丝所需能量。

在图 7-18 所示的 FDM 中,两卷热塑性材料丝(一卷为成型材料,一卷为支撑材料)被送入到移动的、由计算机控制的加热喷头中。喷头内的材料被加热至稍高于熔点 0.5~1℃,以便于材料可在挤出后很短时间内(约 0.1s)固化并粘附在上一层材料之上。喷头准确地按照零件每一截面层的形状移动。当喷头沿 X 和 Y 方向移动时,熔融的热塑性材料由恒流泵挤出。

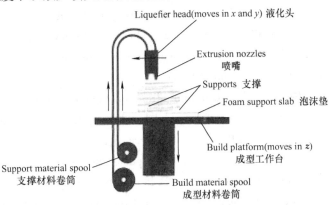

Fig. 7-18 Working principle of FDM 熔融沉积造型工作原理

第一层完成后,工作台下降一个层厚以成型第二层。该过程重复进行直到完成整个零件。之后,拿出零件并去掉支撑。可手工去除支撑,但当支撑为水溶性材料时,也可将零件放入某种化学溶液中将支撑材料溶解掉。

FDM 中常用的热塑性材料包括 ABS 工程塑料、PC 树脂、PPSF 和熔模铸造用蜡。FDM 工艺的主要优点在于可用各种工程塑料加工出非常耐用的零件。然而采用 FDM 工艺,零件成型时间长,且由于采用挤压成型原理,零件表面分层清晰可见。

(4) 叠层制造 在图 7-19 所示的 LOM 工艺中,首先在工作台上铺上一种涂有热敏胶的片材,加热辊 8 热压片材 9,片材上的热敏胶将其粘结在下层材料上。然后,激光头按照由 CAD 模型生成的零件截面轮廓运动,将片材切割成所需形状,并在截面轮廓和外框之间的多余区域切割出网格。这些网格可将轮廓截面与多余材料分开,同时便于后处理时多余材料

traces the outline of the layer so that the laser beam could cut sheet material into the required shape according to the contour of the part cross section generated by CAD. The excess area is also cross-hatched. Cross-hatching breaks up the extra material, making it easier to be removed during post-processing. During the build, the excess material provides excellent support for overhangs and thin-walled sections. Next, the platform drops by the thickness of the sheet material. The sheet material can be advanced by feed mechanism, which winds the excess material onto the take-up roller. Then another layer is laid down, rolled and cut. The process continues as needed until the part is complete. In many cases, parts built in this way are made of paper. They must be sealed and finished with paint or varnish to prevent moisture damage. In addition to paper material, other materials including composite sheet, ceramic and metal powder tapes are also used.

The forming speed of LOM process is 5 – 10 times as fast as that of other RP processes because the laser only cuts the outline of component and cross-hatching instead of scanning all the internal area of the part cross section. Parts can be made quite large because there is no chemical reaction involved. Also, no predesigned supports are needed because of the usage of solid material. The limitations of LOM process are that it would produce smoke in forming component; it would damage the delicate parts when removing excess materials, and the properties of molding parts have specific direction.

4. Applications of RP&M

Although RP models have poor surface finish, limited strength and accuracy, RP technologies are successfully used in various industries like automotive, aerospace, and medical etc. Nearly all the applications of RP technology are fall into one of the following aspects: rapid product development and rapid manufacturing.

RP technology can be used to fabricate concept models, functional models and models for engineering analysis etc. These models are necessary and useful for designers and engineers to explore alternatives, test theories and confirm performance prior to starting production of a new product because new designs often have unexpected problems.

Rapid manufacturing refers to rapidly develop molds or tools for moderate volume parts or products. Rapid production of molds or tools can be accomplished through the use of some rapid prototyping processes followed by some subsequent processes. For example, a RP model of the part needed can be produced and subsequently used as a sacrificial pattern to investment cast the part. Alternatively, a mold can be designed and the patterns for making the mold can be produced in plastic or wax using an RP technology. These RP pieces can be used sacrificially in the investment-casting process to form mold inserts in metal. Compared with conventional tooling and molds manufacturing methods, rapid manufacturing processes can reduce tooling costs and development time greatly.

的剥离。进给机构带动片材前进并将多余材料卷在收料辊上。接着，铺上另一层片材，再进行辊压、切割。该过程重复进行，直到完成整个零件。叠层制造常用材料为纸。生成的原型件需进行喷漆或涂色处理以防止受潮。除了以纸为原料外，也可使用复合材料、陶瓷或金属粉制成的片材。

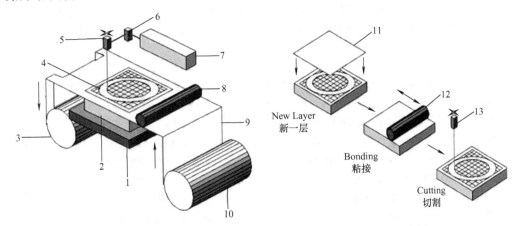

Fig. 7-19　Schematic diagram of LOM process 叠层制造工艺示意图
1—Platform 工作台　2—Part block 块状零件　3—Take-up roll 收料滚筒
4—Layer outline and crosshatch 片材轮廓及网格　5—X-Y positioning device X-Y 定位装置
6—Optics 光学装置　7—Laser 激光器　8、12—Heating roller 加热辊　9—Sheet material 片材
10—Material supply roll 供料滚筒　11—Laminate 层压材料　13—Laser head 激光头

LOM 工艺的成型速度是其他快速原型工艺的 5~10 倍，因为激光器仅切割出零件轮廓及网格，而不需要扫描零件截面内的所有区域。LOM 工艺不涉及化学反应，可以成型较大零件。此外，该工艺采用固体材料，因此不需要设计支撑结构。其局限性包括成型时有烟雾、去除多余材料时会损坏精致的零件以及成型件性能具有方向性。

4. 快速原型制造技术的应用

尽管快速原型件表面质量较差，强度和精度较低，快速原型制造技术仍成功应用于汽车、航空、医疗等行业。其应用领域包括产品快速开发和快速制造两个方面。

快速原型技术可用于生产概念模型、功能模型以及工程分析模型。因为新设计总是存在着不可预计的问题，所以快速原型对于工程技术人员在新产品投入生产之前进行方案选择、理论测试和性能确认非常必要。

快速制造是指可进行一定批量零件或产品生产的工、模具的快速开发。快速制造可采用快速原型工艺与其他加工工艺相结合的方法。例如，首先加工出所需零件的快速原型，然后将该原型件作为熔模，铸造出零件。此外，也可以在模具设计完成后，采用快速原型技术生产出以塑料或蜡为材料的铸型，进而加工出模具。这些快速原型件在熔模铸造中将被融化掉，最终获得金属模具镶件。与传统工、模具加工方法相比，快速制造可大大降低生产成本，缩短开发周期。

References 参 考 文 献

[1] 任小中，康红艳，杨丙乾. 机械制造技术基础［M］. 北京：科学出版社，2012.
[2] 赵雪松，任小中，赵晓芬. 机械制造技术基础［M］. 2版. 武汉：华中科技大学出版社，2010.
[3] Kalpakjian S, Schmid S R. Manufacturing Engineering and Technology［M］. New Jersey：Prentice Hall, Inc., 2001.
[4] P N Rao. Manufacturing Technology［M］. 北京：机械工业出版社，2003.
[5] 曾志新，吕明，轧刚. 机械制造技术基础［M］. 武汉：武汉理工大学出版社，2004.
[6] Donald H Nelson, George Schneider, Jr. Applied Manufacturing Process Planning with Emphasis on Metal Forming and Machining［M］. New Jersey：Prentice Hall, Inc., 2001.
[7] George Tlusty. Manufacturing Processes and Equipment［M］. New Jersey：Prentice Hall, Inc., 2000.
[8] 唐一平. Advanced Manufacturing Technology［M］. 北京：机械工业出版社，2004.
[9] Paul Kenneth Wright. 21 Century Manufacturing［M］. 北京：清华大学出版社，2002.
[10] 于骏一，邹青. 机械制造技术基础［M］. 2版. 北京：机械工业出版社，2009.
[11] 韩秋实，等. 机械制造技术基础［M］. 2版. 北京：机械工业出版社，2005.
[12] 熊良山，严晓光，张福润. 机械制造技术基础［M］. 武汉：华中科技大学出版社，2007.
[13] 卢秉恒，等. 机械制造技术基础［M］. 2版. 北京：机械工业出版社，2005.
[14] 张世昌，李旦，高航. 机械制造技术基础［M］. 北京：高等教育出版社，2003.
[15] 赵雪松，任小中，于华. 机械制造装备设计［M］. 武汉：华中科技大学出版社，2009.
[16] 周宏甫. 机械制造技术基础［M］. 北京：高等教育出版社，2004.
[17] 吉卫喜，等. 机械制造技术［M］. 北京：机械工业出版社，2004.
[18] 郑修本. 机械制造工艺学［M］. 2版. 北京：机械工业出版社，2011.
[19] 曾志新，吕明. 机械制造技术基础［M］. 武汉：武汉理工大学出版社，2001.
[20] 王先逵. 机械制造工艺学［M］. 北京：机械工业出版社，2003.
[21] 陈立德，李晓晖. 机械制造技术［M］. 上海：上海交通大学出版社，2004.
[22] 肖继德，陈宁平. 机床夹具设计［M］. 北京：机械工业出版社，2000.
[23] 任小中. 先进制造技术［M］. 武汉：华中科技大学出版社，2009.
[24] 盛晓敏，邓朝晖. 先进制造技术［M］. 北京：机械工业出版社，2000.
[25] 李伟. 先进制造技术［M］. 北京：机械工业出版社，2005.
[26] 陈宏钧. 实用机械加工工艺手册［M］. 3版. 北京：机械工业出版社，2009.